"This latest work of Sumudu Atapattu is a much-needed contribution to the field of study of climate change and human rights. The devastating impacts of climate change are worsening as actors around the world engage to safeguard human life and protect the rights of all, including those most vulnerable and exposed. Professor Atapattu helps the academic as well as the policy and activist communities better understand how a human rights framing helps victims of environmental degradation. In whatever environmental arena one finds oneself, this book is a required tool and an important asset."

—Ambassador Dessima Williams,
Formerly, Special Advisor on Sustainable Development Goals and Director, Office of the President of the UN General Assembly, Ambassador and Permanent Representative of Grenada to the UN & Chair, Alliance of Small Island Developing States

"UN Human rights institutions are increasingly addressing the human rights impact of environmental degradation including climate change and unsustainable development, despite their mandates' silence on these issues. In this volume, Sumudu Atapattu lends her expertise to this emerging field and provides a comprehensive overview of how these human rights bodies have addressed these interlinkages. This volume fills a gap in the scholarly literature and should be of interest to scholars, students, activists, government officials, think tanks, and practicing lawyers."

—Professor Carmen G. Gonzalez, *Morris I. Leibman, Professor of Law, Loyola University Chicago School of Law*

"This brilliant survey by Sri Lankan international lawyer Sumudu Atapattu comes at a propitious moment – viz., the adoption by the UN General Assembly of Resolution on 'the human right to a clean, healthy and sustainable environment', preceded by a UN Human Rights Council Resolution. As explained by the author, it is not the intention of this study to engage in a discussion of scholarly writing or a literature review of the multiple global and regional human rights institutions so surveyed, but to examine *the practice* of these institutions in relation to environmental issues and whether there are any principles, standards and consistencies, giving birth to a new interdisciplinary area, and concluding with a critical analysis of the potential fragmentation and cross-fertilization in this emerging field."

—Professor Peter H. Sand,
Institute of International Law, University of Munich, Formerly Legal Adviser for Environmental Affairs, World Bank, Commissioner for environmental damage, UN Compensation Commission, Geneva

UN Human Rights Institutions and the Environment

This book presents an in-depth analysis of how UN human rights institutions and mechanisms have addressed environmental protection, sustainable development, and climate change.

Despite the increasing involvement of UN human rights bodies in addressing environmental degradation and climate change, a systematic review of the convergence between human rights and the environment in these bodies has not been carried out. Filing this lacuna, this book surveys the resolutions, general comments, concluding observations, decisions on individual communications and press releases. It identifies principles that have emerged, explores the ways in which human rights charter-based and treaty-based institutions are interpreting environmental principles and examines how they contribute to the emerging field of human rights and the environment. Given the disproportionate effect that polluting activities have on marginalized and vulnerable groups, Atapattu also discusses how these human rights mechanisms have addressed the impact on women, children, indigenous peoples, people with disabilities, and racial minorities.

Written by a world-renowned expert on human rights and the environment, this book will be of great interest to students and scholars researching and teaching in this important field of study.

Sumudu Atapattu is Teaching Professor and Director of the Global Legal Studies Center at the University of Wisconsin Law School. She is also the Executive Director of the campus-wide interdisciplinary Human Rights Program. She serves as the Lead Counsel for Human Rights at the Center for International Sustainable Development Law based in Montreal, Canada, is on the advisory board of the *McGill International Journal of Sustainable Development Law and Policy*, and is affiliated faculty at the Raoul Wallenberg Institute for Human Rights and Humanitarian Law, Sweden. She has published widely in the fields of international environmental law, climate change and human rights, environmental rights, and sustainable development.

Routledge Explorations in Environmental Studies

Sustainable Places
Addressing Social Inequality and Environmental Crisis
David Adamson, Lorena Axinte, Mark Lang and Terry Marsden

Urban Aerial Pesticide Spraying Campaigns
Government Disinformation, Industry Profits, and Public Harm
Manuel Vallée

Addicted to Growth
Societal Therapy for a Sustainable Wellbeing Future
Robert Costanza

Coastal Disaster Risk Management in Bangladesh
Vulnerability and Resilience
Edited by Mahbuba Nasreen, Khondoker Mokaddem Hossain and Mohammed Moniruzzaman Khan

A History of Radioecology
Patrick C. Kangas

Satire, Humor, and Environmental Crises
Massih Zekavat and Tabea Scheel

UN Human Rights Institutions and the Environment
Synergies, Challenges, Trajectories
Sumudu Atapattu

For more information about this series, please visit: www.routledge.com/Routledge-Explorations-in-Environmental-Studies/book-series/REES

UN Human Rights Institutions and the Environment

Synergies, Challenges, Trajectories

Sumudu Atapattu

Routledge
Taylor & Francis Group
LONDON AND NEW YORK

earthscan
from Routledge

First published 2023
by Routledge
4 Park Square, Milton Park, Abingdon, Oxon OX14 4RN

and by Routledge
605 Third Avenue, New York, NY 10158

Routledge is an imprint of the Taylor & Francis Group, an informa business

British Library Cataloguing-in-Publication Data
A catalogue record for this book is available from the British Library

ISBN: 978-0-367-65310-1 (hbk)
ISBN: 978-0-367-65313-2 (pbk)
ISBN: 978-1-003-12884-7 (ebk)

DOI: 10.4324/9781003128847

Typeset in Goudy
by codeMantra

To
My grandson, Charlie, and children of his generation who must bear the brunt of inaction of our generation

Contents

Foreword xv
Acknowledgments xvii
Abbreviations xix
Table of Cases xxi
Table of Treaties and Other Instruments xxiii

PART I
Introduction & Overview 1

1 UN Human Rights Institutions and the Environment: Framing the Issues 3
1.1 Introduction 3
1.2 Environmental Issues Covered 6
1.2.1 Environmental Pollution, Environmental Degradation and Environmental Protection 6
1.2.2 Climate Change 6
1.2.3 Sustainable Development and Sustainable Development Goals 7
1.2.4 Other Issues Covered 8
1.3 Methodology 8
1.4 Outline of the Book and Overview of Chapters 8

2 An Overview of International Human Rights Law and the Emergence of Environmental Rights 15
2.1 Introduction 15
2.2 An Overview of International Human Rights Law 15
2.2.1 International Bill of Rights 17
2.2.2 Typology of Rights 17
2.3 An Overview of Human Rights Institutions 18
2.3.1 Treaty-Based Mechanisms 18
2.3.2 Charter-Based Mechanisms 19

2.3.3 Office of the High Commissioner for Human Rights 20
2.4 Rights Affected by Environmental Degradation 20
2.4.1 Right to Life 21
2.4.2 Right to Health 21
2.4.3 Rights to Food, Water and Sanitation 22
2.4.4 Right of Self-Determination 22
2.5 Specialized Treaties on Vulnerable Groups 23
2.6 Emergence of Environmental Rights and the Right to a Healthy Environment 23
2.6.1 Articulation of Environmental Rights 24
2.6.2 Developments at the UN: The Creation of Special Mandates 26
2.7 Other Global Efforts to Recognize Environmental Rights 27
2.7.1 Draft International Covenant on the Human Right to the Environment 27
2.7.2 Global Pact for the Environment 27
2.8 Regional Developments 28
2.9 Developments at the National Level 29
2.10 Conclusion 30

PART II
Charter-Based Mechanisms 39

3 UN Human Rights Council and the Office of the High Commissioner for Human Rights 41
3.1 Human Rights Council 41
3.1.1 Introduction and Mandate 41
3.1.2 Universal Periodic Review and Working Group Recommendations 42
3.1.2.1 Environmental Protection 43
3.1.2.2 Climate Change 45
3.1.2.3 Sustainable Development and SDGs 49
3.1.2.4 Indigenous Rights 49
3.1.2.5 Business Enterprises and Transnational Corporations 51
3.1.3 Resolutions of the HRC 51
3.1.3.1 Environmental Protection/Degradation 51
3.1.3.2 Climate Change 53
3.1.3.3 Sustainable Development and SDGs 54
3.2 Office of the High Commissioner for Human Rights 55
3.2.1 Environmental Protection/Degradation 55
3.2.2 Climate Change 56
3.2.3 Sustainable Development and SDGs 61
3.3 Conclusion 62

4 Special Procedures Mandate Holders 75

4.1 Introduction 75

4.2 Appointment of a Special Rapporteur on Human Rights and the Environment 75

4.3 Special Rapporteur on the Implications for Human Rights of the Environmentally Sound Management and Disposal of Hazardous Substances and Wastes 77

4.4 Independent Expert on Human Rights and the Environment 78

4.5 Special Rapporteur on Human Rights and Environment 79

4.5.1 Framework Principles on Human Rights and Environment 80

4.5.2 David Boyd as Special Rapporteur 81

4.6 Special Rapporteur on Cultural Rights 86

4.7 Special Rapporteur on the Right to Food 89

4.8 Special Rapporteur on Extreme Poverty and Human Rights 95

4.9 The Working Group on Human Rights and Transnational Corporations and Other Business Enterprises 97

4.10 Special Rapporteur on Contemporary Forms of Racism, Racial Discrimination, Xenophobia, and Related Intolerance 98

4.11 Special Rapporteur on Indigenous Peoples 100

4.12 Special Rapporteur on Climate Change and Human Rights 103

4.13 Complaint Submitted to Special Mandate Holders on Rights of Indigenous Peoples in Addressing Climate-Forced Displacement 103

4.14 Conclusion 105

PART III

Treaty-Based Mechanisms 117

5 UN Human Rights Committee 119

5.1 Introduction and Mandate 119

5.2 General Comments 120

5.3 Concluding Observations 123

5.3.1 Environmental Pollution, Environmental Management and Natural Resource Management 123

5.3.2 Climate Change 124

5.3.3 Indigenous Peoples 125

5.4 Individual Communications 127

5.5 Conclusion 137

6 Committee on Economic, Social and Cultural Rights 147

6.1 Introduction 147

6.2 General Comments 148

6.3 *Concluding Observations 154*
6.3.1 Climate Change 154
6.3.2 Environmental Pollution, Environmental Degradation, and Environmental Impact of Mining Operations and Extractive Industries 158
6.3.3 Sustainable Development and Sustainable Development Goals 167
6.3.4 Indigenous Rights 168
6.3.5 Human Rights Defenders 172
6.4 *Resolutions, Statements, and Press Releases 173*
6.5 *Conclusion 174*

7 Committee on the Elimination of Discrimination against Women 191
7.1 *Introduction and Mandate 191*
7.2 *General Recommendations 192*
7.2.1 Sustainable Development and SDGs 193
7.2.2 Environmental Impact Assessments and Free, Prior, and Informed Consent 194
7.2.3 Climate Change 195
7.3 *Concluding Observations 199*
7.3.1 Climate Change 199
7.3.2 Sustainable Development Goals 206
7.3.3 Environmental Pollution 207
7.3.4 Renewable Energy 210
7.3.5 Women Human Rights Defenders 211
7.4 *Conclusion 211*

8 Committee on the Rights of the Child 225
8.1 *Introduction and Mandate 225*
8.2 *General Comments 227*
8.2.1 Proposed General Comment on Environment and Climate Change 229
8.3 *Concluding Observations 230*
8.3.1 Climate Change 230
8.3.2 Environmental Health/Pollution/Degradation 236
8.3.3 Children's Rights and the Business Sector 240
8.3.4 Sustainable Development/SDGs 243
8.3.5 Other Issues Discussed/Recommendations Made 244
8.4 *Individual Complaints 244*
8.4.1 Amici Curiae Brief of Special Rapporteurs 246
8.4.2 CRC Committee Decision 247
8.5 *Conclusion 252*

9 Other Treaty Bodies 268

9.1 The Committee on the Rights of Persons with Disabilities 268

9.1.1 General Comments 268

9.1.2 Concluding Observations 270

9.2 Committee on the Protection of Rights of All Migrant Workers and Members of Their Families 272

9.2.1 General Comments 272

9.2.2 Concluding Observations 273

9.3 Committee on the Elimination of Racial Discrimination 273

9.3.1 General Recommendations 274

9.3.2 Concluding Observations 275

9.4 Conclusion 285

10 UN Human Rights Institutions and the Environment: Reflections, Challenges, Trajectories 297

10.1 Introduction 297

10.2 Principles, Guidelines and Standards 298

10.3 General Comments 299

10.4 Individual Communications 299

10.5 Critiques 300

10.6 Challenges 300

10.7 Fragmentation, Cohesion, or Cross-Fertilization? 301

10.8 A Research Agenda for the Future 303

Index 305

Foreword

This comprehensive volume is the first to undertake a systematic study of the practice of the major human rights treaty and charter bodies as they have dealt with environmental matters. These bodies have opted to take a major role in relation to environmental protection, one that extends well beyond their original mandates. This leads to many questions that are important not only for environmental protection and human rights but also for international organizations and international law generally. Author Sumudu Atapattu expertly addresses them in detail. They include considering how human rights mechanisms address environmental degradation, sustainability, and climate change; also, how often do treaty bodies admit and decide individual communications that touch on environmental issues. Similarly, looking at another function of treaty bodies, how do they evaluate country reports when environmental issues impact human rights. More generally, how effective are their decisions, resolutions, and General Comments? As she notes, many treaty bodies have referred to states' obligations in relation to climate change and SDGs in their Concluding Observations.

To begin with, she reviews how during the last three decades, the topic "human rights and the environment" has received the attention of scholars and activists, with the publication of books, articles, and reports on the issue. In addition, a substantial body of jurisprudence has also emerged, especially at regional and domestic levels with many states adopting constitutional provisions and/or legislation on environmental rights. Some states have even recognized rights of nature or inanimate objects such as rivers and forests. Due to these developments, many believe that a customary international law principle on the right to a healthy environment is now emerging.

However, despite the increasing involvement of UN human rights bodies in addressing issues of environmental degradation and climate change, a systematic review of their work has not been carried out. This book fills that lacuna by surveying the resolutions, General Comments, Concluding Observations, and decisions of treaty bodies on individual communications. It aims to identify relevant emerging principles and looks at how human rights treaty bodies have interpreted environmental principles as well as the extent to which they contribute to the emerging field of environmental rights.

Key areas discussed are environmental degradation, climate change, and sustainable development, including the Sustainable Development Goals. Given the disproportionate impact of polluting activities on marginalized and vulnerable communities, the book also discusses how human rights mechanisms for women, children, indigenous peoples, people with disability, and racial minorities have addressed these issues. The book refers to the scientific reports that human rights bodies use in their reports and decisions.

Each chapter begins with an introduction to the human rights institution under discussion, how it is constituted and its mandate. The author correctly notes her inability to analyze the actual impact on the ground of the measures studied, as a full systematic analysis of their impact demands a much larger undertaking. Moreover, actual impacts are usually a long-term matter to measure and evaluate. The concluding chapter takes up the more wide-ranging issues about the fragmentation of international law as well as cross-fertilization.

The author is to be thoroughly congratulated on successfully undertaking such a huge project and doing such a masterful job in addressing the complexities involved in overlapping, if not merging, the fields of environmental law and human rights law.

Dinah Shelton
Manatt/Ahn Professor of International Law Emeritus
George Washington University School of Law

Acknowledgments

This book is a "pandemic project." It started as a proposal to the Law & Society Association's Annual Conference in 2020 as a paper presentation examining how the UN Committee on Economic, Social and Cultural Rights has addressed climate change. What caught my attention were the extensive comments made in relation to climate change in its concluding observations on Australia in 2017. However, I soon realized that there was too much material for an article. Also, a systematic review of how UN Human Rights institutions have addressed environmental issues had not been done before. This realization led to a book proposal to Routledge! I optimistically thought that the pandemic was a good time to write the book as I will have so much "free time" to devote to it. Not only did I *not* have free time to write it, it was also hard to separate "Law School work" (done virtually) from "research work" (also done virtually) and carve out time for writing. Even several months after we returned to in-person activities, here I am, still writing it! In my defense I must say that there was more material than I expected (which, of course, is true!)

A project of this nature truly takes a village. I could not have completed the task without the help of many individuals. I owe a huge debt of gratitude to my research assistants without whom this project would not have taken off the ground. I want to sincerely thank Charlotte (Charli) Wynes, Samuel (Sam) Bennington, Alexander (Alex) Duym and Mehak Qureshi for their dedication, diligence, and hard work. To Vera Leone Brown, I owe a special debt of gratitude. She undertook the extremely important yet thankless task of preparing the manuscript for publication. Her attention to detail, consistency, and dedication is much appreciated. Without her tireless work, this book would not have seen the light of day.

I would like to thank the University of Wisconsin Law School for its financial support which enabled me to hire outstanding law students as research assistants. I am especially grateful to Dean Dan Tokaji and Associate Dean Susannah Tahk for their support and encouragement. I am thankful for the incredible opportunity to teach novel courses there and for the excellent body of students who constantly challenge us to think creatively. A special word of thanks is due to the amazing staff at the Law Library, especially Kris Turner and Sunil Rao, who are extremely generous with their time when it comes to helping with our research.

I am especially thankful to Professor Dinah Shelton for writing the foreword. Her pioneering work on environmental rights is truly inspiring.

I would also like to thank Annabelle Harris, senior editor – Environment and Sustainability at Routledge for her support over the years and Jyotsna Gurung, editorial assistant, for her patience with my never-ending requests for extensions and the equally never-ending list of questions about the manuscript. A big thank you to the entire production team at Routledge.

Finally, I would like to thank my family for their constant love and support and for saying, "Mom, you can do it"!!

I have tried to include material up to mid-2022. Just as I was completing the manuscript the decision in the Torres Strait Islanders complaint was issued so I was able to include it in Chapter 5. I am happy that this book is being published at a historic moment when the international community has finally recognized a human right to a healthy environment. This recognition could not have come at a more opportune time, as we experience worsening consequences of environmental degradation and climate change. I hope this publication will contribute in a small way to our understanding of how UN Human Rights institutions are developing environmental rights and pushing boundaries in their own way.

Of course, the errors are all mine!

Sumudu Atapattu
Madison, WI, USA
October 2022

Abbreviations

ACPHR	African Commission on Human and Peoples' Rights
AICHR	ASEAN Intergovernmental Commission on Human Rights
ASEAN	Association of Southeast Asian Nations
CAFTA	Central America Free Trade Agreement
CAT	Convention Against Torture and Other Cruel, Inhuman, or Degrading Treatment or Punishment
CBDR	Common But Differentiated Responsibilities
CED	International Convention for the Protection of All Persons from Enforced Disappearance
CEDAW	Convention on the Elimination of All Forms of Discrimination Against Women
CERD	International Convention on the Elimination of All Forms of Racial Discrimination
CMW	International Convention on the Protection of the Rights of All Migrant Workers and Members of their Families
CRC	Convention on the Rights of the Child
CRPD	Convention on the Rights of Persons with Disabilities
CSOs	Civil Society Organizations
CSW	Commission on the Status of Women
DGD	Day of General Discussion
ECE	United Nations Economic Commission for Europe
ECLAC	United Nations Economic Commission for Latin America and the Caribbean
ECOSOC	United Nations Economic and Social Council
EITI	Extractive Industries Transparency Initiative
ESC	Economic, Social and Cultural
FAO	Food and Agriculture Organization of the United Nations
FPIC	Free, Prior and Informed Consent
GC	General Comment
GHGs	Greenhouse Gases
GPBHR	United Nations Guiding Principles on Business and Human Rights
GR	General Recommendation
HEPA	High Efficiency Particulate Air

HRC	Human Rights Council
HCHR	High Commissioner for Human Rights
IACHR	Inter-American Commission of Human Rights
IACtHR	Inter-American Court of Human Rights
ICCPR	International Covenant on Civil and Political Rights
ICESCR	International Covenant on Economic, Social and Cultural Rights
IDPs	Internally Displaced Persons
IE	Independent Expert
ILC	International Law Commission
ILO	International Labour Organization
INDCs	Intended Nationally Determined Contributions
IPBES	Intergovernmental Science-Policy Platform on Biodiversity and Ecosystem Services
IPCC	Intergovernmental Panel on Climate Change
ISDL	International Sustainable Development Law
LGBTQ+	Lesbian, Gay, Bisexual, Transgender, Queer, and Others
MDGs	United Nations Millennium Development Goals
NAPA	National Adaptation Programme of Action
NDCs	Nationally Determined Contributions
NHRIs	National Human Rights Institutions
OAS	Organization of American States
OECD	Organisation for Economic Co-operation and Development
OHCHR	Office of the High Commissioner for Human Rights
OPIC	Optional Protocol to the Convention on the Rights of the Child on a Communications Procedure
PNPCT	National Policy for Sustainable Development of Traditional Peoples and Communities
REDD+	Reducing Emissions from Deforestation and Forest Degradation in Developing Countries
SDGs	Sustainable Development Goals
SPT	Subcommittee on Prevention of Torture and Other Cruel, Inhuman or Degrading Treatment or Punishment
SR	Special Rapporteur
UDHR	Universal Declaration of Human Rights
UNCTs	United Nations Country Teams
UNDRIP	United Nations Declaration on the Rights of Indigenous Peoples
UNEP	United Nations Environment Programme
UNECE	United Nations Economic Commission for Europe
UNESCO	United Nations Educational, Scientific and Cultural Organization
UNFCCC	United Nations Framework Convention on Climate Change
UNGA	United Nations General Assembly
UNICEF	United Nations International Children's Emergency Fund
UNSG	United Nations Secretary General
UPR	Universal Periodic Review
WCED	World Commission on Environment and Development
WHO	World Health Organization
WMO	World Meteorological Organization

Table of Cases

International Cases and Complaints to Human Rights Treaty Bodies

- André Brun v. France, Communication No. 1453/2006, U.N. Doc. CCPR/C/88/D/1453/2006 (2006).
- Ángela Poma Poma v. Peru, Communication No. 1457/2006 U.N. Doc. CCPR/C/95/D/1457/2006 (2009).
- Apirana Mahuika et al. v. New Zealand, Communication No. 547/1993, U.N. Doc. CCPR/C/70/D/547/1993 (2000).
- Case Concerning Gabcikovo-Nagymaros Project (Hungary v. Slovakia), Judgment, 1997 I.C.J. Rep. 7 (September 25).
- E.H.P. v. Canada, Communication No. 67/1980, U.N. Doc. CCPR/C/OP/1 (1984).
- E.P. et al. v. Colombia, Communication No. 318/1988, U.N. Doc. CCPR/C/39/D/318/1988 (1990).
- Jouni E. Länsman et al. v. Finland, Communication No. 671/1995, U.N. Doc. CCPR/C/58/D/671/1995 (1996).
- Länsman et al. v. Finland, Communication No. 511/1992, U.N. Doc. CCPR/C/52/D/511/1992 (1994).
- Lubicon Lake Band v. Canada, Communication No 167/1984, U.N. Doc. Supp. No. 40 (A/45/40) (1990).
- Petition of Torres Strait Islanders to the United Nations Human Rights Committee Alleging Violations Stemming from Australia's Inaction on Climate Change
- Rights of Indigenous People in Addressing Climate-Forced Displacement, USA 16/2020, complaint (January 15, 2020).
- Sacchi, et al. v. Argentina, et al., Communication No. 104/2019 (Argentina), Communication No. 105/2019 (Brazil), Communication No. 106/2019 (France), Communication No. 107/2019 (Germany), Communication No. 108/2019 (Turkey), U.N. Doc. CRC/C/88/D/104/2019 (2021).
- Susila Malani Dahanayake and 41 other Sri Lankan citizens v. Sri Lanka, Communication No. 1331/2004, U.N. Doc. CCPR/C/87/D/1331/2004 (2006).
- Ioane Teitiota v. New Zealand, Communication No. 2728/2016, CCPR/C/127/D/2728/2016, UN Human Rights Committee (7 January 2020).

- Vaihere Bordes and John Temeharo v. France, Communication 645/1995, U.N. Doc. CCPR/C/57/D/645/1995 (1996).

Regional Cases

- Catan and others v. the Republic of Moldova and Russia, European Court of Human Rights, Application Nos. 43370/04, 8252/05 and 18454/06.
- Kichwa v. Ecuador, Merits, Reparations and Costs, Judgment, Inter-Am. Ct. H.R. (ser. C) No. 245 (June 27, 2012).
- Mossville Environmental Action Now v. United States, Inter-Am. Comm'n H.R., Report No. 43/10 (2010).
- The Environment and Human Rights, Inter-Am. Ct. of H. R. Advisory Opinion, OC-23/17 (15 Nov 2017).
- The Social and Economic Rights Action Centre and the Centre for Economic and Social Rights v. Nigeria, Communication 155/96, UN. Doc. ACHPR/COMM/A044/1 (22 May 2002).
- Watt-Cloutier, et al. v. United States, Inter-Am. Comm'n H.R. (Inuit Petition), Petition No. P-1413–05 (2005).

Domestic Cases

- Oposa v. Factoran, G.R. No. 101083, 224 S.C.R.A. 792 (1993) (Phil.).

Table of Treaties and Other Instruments

1945

United Nations Charter, signed June 26 1945, entered into force October 24 1945, https://www.un.org/en/about-us/un-charter

1948

Universal Declaration of Human Rights, United Nations General Assembly Resolution 217 A, signed December 10 1948, https://www.un.org/en/about-us/universal-declaration-of-human-rights

1950

European Convention on Human Rights, signed November 4 1950, entered into force September 3 1953, https://echr.coe.int/Pages/home.aspx?p=basictexts&c=

1951

Convention Relating to the Status of Refugees, signed July 28 1951, entered into force April 22 1954, https://www.unhcr.org/en-us/1951-refugee-convention.html

1965

International Convention on the Elimination of All Forms of Racial Discrimination, adopted December 21 1965, entered into force January 4 1969, https://www.ohchr.org/en/instruments-mechanisms/instruments/international-convention-elimination-all-forms-racial

1966

International Covenant on Civil and Political Rights, United Nations General Assembly resolution 2200A (XXI), adopted December 16 1966, entered into force March 23 1976, https://www.ohchr.org/en/instruments-mechanisms/instruments/international-covenant-civil-and-political-rights

International Covenant on Economic, Social, and Cultural Rights, United Nations General Assembly resolution 2200A (XXI), adopted December 16 1966, entered into force January 3 1976, https://www.ohchr.org/en/instruments-mechanisms/instruments/international-covenant-economic-social-and-cultural-rights

1979

Convention on the Elimination of All Forms of Discrimination Against Women, signed December 18 1979, entered into force September 3 1981, https://www.ohchr.org/en/instruments-mechanisms/instruments/convention-elimination-all-forms-discrimination-against-women

1981

African Charter on Human and Peoples' Rights, adopted June 27 1981, entered into force October 21 1986, https://www.achpr.org/legalinstruments/detail?id=49

1988

Additional Protocol to the American Convention on Human Rights in the Area of Economic, Social, and Cultural Rights "Protocol of San Salvador," OAS Treaty Series No. 69, adopted November 17 1988, https://www.oas.org/juridico/english/treaties/a-52.html

1989

Indigenous and Tribal Peoples Convention, No. 169, adopted June 27 1989, entered into force September 5 1991

Convention on the Rights of the Child, signed November 20 1989, entered into force September 2 1990, https://www.ohchr.org/en/instruments-mechanisms/instruments/convention-rights-child

1990

International Convention on the Protection of the Rights of All Migrant Workers and Members of Their Families, adopted December 18 1990, entered into force July 1 2003, https://www.ohchr.org/en/instruments-mechanisms/instruments/international-convention-protection-rights-all-migrant-workers

1992

Rio Declaration on Environment and Development, A/CONF.151/26 (Vol. I), June 3–14 1992, https://www.un.org/en/conferences/environment/rio1992

Convention on Biological Diversity, signed June 5 1992, entered into force December 29 1993, https://www.cbd.int/convention

UN Framework Convention on Climate Change, signed June 4–14 1992, entered into force March 21 1994, https://unfccc.int

1998

UN Guiding Principles on Internal Displacement, adopted February 11 1998, https://www.ohchr.org/en/special-procedures/sr-internally-displaced-persons/international-standards

Asian Human Rights Charter, signed March 17 1998, http://www.refworld.org/pdfid/452678304.pdf

Convention on Access to Information, Public Participation in Decision-Making and Access to Justice in Environmental Matters (Aarhus Convention), adopted June 25 1998, https://unece.org/DAM/env/pp/documents/cep43e.pdf

2001

UNESCO Universal Declaration on Cultural Diversity, adopted November 2 2001, https://www.ohchr.org/en/instruments-mechanisms/instruments/universal-declaration-cultural-diversity

2003

Extractive Industries Transparency Initiative, June 2003, https://eiti.org

2004

Arab Charter on Human Rights, 22 May 2004, reprinted in 12 Int'l Hum. Rts. Rep. 893 (2005), entered into force 15 March 2008

2006

Convention on the Rights of Persons with Disabilities, adopted December 13 2006, entered into force May 3 2008, https://www.un.org/development/desa/disabilities/convention-on-the-rights-of-persons-with-disabilities.html

2007

United Nations Declaration on the Rights of Indigenous Peoples, adopted September 13 2007, https://www.ohchr.org/en/indigenous-peoples/un-declaration-rights-indigenous-peoples

2011

UN Guiding Principles on Business and Human Rights, June 2011, https://www.ohchr.org/documents/publications/guidingprinciplesbusinesshr_en.pdf

2012

ASEAN Human Rights Declaration, signed November 19 2012, https://asean.org/asean-human-rights-declaration

2015

Transforming our World: the 2030 Agenda for Sustainable Development, adopted September 25 2015, https://sdgs.un.org/2030agenda

Paris Agreement, signed December 12 2015, entered into force November 4 2016, https://unfccc.int/process-and-meetings/the-paris-agreement/the-paris-agreement

2017

Draft Global Pact for the Environment, presented June 24 2017, https://globalpactenvironment.org/en

Draft International Covenant on the Human Right to the Environment, drafted by The Centre International de Droit Compare de l'Environnement

(International Centre of Comparative Environmental Law), https://cidce.org/wp-content/uploads/2016/08/Draft-of-the-International-Covenant-on-the-Human-Right-to-the-Environment_15.II_.2017_EN.pdf

2018

Framework Principles on Human Rights and the Environment (not formally adopted), January 24 2018, https://www.ohchr.org/sites/default/files/Documents/Issues/Environment/SREnvironment/FrameworkPrinciplesUserFriendlyVersion.pdf

Regional Agreement on Access to Information, Public Participation and Justice in Environmental Matters in Latin America and the Caribbean (Escazú Convention), adopted March 4 2018, entered into force April 22 2021, https://www.cepal.org/en/escazuagreement

Global Compact for Safe, Orderly and Regular Migration, signed December 19 2018, https://refugeesmigrants.un.org/migration-compact

Part I

Introduction & Overview

1 UN Human Rights Institutions and the Environment

Framing the Issues

1.1 Introduction

The atrocities committed during World War II were the impetus for the adoption of the UN Charter with its strong emphasis on human rights.[1] Its preamble specifically reaffirms "faith in fundamental human rights, in the dignity and worth of the human person, in the equal rights of men and women and of nations large and small...."[2] The international community also sought to promote social progress and a better standard of living in larger freedom,[3]

> To achieve international co-operation in solving international problems of an economic, social, cultural, or humanitarian character, and in promoting and encouraging respect for human rights and for fundamental freedoms for all without distinction as to race, sex, language, or religion.[4]

Chapter XI of the Charter is devoted to international economic and social co-operation. Article 55 emphasizes the need to promote, *inter alia*, higher standards of living, full employment, and economic and social progress and development, and universal respect for human rights and fundamental freedoms. All members have pledged to achieve the purposes in Article 55.[5] The Economic and Social Council (ECOSOC), established under Chapter X, was tasked with, *inter alia*, making recommendations to promote human rights and fundamental freedoms, and preparing draft conventions.[6]

While the promotion of human rights, raising the living standards of people, and promoting social progress were central to the purposes of the UN Charter, environmental protection was not. It was not until the late 1960s that the negative consequences of economic development in the form of environmental externalities were beginning to surface which led to the convening of the Stockholm Conference on the Human Environment in 1972.[7] Neither the Universal Declaration of Human Rights (UDHR) nor the two human rights covenants that followed include a reference to environmental protection.

Human rights law has flourished since the adoption of the UN Charter with an enviable body of courts, institutions, and treaties. Despite these developments, and the clear link between the enjoyment of human rights and environmental

DOI: 10.4324/9781003128847-2

degradation, international human rights law took almost 50 years since the Stockholm Conference to adopt a right to a healthy environment as a distinct right.[8] This is certainly not due to the lack of action on this right. Scholars have been writing on this link for decades[9] while over 80% of states have included some form of environmental rights in their constitutions.[10] Likewise, national judiciaries have interpreted existing rights expansively to include environmental rights[11] and regional human rights treaties and institutions have specifically incorporated or pronounced on this right.[12] The UN itself recognized the link between human rights and environmental protection when it appointed a special rapporteur to study the link in 1990[13] and, more recently, created another special mandate to study the link.[14] This mandate has been alive for the past ten years and both the past mandate holder (John Knox) and the current holder (David Boyd) have called upon the UN General Assembly to recognize a right to a healthy environment.[15] The Framework Principles on Human Rights and the Environment, appended to the final report of the former Special Rapporteur contain a set of human rights obligations relating to the environment, including the obligation of states to ensure a safe, clean, healthy, and sustainable environment.[16] Finally, on the eve of the World Environment Day in 2021, a group of over 50 UN experts issued a joint statement calling on the UN to recognize a right to a healthy environment:

> The world is currently facing a climate emergency, pervasive toxic pollution, dramatic loss of biodiversity, and a surge in emerging infectious diseases of zoonotic origin, such as COVID-19. The environmental crisis has negative impacts on a wide range of human rights including the rights to life, health, water, sanitation, food, decent work, development, education, peaceful assembly and cultural rights, as well as the right to live in a healthy environment.
>
> The adverse effects have a disproportionate impact on women and girls and the rights of billions of people, especially those who are already vulnerable to environmental harm including people living in poverty, minorities, older persons, LGBT persons, racially and ethnically marginalized groups, indigenous peoples, people of African descent, persons with disabilities, migrants, internally displaced persons, and children.[17]

Although environmental protection is not mentioned in any of the human rights treaties, the institutions established under them are increasingly confronted with human rights violations caused by environmental degradation. Moreover, human rights provide the overarching framework for Agenda 2030 and Sustainable Development Goals (SDGs) adopted by the international community in 2015.[18] These human rights bodies have undertaken an oversight role in relation to environmental protection that was not part of their original mandate. How do these human rights mechanisms address environmental degradation, sustainability and climate change? How do they decide individual communications that impinge upon environmental issues? How do they evaluate country reports when environmental issues impact human rights? How effective are their decisions, resolutions, and General Comments? Many treaty bodies have referred to states' obligations in relation to climate change and SDGs in their Concluding Observations. For example, in its Concluding Observations on the country report of Australia in 2017, the UN Committee on Economic, Social and Cultural Rights expressed concern about the increase of carbon dioxide emissions in the state party despite its commitments under the United Nations Framework Convention on Climate Change (UNFCCC) and the Kyoto Protocol, and its nationally determined contribution (NDC) under the Paris Agreement. The Committee also noted that climate change is disproportionately affecting Covenant rights of indigenous peoples.[19]

The Committee recommended, *inter alia*, that Australia: revise its climate change and energy policies; take immediate measures to reverse the current trend of increasing greenhouse gas emissions and pursue alternative and renewable energy production; review its position in support of coal mines and coal exports; and address the impact of climate change on indigenous peoples, with their participation. Except for the last recommendation, none of the other recommendations directly impact human rights. Furthermore, the Committee called upon Australia to take steps to ensure that when implementing SDGs its obligations under the Covenant are fully taken into account.[20]

The growing trend of human rights bodies addressing environmental protection and climate change is a fairly recent phenomenon. The first comprehensive report on the link between climate change and human rights was compiled by the Office of the High Commissioner for Human Rights (OHCHR) in 2009[21] pursuant to a request made by the UN Human Rights Council (HRC).[22] Since then, the HRC has adopted several resolutions on climate change and human rights.[23]

Of course, the work of the Independent Expert (later Special Rapporteur) on Human Rights and the Environment appointed in 2012 and that of his successor is directly relevant here.[24] Their work elevated the status of environmental rights and culminated in the adoption of a distinct right to a healthy environment by the HRC in October 2021 followed by the UN General Assembly in July 2022.[25]

In the last three decades, the topic "human rights and the environment" has received considerable attention with books, articles, and reports published on the topic, including textbooks. A substantial body of jurisprudence has also emerged, especially at regional and domestic levels and many states have adopted constitutional provisions on environmental rights. Some states have gone to the extent of recognizing rights of nature and some national courts have accorded rights to inanimate objects such as rivers and forests. Due to these developments, many believe that a customary international law principle on the right to a healthy environment is now emerging.

Despite the increasing involvement of UN human rights bodies in environmental degradation and climate change, a systematic review of their work has not been carried out. This book seeks to fill that void. It will survey the resolutions, General Comments, Concluding Observations, and decisions on individual communications (where applicable). It will identify any relevant principles that have emerged, how human rights treaty bodies are interpreting environmental principles and how they contribute to the emerging field of environmental rights. Key areas that will be discussed are environmental degradation, climate change, and sustainable development including SDGs. Given the disproportionate impact of polluting activities on marginalized and vulnerable communities, the book will discuss how the human rights mechanisms on women, children, indigenous peoples, people with disability and racial minorities have addressed these issues. Moreover, as environmental issues are based on science, the book will refer to the scientific reports that these human rights bodies use in their reports and decisions. Each chapter will start with an introduction to the human rights institution under discussion, how it is constituted and its mandate.

One of the drawbacks of the book is the inability to analyze the actual impact of these works on the ground as a full, systematic analysis of their impact will be a much bigger undertaking. It often takes time to see the actual impact on the ground. In addition, given that these issues are intrinsically intertwined with politics, some of the critiques of non-compliance go to the very heart of the international legal system itself. The concluding chapter will discuss fragmentation of international law as well as cross-fertilization.

However, one caveat should be noted at the outset: it is not the intention of the book to engage in a discussion of scholarly writings or do a literature review of the human rights institutions surveyed in the book. Where possible such writings are cited, but the intention of the book is to examine *the practice* of these institutions in relation to environmental issues and whether there are any principles, standards, and consistencies or even contradictions or inconsistencies across the different institutions surveyed in the book.

1.2 Environmental Issues Covered

1.2.1 Environmental Pollution, Environmental Degradation and Environmental Protection

The book adopts a wide definition of environmental pollution, environmental degradation, and environmental protection to encompass all aspects of environmental deterioration, including the pollution of air, water, soil, and land as well as the impact of mining, deforestation, depletion of natural resources, availability of clean water, sanitation, and conservation efforts in general.

1.2.2 Climate Change

As one of the biggest challenges facing the international community (if not *the biggest challenge*), the book considers climate change as the phenomenon that has resulted from the human interference with the climate system, mainly due to the accumulation of greenhouse gases (GHGs) in the atmosphere emitted by countries since the industrial revolution. The UNFCCC defines climate change as "a change of climate which is *attributed directly or indirectly to human activity* that alters the composition of the global atmosphere and which is in addition to natural climate variability observed over comparable time periods."[26] The objective of the UNFCCC is the "stabilization of greenhouse gas concentrations in the atmosphere at a level that would prevent dangerous anthropogenic interference with the climate system."[27] While the UNFCCC does not define what "dangerous anthropogenic interference with the climate system" is, the Copenhagen Declaration set this level as not exceeding the temperature increase by 2 degrees Celsius. At the insistence of small island states, the Paris Agreement lowered this to 1.5 degrees Celsius[28] as a 2-degree increase would mean the extinction of most small island states. It is important to remember that the world's temperature has *already* increased by 1.1 degrees Celsius since the industrial revolution[29] and, therefore, we do not have much wiggle room here.

The consequences of climate change are far reaching and range from increased severe weather events to water and food scarcity, mass displacement of people, melting of glaciers and sea level rise. The world is already witnessing the consequences of scorching temperatures, droughts and water scarcity.[30] Two of the most serious consequences of climate change are forced migration[31] and inundation of small island states with serious ramifications for human rights.[32] Reference will also be made to disasters whether in the context of climate change or generally.

1.2.3 *Sustainable Development and Sustainable Development Goals*

Sustainable development has been on the global agenda since it was first popularized by the World Commission on Environment and Development (WCED) in its seminal report in 1987. Called *Our Common Future*, the report called the international community to action to change its destructive development practices and to adopt a more sustainable development pathway. It defined sustainable development as "development that meets the needs of the present without compromising the ability of future generations to meet their needs."[33] Although roundly criticized for being vague, sustainable development has had a huge impact on both environmental protection and human rights to the extent of potentially giving birth to a new area of law called International Sustainable Development Law (ISDL).[34]

Despite its critiques, sustainable development has developed into an umbrella term encompassing both substantive and procedural components. Some consider it as "a contestable concept" which has meaning at two levels – at the first level, its meaning is amorphous, similar to concepts like democracy and justice. At the second level, it depends on other principles to be meaningful. Moreover, sustainable development consists of three dimensions – society, ecology, and the economy – and requires their balancing in a given situation. However, critics argue that the environment is not a component or pillar of sustainable development but is *the floor* upon which both society and economy rest. Without a stable environment, neither society nor the economy can flourish.[35] Sustainable development received a new lease of life with the adoption of Agenda 2030 and the SDGs in 2015. Agenda 2030 makes it clear that human rights underlie the entire agenda and the SDGs.[36]

> We reaffirm the importance of the Universal Declaration of Human Rights, as well as other international instruments relating to human rights and

international law. We emphasize the responsibilities of all States, in conformity with the Charter of the United Nations, to respect, protect and promote human rights and fundamental freedoms for all, without distinction of any kind as to race, colour, sex, language, religion, political or other opinion, national or social origin, property, birth, disability or other status.[37]

With 17 goals and 169 targets, the international community adopted an ambitious agenda which has seen a setback due to the COVID-19 pandemic. However, Agenda 2030 clearly linked human rights, environmental protection, and sustainable development and is the first time that all three dimensions of sustainable development were included in one global agenda.

1.2.4 *Other Issues Covered*

Other issues covered include indigenous rights, including the Free, Prior and Informed Consent (FPIC) principle, ancestral lands and land rights, renewable energy, references to environmental rights, environmental health, food security, healthy environment, environmental impact assessments, and (environmental) disasters.[38]

1.3 Methodology

Each of the chapters will survey and analyze the following foundational documents to ascertain how the three categories identified above are addressed: (a) General Comments; (b) Concluding Observations; (c) individual complaints (if any) and their decisions; and (d) resolutions and press releases. In addition, each chapter includes a summary of the mandate of the human rights body discussed in that chapter.

These documents were surveyed utilizing the search function for the four key words/phrases identified above. Other words/phrases searched include: indigenous rights including FPIC, right to water and sanitation, right to food, extractive industries, natural resources, and transnational corporations. OHCHR website was the main source for the documents surveyed. The mapping reports prepared for the former Special Rapporteur on Human Rights and the Environment were used as background reports.

1.4 Outline of the Book and Overview of Chapters

The typology of rights – respect, protect, and fulfill – is used as the framing for the book. The proposed chapters and an overview of each are given below. Of course, it is impossible to cover the work of each and every UN human rights mechanism in a volume of this nature so only the most relevant bodies were selected for this analysis. Each chapter will have an overview of the human rights body discussed, and its structure and mandate under the relevant human rights treaty. The book is divided into three parts – Part I consists of the Introduction and an overview of international human rights law and charter-based and treaty-based mechanisms. Part II discusses the charter-based mechanisms and comprises two chapters: Chapter 3 which discusses the HRC and the OHCHR and Chapter 4 which discusses the special procedure mandate holders, especially the Special Rapporteur on Human Rights and the Environment, and special mandate holders on water, food, poverty, housing, and toxics. Part III discusses the treaty-based mechanisms – the UN Human Rights Committee; Committee on Economic, Social and Cultural Rights; Committee on the Elimination of all Forms of Discrimination Against Women (CEDAW Committee); Committee on the Rights of the Child; and a "catch-all" chapter surveying the Committee on Racial Discrimination, Committee on Migrant Workers, and the Committee on People with Disabilities – and consists of Chapters 5–9. It also includes the concluding chapter, Chapter 10. Below is an overview of each chapter.

Chapter 2 provides an overview of international human rights law and charter-based and treaty-based mechanisms and the emergence of environmental rights. It introduces the reader to various human rights mechanisms at the UN level including individual communications (where relevant), Concluding Observations, General Comments, and resolutions. It also introduces the reader to the special procedures such as independent experts and special rapporteurs. While, technically, the UN charter-based mechanisms include the Security Council and the General Assembly, their work is not examined in this volume as their mandate is not limited to human rights issues.

The HRC is the principal UN body dealing with human rights. It is the successor to the UN Human Rights Commission and was established in 2006. While its mandate does not include environmental issues, its first resolution on climate change was adopted in 2008. Since then, it has adopted a series of resolutions on the topic. In addition, the Universal Periodic Review (UPR) procedure has been used to discuss environmental degradation and climate change and make recommendations. Thus, Chapter 3 will discuss HRC's and OHCHR's involvement in environmental protection, climate change and sustainable development. Apart from referring to the mechanism of appointing special mandate holders by the UN HRC, this chapter will not survey their work. Chapter 3 differs from other chapters because there are no Concluding Observations or General Comments to discuss. Thus, the main focus of the chapter is the UPR process. Since it is impossible to cover all the state reports submitted, we will survey the recommendations of working groups.

Chapter 4 goes on to examine the work of special mandate holders appointed by the HRC. The work of the Independent Expert on Human Rights and the Environment (later special rapporteur) and that of the current special rapporteur is particularly important. The Framework Principles on Human Rights and the Environment submitted by the former Special Rapporteur that consolidated human rights principles applicable to environmental protection merit mention here. In addition, several other mandate holders have discussed environmental protection, climate change, and sustainable development in their reports or devoted reports to these issues and their relationship to their own mandates.

Chapter 5 discusses the work of the UN Human Rights Committee which was established under Article 28 of the International Covenant on Civil and Political Rights (ICCPR). Its main function is to oversee the implementation of the provisions of the Covenant. Toward that end, states are required to submit country reports and the Committee issues Concluding Observations. It also adopts General Comments clarifying provisions of the Covenant. For example, its General Comment No 36 on the Right to Life refers to the need to address environmental degradation in the context of the duty to protect life.

Under the Optional Protocol to the Covenant, individuals have the right to refer communications to the Committee against their state if they are victims of a violation of any of the rights embodied in the Covenant. Several environmental cases have been brought before the Committee by individuals. A 2020 decision by the Committee has implications for refugee status and the principle of *non-refoulement* in the context of climate change. Its latest decision on Torres Strait Islanders case held Australia accountable for damage caused to the Islanders by climate change. When evaluating individual communications, the Committee exercises a quasi-judicial function.

Chapter 6 is devoted to the work of the UN Committee on Economic, Social and Cultural Rights (ESCR Committee) and discusses how it has discussed environmental protection (degradation), climate change, and sustainable development in its work. While the number of individual communications to the Committee is low in number as the Optional Protocol to the Covenant came into effect only in 2013, there are several noteworthy Concluding Observations and General Comments that will be discussed in this chapter.

Chapter 7 surveys the work of the Committee established under the Convention on the Elimination of All Forms of Discrimination against Women (CEDAW). For example, in 2018, the CEDAW Committee adopted General Recommendation 37 titled "Gender-related dimensions of disaster risk reduction in the context of climate change."[39] It recognized the need to take gender differences into consideration when addressing disasters associated with climate change.

Chapter 8 examines the work of the Committee on the Rights of the Child established under the Convention on the Rights of the Child. The petitions brought before the CRC Committee in 2019 by a group of 16 children led by activist Greta Thunberg drew considerable publicity.[40] The chapter examines in detail the decision of the Committee in *Sacchi v. Argentina* (the petitions were taken together as the subject matter was identical). It also discusses the proposed General Comment on climate change and the expert meeting held to discuss it in which the author was invited to participate.

Chapter 9 is a "catch-all" chapter to discuss the work of other treaty bodies and mechanisms that have discussed these three issues within their mandates. It discusses the Committee on Racial Discrimination, Committee on Migrant Workers, and the Committee on People with Disabilities.

The concluding chapter – Chapter 10 – identifies principles, standards, and frameworks that have emerged from the work of the human rights bodies discussed in this book. It discusses to what extent they draw on each other's jurisprudence and pronouncements and establish a more coherent body of law relating to the convergence of human rights and the environment. It highlights a few significant general comments, individual communications as well as critiques and challenges. It further offers some thoughts on fragmentation and cross-fertilization of international law and potential areas for future research.

Notes

1 For an historical overview, see Shelton, D. (2014) *Advanced Introduction to International Human Rights Law*, Cheltenham: Edward Elgar Publishing Ltd., Chapter 2.
2 UN Charter, preamble, ¶ 1, available at: https://treaties.un.org/doc/publication/ctc/uncharter.pdf.
3 Ibid.
4 Ibid., art. 1, ¶ 3.
5 Ibid., art. 56.
6 Ibid., art. 62.
7 Boudes, P. (2014) *United Nations Conference on the Human Environment*. Encyclopedia Britannica, available at: https://www.britannica.com/topic/United-Nations-Conference-on-the-Human-Environment; and see Sands, P. et al. (2018) *Principles of International Environmental Law* (4th edn.), Cambridge: Cambridge University Press, p. 29, doi: 10.1017/9781108355728.
8 Human Rights Council adopted resolution 48/13, A/HRC/RES/48/13 (8 October 2021), recognizing the human right to a clean, healthy and sustainable environment; followed by UN General Assembly which adopted an almost identical resolution on 26 July 2022 recognizing a human right to a clean, healthy and sustainable environment: *The human right to a clean, healthy and sustainable environment*, A/76/L.75.
9 For one of the early works, see Boyle, A. & Anderson, M., eds. (1996) *Human Rights Approaches to Environmental Protection*, Oxford: Clarendon Press. There are even books on the topic now: Anton, D. & Shelton, D. (2011) *Environmental Protection and Human Rights*, Cambridge: Cambridge University Press, doi: 10.1017/CBO9780511974571; Kravchenko, S. & Bonine, J. (2008) *Human Rights and the Environment: Cases, Law & Policy*, Durham, NC: Carolina Academic Press; Atapattu, S. (2016) *Human Rights Approaches to Climate Change, Challenges and Opportunities*, London: Routledge, doi: 10.4324/9781315849683; Boer, B., ed. (2015) *Environmental Law Dimensions of Human Rights*, Oxford: Oxford University Press, doi: 10.1093/acprof:oso/9780198736141.001.0001.
10 Boyd, D. (2012) *The Environmental Rights Revolution: A Global Study of Constitutions, Human Rights, and the Environment*, Vancouver: UBC Press, doi: 10.1353/hrq.2013.0061.
11 See Atapattu, S. & Schapper, A. (2019) *Human Rights and the Environment: Key Issues*, London: Routledge, Chapter 8, for a discussion of cases at the national level.
12 Ibid., Chapter 4. African Charter on Human and Peoples' Rights (1981) was the first regional treaty to incorporate a right to a healthy environment as a collective right. It was followed by the Additional Protocol to the American Convention on Human Rights in the Area of Economic, Social and Cultural Rights (Protocol of San Salvador), OAS Treaty Series No. 69, which incorporated a right to a healthy environment as an individual right but was considered unjusticiable until the Advisory Opinion of the Inter-American Court of Human Rights in 2017. Two regional environmental treaties also incorporate this right: Aarhus Convention (1998), and the Escazú Convention (2018).
13 See Chapter 2 for a discussion of Ksentini's reports.
14 *Human Rights and the Environment*, A/HRC/19/L.8/rev.1 (20 March 2012).

15 When John Knox submitted his final report to the Human Rights Council, he called upon it to recognize a stand-alone right to a healthy environment stating that

> I hope that all of you will support the movement for global recognition of a human right to a healthy environment. As I told the Council, there is nothing so powerful as an idea whose time has come, and this is an idea whose moment is here.

See http://srenvironment.org/2018/07/30/newsletter-no-28. David Boyd, the current Special Rapporteur, spearheaded efforts to get this right recognized by the UNGA. See https://news.un.org/en/story/2022/07/1123142.

16 *Framework Principles on Human Rights and the Environment*, art. 1 (2018), available at: https://www.ohchr.org/sites/default/files/Documents/Issues/Environment/SREnvironment/FrameworkPrinciplesUserFriendlyVersion.pdf.

17 *Joint Statement by UN human rights experts for World Environment Day* (4 June 2021), available at: https://www.ohchr.org/EN/NewsEvents/Pages/DisplayNews.aspx?NewsID=27130&LangID=E.

18 Ibid.

19 *Concluding Observations on the Fifth Periodic Report of Australia*, E/C.12/AUS/CO/5 (11 July 2017).

20 Ibid, ¶ 61.

21 *Report of the Office of the United Nations High Commissioner for Human Rights on the Relationship between Climate Change and Human Rights*, A/HRC/10/61 (15 January 2009).

22 *Resolution 7/23*, (28 March 2008).

23 See *OHCHR and Climate Change*, https://www.ohchr.org/EN/Issues/HRAndClimateChange/Pages/HRClimateChangeIndex.aspx (2022); and Chapter 3.

24 See Chapter 4.

25 See Chapter 3.

26 UNFCCC (1992) art. 1, ¶ 2 (emphasis added), available at: https://unfccc.int/resource/docs/convkp/conveng.pdf.

27 Ibid., art. 2.

28 UN Paris Agreement, art. 2, available at: https://unfccc.int/sites/default/files/english_paris_agreement.pdf.

29 See Lindsey R. & Dahlman, L., *Climate Change: Global Temperature* (28 June 2022), available at: https://www.climate.gov/news-features/understanding-climate/climate-change-global-temperature.

30 England experienced its highest ever temperature in July 2022 which saw runways melting, parched gardens and water rationing.

31 This category of people is generally referred to as "climate refugees" even though it is not a legally accepted term. See Behrman S. & Kent, A., eds. (2022) *Climate Refugees: Global, Local and Critical Approaches*, Cambridge: Cambridge University Press, doi: 10.1017/9781108902991.

32 In the Teitiota case, discussed in Chapter 5, the UN Human Rights Committee accepted that Kiribati (and by implication other small island states) could become uninhabitable due to consequences associated with climate change in 10–15 years' time. Disappearance of small island states raises legal as well as humanitarian questions – among the legal questions are those relating to statehood and nationality while the biggest humanitarian question is about the fate of the populations.

33 World Commission on Environment and Development (1987) *Our Common Future*, Oxford: Oxford University Press, p. 43, available at: https://sustainabledevelopment.un.org/content/documents/5987our-common-future.pdf.

34 In fact, Principle 27 of the Rio Declaration on Environment and Development (1992) called upon states to cooperate to further develop *international law in the field of sustainable development*, not international environmental law.

35 Dawe, N. & Ryan, K. (2003) "The Faulty Three-Legged Stool Model of Sustainable Development," *Conservation Biology*, vol. 17, p. 1458, doi: 10.1046/j.1523–1739.2003.02471.x.
36 Transforming our World: Agenda 2030 for Sustainable Development, A/RES/70/1 (21 October 2015), ¶ 10.
37 Ibid., ¶ 19.
38 References to "natural disasters" is also covered as some of these bodies still refer to climate-induced disasters as "natural disasters."
39 *General Recommendation No. 37,* CEDAW/C/GC/37 (13 March 2018).
40 Earthjustice, *16 Young People File UN Human Rights Complaint on Climate Change* (23 September 2019), available at: https://earthjustice.org/news/press/2019/un-committee-on-the-rights-of-the-child-receives-first-ever-human-rights-complaint-on-climate-change.

References

ACPHR (1981) *African Charter on Human and Peoples' Rights*, https://www.achpr.org/legalinstruments/detail?id=49.

Anton, D. & Shelton, D. (2011) *Environmental Protection and Human Rights*, Cambridge: Cambridge University Press, doi: 10.1017/CBO9780511974571.

Atapattu, S. (2016) *Human Rights Approaches to Climate Change, Challenges and Opportunities*, London: Routledge, doi: 10.4324/9781315849683.

Atapattu, S. & Schapper, A. (2019) *Human Rights and the Environment: Key Issues*, London: Routledge.

Behrman S. & Kent, A., eds. (2022) *Climate Refugees: Global, Local and Critical Approaches*, Cambridge: Cambridge University Press, doi: 10.1017/9781108902991.

Boer, B. (ed.) (2015) *Environmental Law Dimensions of Human Rights*, Oxford: Oxford University Press, doi: 10.1093/acprof:oso/9780198736141.001.0001.

Boudes, P. (2014) *United Nations Conference on the Human Environment*. Encyclopedia Britannica, https://www.britannica.com/topic/United-Nations-Conference-on-the-Human-Environment.

Boyd, D. (2012) *The Environmental Rights Revolution: A Global Study of Constitutions, Human Rights, and the Environment*, Vancouver: UBC Press, doi: 10.1353/hrq.2013.0061.

Boyd, D. (2022) *Why the UN General Assembly Must Back the Right to a Healthy Environment*, https://news.un.org/en/story/2022/07/1123142.

Boyle, A. & Anderson, M., eds. (1996) *Human Rights Approaches to Environmental Protection*, Oxford: Clarendon Press.

Concluding Observations on the Fifth Periodic Report of Australia (2017), E/C.12/AUS/CO/5.

Dawe, N. & Ryan, K. (2003) "The Faulty Three-Legged Stool Model of Sustainable Development," *Conservation Biology*, vol. 17, p. 1458, doi: 10.1046/j.1523–1739.2003.02471.x.

ECLAC (2018) Escazú Convention, https://repositorio.cepal.org/bitstream/handle/11362/43583/1/S1800428_en.pdf.

Earthjustice (2019) *16 Young People File UN Human Rights Complaint on Climate Change*, https://earthjustice.org/news/press/2019/un-committee-on-the-rights-of-the-child-receives-first-ever-human-rights-complaint-on-climate-change.

Framework Principles on Human Rights and the Environment (2018) https://www.ohchr.org/sites/default/files/Documents/Issues/Environment/SREnvironment/FrameworkPrinciplesUserFriendlyVersion.pdf.

General recommendation No. 37 (2018), CEDAW/C/GC/37.

Joint Statement by UN human rights experts for World Environment Day (2021) https://www.ohchr.org/EN/NewsEvents/Pages/DisplayNews.aspx?NewsID=27130&LangID=E.

Knox, J. (2018) *Newsletter No. 28*, http://srenvironment.org/2018/07/30/newsletter-no-28.

Kravchenko, S. & Bonine, J. (2008) *Human Rights and the Environment: Cases, Law & Policy*, Durham, NC: Carolina Academic Press.

Lindsey, R. & Dahlman, L. (2022) *Climate Change: Global Temperature*, https://www.climate.gov/news-features/understanding-climate/climate-change-global-temperature.

OAS (1988) Additional Protocol to the American Convention on Human Rights in the Area of Economic, Social and Cultural Rights (Protocol of San Salvador), OAS Treaty Series No. 69, https://www.oas.org/juridico/english/treaties/a-52.html.

OHCHR and Climate Change (2022) https://www.ohchr.org/EN/Issues/HRAndClimateChange/Pages/HRClimateChangeIndex.aspx.

OHCHR, Report of the Office of the United Nations High Commissioner for Human Rights on the Relationship between Climate Change and Human Rights (2009) A/HRC/10/61.

Rio Declaration on Environment and Development (1992) A/CONF.151/26 (Vol. I), https://www.un.org/en/development/desa/population/migration/generalassembly/docs/globalcompact/A_CONF.151_26_Vol.I_Declaration.pdf.

Sands, P. et al. (2018) *Principles of International Environmental Law* (4th edn.), Cambridge: Cambridge University Press, doi: 10.1017/9781108355728.

Shelton, D. (2014) *Advanced Introduction to International Human Rights Law*, Cheltenham: Edward Elgar Publishing Ltd.

UN Charter (1945) https://treaties.un.org/doc/publication/ctc/uncharter.pdf.

UNFCCC (1992) https://unfccc.int/resource/docs/convkp/conveng.pdf.

UNGA, The Human Right to a Clean, Healthy and Sustainable Environment (2022), A/76/L.75.

UNGA, Transforming our World: Agenda 2030 for Sustainable Development (2015), A/RES/70/1.

UN Paris Agreement (2015), https://unfccc.int/sites/default/files/english_paris_agreement.pdf.

UNECE (1998) *Convention on Access to Information, Public Participation in Decision-Making and Access to Justice in Environmental Matters*, https://unece.org/DAM/env/pp/documents/cep43e.pdf.

World Commission on Environment and Development (1987) *Our Common Future*, Oxford: Oxford University Press, https://sustainabledevelopment.un.org/content/documents/5987our-common-future.pdf.

2 An Overview of International Human Rights Law and the Emergence of Environmental Rights

2.1 Introduction

International human rights law has developed quite remarkably since the adoption of the UN Charter and especially the UDHR in 1948.[1] The number of human rights instruments as well as institutions have proliferated internationally and regionally, and a large number of states have ratified these treaties. States' human rights records have come under international scrutiny pursuant to these international and regional efforts and these institutions and courts have generated an impressive body of jurisprudence. This chapter provides an overview of these developments and the emergence of environmental rights.[2]

According to the OHCHR:

> Human rights are rights we have simply because we exist as human beings – they are not granted by any state. These universal rights are **inherent** to us all, regardless of nationality, sex, national or ethnic origin, color, religion, language, or any other status. They range from the most fundamental – the right to life – to those that make life worth living, such as the rights to food, education, work, health, and liberty.[3]

While not engaging in a debate on what human rights are, it is nonetheless important to point out that even the word "rights" is contested.[4] Rights can mean different things to different people and range from legal rights to customary rights. In this volume, we adopt the human rights framework as incorporated in human rights treaties and interpreted by various institutions and courts. As the UDHR emphasized, the notion of dignity underlies all human rights.[5]

2.2 An Overview of International Human Rights Law

Human rights law is one of the few areas of international law that does not govern relations between states. Rather, it regulates the relationship between the State and individuals under its jurisdiction or control. Thus, human rights law operates vertically between states as *duty bearers* and individuals as *rights holders* usually bound by the link of nationality and typically operate territorially.[6] Prior to the

DOI: 10.4324/9781003128847-3

advent of human rights law, the only area of international law that governed individuals, with the exception of humanitarian law, related to the treatment of aliens.[7] How states treated its own nationals was not governed by international law and was considered an internal matter subject to state's sovereignty.[8] However, this situation has changed dramatically since the adoption of the UN Charter and the UDHR.[9]

While historically individual rights may be traced back to the Magna Carta of 1215 and other Bills of Rights have influenced the development of human rights,[10] the UN Charter is considered as the modern foundation of human rights.[11] The Preamble to the UN Charter is enlightening: "We the Peoples of the United Nations, determined… to reaffirm faith in fundamental human rights, in the dignity and worth of the human person, in the equal rights of men and women and of nations large and small…."[12] In addition, Article 1 specifically includes "promoting and encouraging respect for human rights and for fundamental freedoms…"[13] as one of the purposes of the UN. The Commission on Human Rights led by Eleanor Roosevelt was established in 1946. One of the first tasks of the Commission was to draft an international bill of rights. From a small body comprising nine members when it was originally established, it grew to 53 members in 2006 when it was replaced by the Human Rights Council. Its functions remained largely limited to drafting until the adoption of Resolutions 1235 and 1503 in 1970.[14]

In addition, ECOSOC established the Commission on the Status of Women in 1946[15] which was originally envisaged as a sub-commission subordinate to the Commission on Human Rights.[16] Additionally, the Sub-Commission on the Discrimination and Protection of Minorities was established in 1946 which was renamed the Sub-Commission on the Promotion and Protection of Human Rights in 1999. A body of independent experts, the Sub-Commission was charged with undertaking studies and making recommendations to the Commission on important issues relating to human rights. The Commission was replaced by the HRC in 2006.[17]

The first task of the Commission was to draft a Declaration of Rights. Three separate working groups were established: one to work on a non-binding declaration; and the other two to work on a text of a convention (binding for those who ratified it). This allowed the working group drafting the non-binding declaration to reach a consensus within a short period of time and the UDHR was adopted by the UN General Assembly (UNGA) with 48 votes to none with eight abstentions.[18] The date of adoption of the UDHR, December 10th, is now celebrated as the World Human Rights Day.

The UDHR which embodies both civil and political rights, and economic, social, and cultural (ESC) rights, was supposed to have been closely followed by a treaty embodying its provisions to denote that all rights are interrelated and interdependent. However, opposition to including both sets of rights in one document grew in ECOSOC on the basis that civil and political rights require abstention by states and can be implemented immediately whereas ESC rights require resources to be allocated for their realization as well as positive action by states.[19] In the end and after almost two decades, two separate treaties were adopted in 1966: the International Covenant on Civil and Political Rights (ICCPR) and the International Covenant on Economic, Social and Cultural Rights (ICESCR). It would take another ten years for the two treaties to enter into force. Despite this rather slow start, international human rights law has evolved into a sophisticated body of law with a plethora of treaties, soft law instruments, and institutions.

2.2.1 International Bill of Rights

Together, the UDHR, the ICCPR, ICESCR and the first optional protocol to the ICCPR form the International Bill of Rights and embody the core, foundational rights that everybody enjoys. Even though civil and political rights and ESC rights were bifurcated and embodied in two separate treaties, the official UN position is that all rights are universal, indivisible, interdependent, and interrelated.[20] Similarly, reference is made to "generations" of rights: *first generation rights* refer to civil and political rights; *second generation rights* refer to ESC rights and *third generation rights* refer to group rights or collective rights reflecting the different stages of their emergence, rather than their hierarchy or significance.[21]

Human rights are based on "the inherent dignity and of the equal and inalienable rights of all members of the human family" and is the foundation of freedom, justice, and peace in the world.[22] Article 1 of the UDHR further endorses the dignity and equality principles: "All human beings are born free and equal in dignity and rights."[23] Principle of non-discrimination is embodied in Article 2 and Article 7 endorses the principle of equality. While most rights embodied in the UDHR found their way into the two covenants, a few rights did not – the right to a nationality (Article 15) and the right to own property (Article 17) are two examples.

2.2.2 Typology of Rights

Human rights are based on the dignity of the human person and the rights that are recognized accrue to them by virtue of being born human. Some rights, such as the right to vote and the freedom of movement, are confined to citizens while all individuals irrespective of nationality enjoy other rights. However, rights are not confined to human beings alone. Animals and corporations have rights, and a more recent development is where nature, including rivers and forests, are accorded legal rights.[24] Even the notion of "a human being" is now undergoing transformation, in light of technological advancements that have ramifications for human rights.

International human rights law adopts a tripartite typology of obligations with regard to human rights – respect, protect, and fulfill. While this typology is usually applied in relation to ESC rights,[25] it can equally be relevant in relation to civil and political rights. The obligation to *respect* requires States to refrain from interfering directly or indirectly with the enjoyment of the right. The obligation to *protect* requires States to take measures that prevent third parties from interfering with the enjoyment of the right. The obligation to *fulfill* contains obligations to facilitate, provide, and promote and requires states to adopt appropriate legislative, administrative, budgetary, judicial, promotional, and other measures toward the full realization of the right.[26] This typology has been widely adopted in relation to ESC rights[27] and is useful in relation to environmental rights as well.

2.3 An Overview of Human Rights Institutions

International human rights law and institutions have largely developed along two tracks: treaty-based mechanisms and charter-based mechanisms. Treaty-based mechanisms are those mechanisms established under human rights treaties as described under Section 2.3.1 and include several specialized regimes governing various vulnerable groups including women, children, migrant workers, and people with disability. Charter-based mechanisms are those established under the UN Charter itself and include special procedure mandate holders; procedure 1503 under ECOSOC; and the Universal Periodic Review (UPR).[28]

In addition to these international mechanisms, there are three regional human rights systems that will not be discussed in this volume: (a) the European Court of Human Rights; (b) the Inter-American Commission and Court of Human Rights; and (c) the African Commission and Court of Human and Peoples Rights.[29] While these regional systems have been at the forefront of developing environmental rights, their jurisprudence falls outside the scope of this book.

2.3.1 Treaty-Based Mechanisms

Ten treaty-based mechanisms currently exist under the main international human rights treaties.[30] These are the mechanisms established under the following treaties: ICCPR; ICESCR; Convention on the Rights of the Child (CRC); Convention on All forms of Discrimination against Women (CEDAW); International Convention on the Elimination of All Forms of Racial Discrimination (CERD); International Convention on the Protection of the Rights of All Migrant Workers and Members of their Families (CMW); Convention against Torture and other Cruel, Inhuman or Degrading Treatment or Punishment (CAT); Convention on Rights of Persons with Disabilities (CRPD); and the International Convention for the Protection of All Persons from Enforced Disappearance (CED). In addition, a Subcommittee on Prevention of Torture and other Cruel, Inhuman or Degrading Treatment or Punishment (SPT) was established in 2007 as "a new kind of treaty body" which has "a preventive mandate focused on an innovative, sustained and proactive approach to the prevention of torture and ill treatment."[31]

Treaty bodies are committees of independent experts that monitor the implementation of human rights treaties. They adopt several mechanisms to ensure compliance with the obligations covered under their respective treaties: submission of country reports;[32] concluding observations on country reports; general comments elaborating on various provisions of the treaty;[33] and an individual complaints procedure if the treaty provides for that.[34] While the scope of the treaty bodies' supervisory functions is defined in each of the human rights treaties, their main responsibility is to examine reports submitted by state parties on how they have fulfilled their treaty obligations[35] as states have an obligation to ensure that everyone in the State enjoys the rights embodied in the treaty.[36] Some treaty bodies can conduct inquiries in the territory of state parties if they have reason to believe that serious human rights violations are taking place. These treaty bodies thus discharge functions of an *administrative or investigative* nature (with regard to examination of state reports), *legislative* nature (with regard to the adoption of general comments) and *judicial* nature (with regard to the determination of individual complaints).

Treaty bodies elaborate on the provisions of their respective treaty through general comments. They play several roles: (a) further interpret a particular provision of the treaty; (b) establish rules of procedure;[37] (c) elaborate on the meaning of a particular provision and its relationship to other aspects;[38] (d) occasionally, existing rights can be interpreted as embodying new rights;[39] and (e) elaborate on an issue that the Committee deems important and relevant.[40]

There are three mechanisms by which complaints can be brought before these human rights bodies: individual communications; state-to-state complaints; and inquiries.[41] Currently, eight treaties accept individual communications: ICCPR, CERD, CAT, CEDAW, CRPD, CED, CRC, and ICESCR.[42] The individual communications procedure under the CMW has not yet entered into force. Together, these reports, general comments, concluding observations and pronouncements on individual petitions have contributed to the development of international human rights law which increasingly involve environmental issues.

2.3.2 Charter-Based Mechanisms

The UN Charter itself provides the legal basis for these mechanisms. Acting pursuant to Article 65 of the UN Charter, the ECOSOC has established various bodies to address human rights. These include the Commission on the Status of Women (CSW), HRC (which succeeded the Commission on Human Rights), the complaints mechanisms, UPR procedure and special procedures whose mandate is either country-specific or thematic. Currently, there are 44 thematic mandates and 11 country-specific mandates.[43]

These special procedures mandate holders are called special rapporteurs or independent experts or can be a working group.[44] Several special mandate holders have discussed environmental issues, especially climate change, in the context of their mandate. These include the Special Rapporteurs on the Right to Food, Health, Water, Indigenous rights, Poverty, Hazardous Waste, etc. Of these, the Special Rapporteur on Human Rights and the Environment and the Independent Expert on Human Rights and the Environment are relevant to the present discussion and will be discussed in Chapter 4. In addition, the UPR is a unique process that has proven to be an important mechanism to evaluate the human rights situation in all UN member countries. This includes environmental obligations to the extent that they impact human rights. It is:

> [a] State-driven process, under the auspices of the Human Rights Council, which provides the opportunity for each State to declare what actions they have taken to improve the human rights situations in their countries and to fulfill their human rights obligations.[45]

The HRC, UPR process and OHCHR in relation to environmental issues are discussed in Chapter 3.

The CSW, "the principal global intergovernmental body exclusively dedicated to the promotion of gender equality and the empowerment of women,"[46] adopts a multi-year program of work[47] to evaluate the progress made under the Beijing Platform for Action and to contribute to the 2030 Agenda for Sustainable Development to accelerate the realization of gender equality and empowerment of women. For 2022, its priority theme was "achieving gender equality and the empowerment of all women and girls in the context of climate change, environmental and disaster risk reduction policies and programmes."[48] The work of the CSW will not be discussed in detail but significant reports and decisions will be highlighted where relevant.

2.3.3 *Office of the High Commissioner for Human Rights*

The High Commissioner for Human Rights is the principal human rights official of the UN and heads the OHCHR and spearheads UN's work on human rights. OHCHR's thematic priorities include strengthening international human rights mechanisms; integrating human rights in development and in the economic sphere; widening the democratic space; and early warning and protection of human rights in situations of conflict, violence, and insecurity.[49] In addition, it acts as the secretariat for the HRC, and supports UN human rights treaty bodies and special procedures, thus bridging the gap between treaty-based mechanisms and charter-based mechanisms. OHCHR has spearheaded several efforts relating to environmental issues, including climate change, water, health, and food and is discussed in Chapter 3.

2.4 Rights Affected by Environmental Degradation

No global human rights treaty embodies a distinct, stand-alone right to a healthy environment. The CRC is the only global human rights treaty that even refer to pollution.[50] The ICESCR refers to improving all aspects of environmental and industrial hygiene in the context of the right to health.[51] However, many protected rights can be infringed as a result of environmental degradation as a healthy environment is necessary to enjoy many rights.[52] In the context of environmental degradation, the rights that have been invoked most are the rights to: life, an adequate standard of living, including food, water, shelter, sanitation, privacy, and family life, health, and the procedural rights of access to information, participation in decision-making, and access to remedies.[53] In addition, two regional human rights treaties embody a right to a healthy environment.[54] After years of vacillating, the HRC finally recognized a distinct right to a clean, healthy, and sustainable environment in 2021[55] which was endorsed almost verbatim by the UNGA in 2022 without a single negative vote.[56] Although soft law, the endorsement of this right by the UN's plenary body is significant[57] as it will give the right to a healthy environment a better footing in international human rights law, similar to its endorsement of the human right to water and sanitation in 2010.[58] The enjoyment of both civil and political rights and ESC rights can be jeopardized by a degraded environment. In addition to the rights identified above, the right of self-determination can be affected by climate change. We will elaborate on a few rights here.

2.4.1 Right to Life

The right to life is the supreme right from which no derogation is permitted even during times of emergencies that threaten the life of the nation. While it was originally envisaged as a negative right to prevent state's interference with the right and to curb arbitrary deprivation, it is now recognized that states have a duty to take positive measures to protect life which includes the obligation to address instances of environmental degradation. The UN Human Rights Committee recognized the link between the right to life and environmental degradation in its General Comment (GC) No 36 which replaced GC Nos 6 and 14: "Environmental degradation, climate change and unsustainable development constitute some of the most pressing and serious threats to the ability of present and future generations to enjoy the right to life."[59] In addition, decisions on several individual communications recognize this link. The work of the Human Rights Committee is examined in Chapter 5.

2.4.2 Right to Health

The link between the right to health and a polluted environment needs no special elaboration. Many cases have established this link as have the World Health Organization (WHO) and other organizations working on health issues. For example, a group of over 200 medical journals issued a joint statement in 2021 urging world leaders to cut GHG emissions to avoid catastrophic harm to health.[60] They issued a stark warning that deaths associated with COVID-19 would pale in comparison to the projected deaths associated with climate change. The WHO points out that 24% of all global deaths are linked to the environment: "Healthier environments could prevent almost one quarter of the global burden of disease. The COVID-19 pandemic is a further reminder of the delicate relationship between people and our planet."[61] It further provides:

> Clean air, stable climate, adequate water, sanitation and hygiene, safe use of chemicals, protection from radiation, healthy and safe workplaces, sound agricultural practices, health-supportive cities and built environments, and a preserved nature are all prerequisites for good health.[62]

2.4.3 Rights to Food, Water and Sanitation

The availability of food and water is very much dependent on environmental conditions, and unsanitary conditions and lack of sanitation facilities also lead to disease. Likewise, the enjoyment of many rights depends on the availability of nutritious food and uncontaminated water. Some statistics will help put this into perspective: every year, air pollution kills more than 7 million people, while lack of access to fresh water and sanitation has been linked to the death of 5 million people per year.[63] About 500,000 children under the age of five years die annually due to diarrheal diseases associated with lack of clean water and sanitation facilities.[64] Globally, there are nearly 1.7 billion cases of childhood diarrheal disease every year and diarrhea is a leading cause of malnutrition in children under the age of five years.[65] In 2016, the world generated 242 million tons of plastic waste.[66] If present trends continue, there will be more plastic in the oceans than fish by 2050[67] affecting global fisheries and the marine ecosystem and the livelihood of millions of people. This adds to the fact that over 90% of fish stocks have been exhausted.[68] Between 1945 and the present, the worldwide generation of hazardous waste increased from 5 million to 400 million tons *per year*.[69]

In this regard, the Sustainable Development Goals (SDGs) are significant. Goal No 2 seeks to end hunger, achieve food security and improved nutrition, and promote sustainable agriculture while Goal No 6 seeks to ensure the availability and sustainable management of water and sanitation for all. Moreover, Goal No 3 on health and well-being, Goal No 4 on quality education, Goal No 5 on gender equality, and Goal No 13 on climate action are also important as lack of food and water or contaminated food and water can lead to ill-health which, in turn, affects education. In many cultures, women and girls engage in household chores and collecting water. Availability of water and agriculture are being affected by climate change which affect women and girls more as they have to walk longer to collect water, reducing the time they have for education and other recreational activities, and also exposing them to sexual harassment and gender-based violence. All these factors affect the rights to food, water, and sanitation as well as education, health, and (gender) equality.

2.4.4 Right of Self-Determination

Common Article 1 to the ICCPR and ICESCR embodies the right of self-determination by virtue of which all peoples "freely determine their political status and freely pursue their economic, social and cultural development."[70] It further provides that all peoples may freely dispose of their natural wealth and resources and they may not be deprived of their own means of subsistence. The right of self-determination becomes relevant in the context of small island states and their inhabitants who stand to lose everything they have including their state due to adverse consequences associated with climate change. Can they rely on the right of self-determination to preserve the statehood of their current state and to retain the rights they are currently enjoying? It must be pointed out that this issue has not been tested yet. In its report on the link between climate change and human rights, the OHCHR recognized that one possible scenario of forcible displacement across borders is the eventual submergence of small island states but noted that human rights law does not provide clear answers to the status of the populations who have been displaced from these islands.[71] It recognized that sea level rise and extreme weather events could threaten the habitability and existence of many low-lying islands and the traditional territories of indigenous peoples and their livelihood implicating the right of self-determination:

> While there is no clear precedence to follow, it is clear that insofar as climate change poses a threat to the right of peoples to self-determination, States have a duty to take positive action, individually and jointly, to address and avert this threat. Equally, States have an obligation to take action to avert climate change impacts which threaten the cultural and social identity of indigenous peoples.[72]

While this recognition is important, it is not clear what is meant by *taking positive action to address and avert the threat* (i.e. disappearance of states). Does this mean that it would be up to each small island state to take positive action to avert the threat? Given that the emissions of major contributors since the industrial revolution have led to this situation and the emissions of small island states are minimal in comparison, putting the onus on the small island states is like adding insult to injury. What are the obligations of major emitters to take mitigation measures? What are the obligations of major emitters vis-à-vis these vulnerable countries and their people? Can we rely on the common but differentiated responsibility (CBDR) principle which forms part of the legal regime governing climate change to address this situation?[73]

2.5 Specialized Treaties on Vulnerable Groups

In addition to the general human rights treaties, there are treaties that protect groups that are vulnerable or have been historically marginalized and therefore need special protection. These include children, women, minorities, indigenous peoples, migrant workers, and people with disabilities. Each group has a specialized instrument giving them additional rights. These are: the CRC, CEDAW, CERD, UN Declaration on the Rights of Indigenous Peoples (UNDRIP), CMW, and the CPD. The work of these treaty bodies vis-à-vis environmental issues (including climate change and sustainable development) is examined in later chapters.[74]

2.6 Emergence of Environmental Rights and the Right to a Healthy Environment[75]

As in my previous work, I use "environmental rights" broadly to refer to both *substantive rights* that could be affected by environmental degradation, and *procedural rights* that lead to better environmental outcomes, as well as a stand-alone right to a healthy environment. Substantive rights that could be infringed by living in a polluted environment range from the right to life (in extreme situations), to rights to food, water and sanitation, as well as the right to culture and self-determination. The procedural rights include the right to information, right to participate in the decision-making process and the right to remedies. To give effect to these rights, states may have to promulgate legislation, put institutions in place, monitor the activities of private individuals including businesses and punish them in the event of wrongdoing, ensure the preparation of (environmental) impact assessment reports,[76] publicize those reports and facilitate public participation including consultation and free, prior, and informed consent in some instances, and provide an adequate remedy when these obligations are violated.[77]

Judge Christopher Weeramantry noted the link between human rights and the environment in the *Case Concerning the Gabcikovo Nagymaros Project*:

> The protection of the environment is likewise a vital part of contemporary human rights doctrine, for it is a *sine qua non* for numerous human rights such as the right to health and the right to life itself. It is scarcely necessary to elaborate on this, as damage to the environment can impair and undermine all the human rights spoken of in the Universal Declaration and other human rights instruments.[78]

Although the enjoyment of many rights can be jeopardized by environmental pollution and a healthy environment is necessary to enjoy many rights, it took the human rights community almost 75 years since the adoption of the UDHR to recognize a distinct right to a healthy environment even though the first seeds of the link between human rights and environmental degradation were sown by the Stockholm Declaration on the Human Environment itself. This document, considered to be the foundation of international environmental law, provides:

> Man has the fundamental right to freedom, equality and adequate conditions of life, in an environment of a quality that permits a life of dignity and well-being, and he bears a solemn responsibility to protect and improve the environment for present and future generations…[79]

While falling short of endorsing a distinct human right to a healthy environment, Principle 1 clearly recognized that an environment of a certain quality is necessary for human beings to lead a life of dignity and well-being. The Paris Agreement was the first environmental treaty that incorporated the need to respect human rights in relation to an environmental issue. However, until the adoption of the UNGA resolution in 2022, specific recognition of a right to a healthy environment in international human rights law was confined to two regional treaties.

2.6.1 *Articulation of Environmental Rights*

Two developments *at the national level* contributed to the convergence of the two fields at the international level. The first is the environmental impact assessment process which is subject to public scrutiny and public participation in many countries. This process has incorporated the three procedural rights that were referred to earlier.[80] Access to environmental information is now firmly entrenched in international law and is closely connected to participatory rights.[81] Principle 10 of the Rio Declaration on Environment and Development, 1992 that incorporated these three pillars in the context of sustainable development formed the basis for two regional environmental treaties: one adopted under the auspices of the Economic Cooperation for Europe (called the Aarhus Convention)[82] and the other adopted by Latin American and Caribbean states (called the Escazú Agreement).[83]

The second development is the environmental justice movement which originated as a response to the practice of locating polluting and hazardous industries in low income and minority communities in the US.[84] These polluting activities had a huge impact on the lives of people and impinged on their rights. This practice is by no means over, despite the Executive Order issued by President Clinton in 1994.[85] Examples include: Hurricane Katrina, Kivalina and Shishmarof in Alaska, and the water crisis in Flint, Michigan. A petition alleging environmental racism is pending before the Inter-American Commission of Human Rights in the Mossville Case.[86]

Indigenous rights is another area where this convergence can be seen. Indigenous peoples' culture is intimately related to land and environmental degradation has a severe impact not only on their protected rights but also on their culture. They have an intimate knowledge of their environment and this knowledge has been passed from generation to generation. In the petition brought by the Inuit of Canada and the US against the United States, the petitioners claimed, inter alia, that they could no longer rely on their traditional knowledge due to the unpredictability of a changing climate.[87]

The articulation of environmental rights falls into three broad categories: (a) "greening" human rights law by using existing rights – both civil and political rights, and ESC rights – to articulate environmental rights. Under this approach, substantive human rights have been invoked in relation to environmental issues; (b) using procedural rights to lead to better outcomes relating to projects that can have a significant impact on the environment or to provide relief to victims; and (c) recognition of a distinct right to a healthy environment as a stand-alone right.[88]

The relationship between human rights and environment is two-fold – the enjoyment of human rights can be directly affected by environmental degradation. Moreover, human beings depend on nature for the services it provides – purifying air, water, decomposition of waste etc. (referred to as "ecosystem services" even though that term has been critiqued as being too anthropocentric). Similarly, there could be ecological limitations to human rights such as planetary boundaries.

The predominate approach has been greening existing rights, rather than adopting a substantive right to a healthy environment. Some have gone to the extent of arguing against its adoption on the ground that it could dilute existing rights.[89] While a stand-alone right to a healthy environment under international human rights law was recognized only in 2021, it is clear that the international community has come a long way since the 1972 Stockholm Conference going to the extent of even recognizing rights of nature.[90]

2.6.2 Developments at the UN: The Creation of Special Mandates

The World Commission on Environment and Development (WCED) appointed by the UN General Assembly issued its report called "Our Common Future" which had a huge impact on the development on international environmental law and its future trajectory. WCED called upon states to integrate environmental protection into the economic development process and advanced the notion of sustainable development as the new policy framework for states. Defined as "development that meets the needs of the present without compromising the ability of future generations to meet their needs"[91] sustainable development made a direct reference to human beings and the need to meet their basic needs. Since then, sustainable development has evolved to encompass three dimensions with a definite human rights focus: environmental protection, economic development, and social development.[92] The social pillar has received less attention but recognizes that without addressing social inclusion, social justice and equality, sustainable development cannot be realized.

The draft legal principles for environmental protection and sustainable development proposed by the WCED Experts Group on Environmental Law articulated that: "All human beings have the fundamental right to an environment adequate for their health, and well-being."[93] These draft principles, unfortunately, did not proceed much further and continue to gather dust in the UN archives.

While the contribution of the special mandate holders will be discussed more fully in Chapter 4, a summary will be made here in order to trace the evolution of environmental rights. When the UN Sub-Commission on Prevention of Discrimination and Protection of Minorities (as it was then called) appointed a special rapporteur in 1990 to study the relationship between human rights and the environment, the UNGA adopted a resolution stating that "all individuals *are entitled* to live in an environment adequate for their health and well-being."[94] It further noted that a healthier environment contributes to the enjoyment of human rights. While this acknowledged the link between a healthy environment and the enjoyment of rights, it fell short of recognizing a human right to a healthy environment.

The first Special Rapporteur on Human Rights and the Environment, Ms. Fatma Zohra Ksentini appointed in 1990, submitted a set of draft articles with her final report in 1995.[95] These draft articles were never officially adopted and met the same fate as the draft principles proposed by the WCED Experts Group on Environmental Law. The topic remained dormant until the adverse consequences of climate change began to surface and the human rights community lobbied for a special mandate to be created to address the human rights impacts of climate change. In response, the HRC created a mandate for an independent expert on human rights and the environment in 2012, whose mandate was extended and converted to a special rapporteur position in 2015. John Knox, who held the independent expert position and later special rapporteur was succeeded by David Boyd who is the current mandate holder. In 2021, the HRC created a separate mandate for a special rapporteur on climate change and human rights and Ian Fry was appointed to that position.[96]

2.7 Other Global Efforts to Recognize Environmental Rights

2.7.1 *Draft International Covenant on the Human Right to the Environment*

The *Centre International de Droit Compare de l'Environnement* (International Center of Comparative Environmental Law) which has consultative status with ECOSOC published a draft International Covenant on the Human Right to the Environment in 2017, modeled after the two covenants. The Preamble reaffirms that all human rights *including the right to environment,* are universal, indivisible, and interdependent and it is the responsibility of states, individuals, and other public and private entities to protect and promote these rights. Draft Article 1 provides that: "Everyone, including future generations, has the right to live in an ecologically balanced environment capable of assuring his or her health, security, and wellbeing."[97]

Conspicuously absent are obligations relating to investing in developing countries and the protection of human rights of people in the host state. While transboundary environmental damage is addressed in the Covenant, it does not address the complex issue of remedies for human rights violations in neighboring states due to transboundary environmental damage. The draft covenant was submitted to the HRC in February 2017 and the outcome remains to be seen.

2.7.2 *Global Pact for the Environment*

In another development, the French President Emmanuel Macron spearheaded efforts to adopt a Global Pact for the Environment.[98] Drafted by *Le Club des Juristes*, the preliminary draft refers to both SDGs and the urgent need to address climate change. The Preamble acknowledges "the growing threats to the environment and the need to act in an ambitious and concerted manner at the global level" and is remarkably similar to the Preamble to the Paris Agreement. Draft article 1 acknowledges that "every person has the right to live in an ecologically sound environment adequate for their health, well-being, dignity, culture and fulfillment." It proposes the establishment of a Committee of Independent Experts to monitor compliance focusing on facilitation in a transparent, non-adversarial, and non-punitive manner.

A summit on the Pact was held in 2017, and in 2018 the UNGA requested the Secretary-General to submit a technical and evidence-based report that identifies possible gaps in international environmental law and environment-related instruments to strengthen their implementation.[99] UNGA decided to establish an *ad hoc* open-ended working group to consider the report and discuss possible options to address these gaps and make recommendations. While the Global Pact recognizes the intrinsic value of the environment especially preserving biodiversity, it falls short of recognizing rights of nature.

2.8 Regional Developments

At the regional level, the proliferation of environmental rights was more rapid and consistent. Four regional treaties embody the right to a healthy environment, the first to do so being the African Charter on Human and Peoples' Rights which endorsed the right to a healthy environment as a collective right.[100] While the European Convention on Human Rights, the oldest regional human rights treaty, does not embody a distinct right to a healthy environment, it has articulated environmental rights via existing rights such as the right to life, and the right to privacy and family life.[101] The right to a healthy environment in the San Salvador Protocol to the American Convention on Human Rights remained unjusticiable until the groundbreaking advisory opinion of the Inter-American Court of Human Rights in 2017. Not only did the Court recognize the "existence of an undeniable relationship between the protection of the environment and the realization of other human rights,"[102] it also endorsed that ESC rights (which the right to a healthy environment forms part of) are enforceable in view of the interdependence and indivisibility of rights:

> The human right to a healthy environment has been understood as a right that has both individual and also collective connotations. In its collective dimension, the right to a healthy environment constitutes a universal value that is owed to both present and future generations. That said, the right to a healthy environment also has an individual dimension insofar as its violation may have a direct and an indirect impact on the individual owing to its connectivity to other rights, such as the rights to health, personal integrity, and life. Environmental degradation may cause irreparable harm to human beings; thus, a healthy environment is a fundamental right for the existence of humankind.[103]

Even more significantly, the Court articulated rights of nature – a first (and unprecedented) for a human rights court:

> The Court considers it important to stress that, as an autonomous right, the right to a healthy environment, unlike other rights, protects the components of the environment, such as forests, rivers and seas, as legal interests in themselves, even in the absence of the certainty or evidence of a risk to individuals. This means that it protects nature and the environment.... because of their importance to other living organisms with which we share the planet that also merit protection in their own right.[104]

Noting a tendency in judgments and Constitutions to recognize legal personality and rights of nature, the Court stated that "the right to a healthy environment as an autonomous right differs from the environmental content that arises from the protection of other rights, such as the right to life or the right to personal integrity."[105]

While there have been other regional efforts, nothing akin to these three regional systems[106] exists in other parts of the world. For example, Asian countries adopted the Asian Human Rights Charter in 1998 and recommended the adoption of a regional convention and the establishment of a commission or court which can receive complaints but this has yet to materialize.[107] The ASEAN Human Rights Declaration[108] embodies ESC rights including the right to a safe, clean, and sustainable environment.[109] ASEAN Intergovernmental Commission on Human Rights (AICHR) is entrusted with the task of promoting and protecting human rights in the ASEAN region but lacks a mechanism to receive individual complaints. The Arab Charter on Human Rights, adopted in 2004 by the League of Arab Nations[110] also recognizes a right to a healthy environment as a component of the right to an adequate standard of living.[111] Under Article 45, a Human Rights Committee is established and while state parties are required to submit reports to the Committee,[112] it does not recognize an individual right of petition.

There have been initiatives to establish other regional mechanisms,[113] but a large portion of the globe currently remains outside regional human rights mechanisms. As with all international tribunals where individual right of petition is recognized, local remedies must be exhausted before they can resort to these regional bodies.

2.9 Developments at the National Level

The convergence between human rights and the environment is seen most prominently at the national level. This happened in two distinct, yet interrelated ways: many national constitutions started embodying environmental rights, including a right to a healthy environment either as justiciable rights or as directive principles;[114] and judiciaries around the world started articulating environmental rights by interpreting existing rights expansively and creatively and by expanding standing (*locus standi*), paving the way for public interest litigation.[115] While the Indian Supreme Court has been at the forefront of articulating environmental rights, the seminal case of *Minors Oposa v. Secretary of the Department of Environment and Natural Resources* from the Philippines is possibly the best known case. It involved a challenge by a group of minors of timber license agreements on the ground that deforestation is causing environmental damage to themselves as well as future generations. The Supreme Court of the Philippines noted:

> While the right to a balanced and healthful ecology is to be found under the Declaration of Principles and State Policies and not under the Bill of Rights, it does not follow that it is less important than any of the civil and political

rights enumerated in the latter. Such a right belongs to a different category of rights altogether for it concerns nothing less than self-preservation and self-perpetuation – aptly and fittingly stressed by the petitioners – the advancement of which may even be said to predate all governments and constitutions. *As a matter of fact, these basic rights need not even be written in the Constitution for they are assumed to exist from the inception of humankind.*[116]

2.10 Conclusion

The topic environmental rights has come a long way since it was first articulated by scholars in the early 1990s, finally culminating in the recognition of a right of every person to a clean, healthy, and sustainable environment. In some respects, the topic has come a full circle – from a special rapporteur on the topic who submitted a set of draft articles which went nowhere to another special rapporteur two decades later who has submitted another set of principles on the topic to the HRC. We urge the UNGA to formally adopt the Framework Principles on Human Rights and the Environment submitted by John Knox in 2018.[117] Now that the UNGA has recognized a right to a healthy environment, there is no excuse for further delay.

Notes

1 See Buergenthal, T. (2006) "The Evolving International Human Rights System," *American Journal of International Law*, vol. 100, no. 4, p. 783.
2 This chapter draws from author's previous work: Atapattu, S. & Schapper, A. (2019) *Human Rights and the Environment: Key Issues*, Abingdon: Routledge, Chapter 2. Updated to include developments since the publication of the book. Reproduced with publisher's permission.
3 OHCHR website: https://www.ohchr.org/en/what-are-human-rights (emphasis in original).
4 See Anton, D. & Shelton, D. (2011) *Environmental Protection and Human Rights*, Cambridge: Cambridge University Press, pp. 120–121, doi: 10.1017/CBO9780511974571.
5 See Daly, E. & May, J. (2020) *Dignity Law: Global Recognition, Cases and Perspectives*, Getzville: William S. Hein & Co. For a discussion of dignity in the context of sustainable development see Daly, E. & May, J. (2021) "The Indivisibility of Human Dignity and Sustainability," in *The Cambridge Handbook of Environmental Justice and Sustainable Development*, Atapattu, S., Gonzalez, C., & Seck, S., eds., Cambridge: Cambridge University Press, p. 23.
6 Whether human rights law is applicable extraterritorially is a hotly debated issue, especially in relation to transboundary environmental damage. See Skogly, S. (2010) "Extraterritoriality: Universal Human Rights without Universal Obligations?" in *Research Handbook on International Human Rights Law*, Joseph, S. & McBeth, A., eds., Cheltenham: Edward Elgar, p. 71.
7 See Shaw, M. (2021) *International Law* (9th edn.), Cambridge: Cambridge University Press, p. 715. There were two theories with regard to treatment of aliens: (a) international minimum standard, regardless how state treats its own citizens; and (b) national treatment standard – treating 'aliens' the same way as its own nations.
8 Ibid., p. 244, with two exceptions – piracy *jure gentium* and slavery.
9 See OHCHR webpage: https://www.ohchr.org/en/universal-declaration-of-human-rights.

10 See Anton & Shelton, supra note 4, Chapter 3, for a discussion of historical foundations of human rights.
11 See Keller, H. & Ulfstein, G., eds. (2012) *UN Human Rights Treaty Bodies: Law and Legitimacy*, Cambridge: Cambridge University Press, p. 1. The editors believe that the adoption of the UN Charter "marked a breakthrough for international human rights, with the following decades dedicated to their codification."
12 UN Charter, Preamble, https://treaties.un.org/doc/publication/ctc/uncharter.pdf.
13 Ibid., art. 1, ¶ 3.
14 Resolution 1235 is available at: https://undocs.org/E/RES/1235(XLII) and Resolution 1503 is available at: https://undocs.org/E/4832/Add.1.
15 Resolution 11 (II) of 21 June 1946, https://undocs.org/en/E/RES/11(II).
16 De Schutter, O. (2019) *International Human Rights Law: Cases, Materials, Commentary* (3rd edn.), Cambridge: Cambridge University Press, p. 17.
17 See Chapter 3 for a discussion of the Human Rights Council.
18 De Schutter, O., supra note 16, p. 17.
19 The distinction between negative rights and positive rights is misleading. Even civil and political rights which are generally considered as negative rights require positive action by states as well as resource allocation.
20 Section 5, *Vienna Declaration and Program of Action*, adopted at the World Conference on Human Rights in Vienna (25 June 1993), https://www.ohchr.org/Documents/ProfessionalInterest/vienna.pdf.
21 Despite official UN position that all rights are universal and indivisible, some states consider civil and political rights to be superior to ESC rights. Group rights or collective rights are emerging – many believe environmental rights to be part of collective rights even though the enjoyment of many individual rights can be jeopardized due to environmental degradation.
22 UDHR, Preamble, https://www.un.org/en/about-us/universal-declaration-of-human-rights.
23 Ibid., art. 1.
24 See Kauffman, C. & Martin, P. (2018) "Constructing Rights of Nature Norms in the US, Ecuador and New Zealand," *Global Environmental Politics*, vol. 18, no. 4, p. 43, doi: 10.1162/glep_a_00481.
25 De Schutter, O., supra note 16, pp. 292, 297.
26 *General Comment No. 14*, E/C.12/2000/4 (11 August 2000); see also Anton & Shelton, supra note 4.
27 De Schutter, O., supra note 16, p. 292. See also Henry Shue's proposal with regard to three sets of duties of states referred to there.
28 There are three main procedures to bring complaints for violations of human rights: individual communications, state-to-state complaints and inquiries. Other ways are complaints to special procedures (discussed in Chapter 4), 1503 procedure and the complaints procedure under the Commission on Status of Women. The oldest complaints mechanism and established by the Commission on Human Rights, 1503 procedure generally addresses situations in countries, rather than individual complaints. See https://www.ohchr.org/en/treaty-bodies/human-rights-bodies-complaints-procedures/complaints-procedures-under-human-rights-treaties
29 For a discussion see Atapattu & Schapper, supra note 2, Chapter 4. There are other regional treaties and instruments such as the Arab Charter and the ASEAN Declaration
30 Instruments and mechanisms: https://www.ohchr.org/EN/HRBodies/Pages/HumanRightsBodies.aspx.
31 Subcommittee on the Prevention of Torture: https://www.ohchr.org/EN/HRBodies/OPCAT/Pages/OPCATIndex.aspx.
32 The practice of submitting "shadow reports" by civil society groups has become quite entrenched now although there is no formal mechanism for their submission.

33 Although Keller & Ulfstein, supra note 11, p. 3, refer to these as authoritative interpretations of treaty obligations, General Comments are not binding on state parties. However, they have persuasive value and treaty bodies often refer to their general comments in concluding observations. At a minimum they provide guidance to states as to how to fulfill their treaty obligations.

34 While these decisions are not legally binding, these are not without legal significance. See Keller & Ulfstein, supra note 11, p. 4.

35 Keller & Ulfstein, supra note 11, p 3.

36 Treaty bodies: https://www.ohchr.org/en/treaty-bodies.

37 Thus, General Comment No. 1 elaborates on reporting by states parties. It provides that the reporting obligations are to principally assist states in fulfilling their obligations under the Covenant and to assist the Council and the Committee to monitor states' compliance with their obligations. It also lays down the objectives of submission of reports.

38 E.g., General Comment No. 7 discusses the right to adequate housing and forced evictions. General Comment No. 8 elaborates on the relationship between economic sanctions and ESC rights.

39 E.g., General Comment No. 15 interpreted Article 11 on an adequate standard of living as embodying a right to water.

40 Thus, the Committee is currently soliciting comments for the proposed General Comment on Sustainable Development and ESC rights. See https://www.ohchr.org/en/treaty-bodies/cescr/general-comment-sustainable-development-and-international-covenant-economic-social-and-cultural.

41 OHCHR webpage is a little out of date as the Optional Protocols under both the ICESCR and the CRC are now in force: https://www.ohchr.org/en/treaty-bodies/what-treaty-bodies-do.

42 Ibid.

43 Special Procedures of the Human Rights Council: https://www.ohchr.org/EN/HRBodies/SP/Pages/Welcomepage.aspx.

44 De Schutter, O., supra note 16, p. 970.

45 HRC webpage: https://www.ohchr.org/en/hrbodies/hrc/home.

46 UN Women, Commission on the Status of Women: https://www.unwomen.org/en/csw#multiyear.

47 Multi-year program of work of the Commission on the Status of Women, E/RES/2020/15 (27 July 2020).

48 https://www.unwomen.org/en/csw#multiyear. An experts group meeting was held in October 2011 on this theme and an agreed conclusions document adopted in 2017.

49 Working for Your Rights: https://www.ohchr.org/sites/default/files/Documents/AboutUs/Leaflet_english_web.pdf.

50 Article 24 refers to the dangers and risks of environmental pollution. *Convention on the Rights of the Child* (20 November 1989), https://www.ohchr.org/en/instruments-mechanisms/instruments/convention-rights-child.

51 ICESCR (16 December 1966), art. 12(2)(b).

52 See Knox, J., *Report of the Independent Expert on the issue of human rights obligations relating to the enjoyment of a safe, clean, healthy and sustainable environment (Mapping Report)*, A/HRC/25/53 (30 December 2013), discussed in Chapter 4.

53 These will be discussed in more detail in this volume.

54 These are the African Charter on Human and People's Rights (1981) and the Additional Protocol to the American Convention on Human Rights (1988). See Atapattu & Schapper, supra note 2, Chapter 4.

55 Resolution 48/13, A/HRC/RES/48/13 (8 October 2021).

56 *The Human Right to a Clean, Healthy and Sustainable Environment*, A/76/L/.75 (26 July 2022).

57 See Badrinarayana, D. (May/June 2022) "The Hard Expectations of Soft Law" and Atapattu, S., "A Human Right Whose Time Has Come," Centerpiece and Sidebar, *Environmental Law Institute's The Environmental Forum*, vol. 39 no. 3, https://www.eli.org/the-environmental-forum/hard-expectations-soft-law.

58 *Resolution 64/292*, A/RES/64/292 (3 August 2010). This followed General Comment No. 15 recognizing a right to water by the Committee on ESC Rights (2003).

59 *General Comment No. 36*, CCPR/C/GC/36 (3 September 2019).

60 See Sommer, L. *Climate Change Is the Greatest Threat to Public Health, Top Medical Journals Warn*, (7 September 2021), https://www.npr.org/2021/09/07/1034670549/climate-change-is-the-greatest-threat-to-public-health-top-medical-journals-warn.

61 WHO, Environmental Health, https://www.who.int/health-topics/environmental-health.

62 Ibid.

63 See Hunter, D., Salzman J., & Zelke, D. (2015) *International Environmental Law and Policy* (5th edn.), St. Paul: West Academic, p. 15.

64 WHO, Diarrhoeal Disease, https://www.who.int/news-room/fact-sheets/detail/diarrhoeal-disease.

65 Ibid.

66 World Bank, *Global Waste to Grow by 70 Percent by 2050 Unless Urgent Action Is Taken: World Bank Report* (September 20, 2018), https://www.worldbank.org/en/news/press-release/2018/09/20/global-waste-to-grow-by-70-percent-by-2050-unless-urgent-action-is-taken-world-bank-report.

67 Harrington, R. "By 2050, The Oceans Could Have More Plastic Than Fish," *Business Insider* (26 January 2017), https://www.businessinsider.com/plastic-in-ocean-outweighs-fish-evidence-report-2017-1.

68 World Economic Forum, "90% of fish stocks are used up – fisheries subsidies must stop emptying the ocean," (13 July 2018), https://www.weforum.org/agenda/2018/07/fish-stocks-are-used-up-fisheries-subsidies-must-stop.

69 The World Counts, *Hazardous Waste Statistics*, https://www.theworldcounts.com/counters/waste_pollution_facts/hazardous_waste_statistics.

70 ICCPR, art. 1, https://www.ohchr.org/sites/default/files/ccpr.pdf and ICESCR, art. 1, https://www.ohchr.org/sites/default/files/cescr.pdf.

71 *Report of the Office of the United Nations High Commissioner for Human Rights on the relationship between climate change and human rights*, A/HRC/10/61 (15 January 2009), ¶ 60.

72 Ibid., ¶ 41 (footnotes omitted).

73 It is noteworthy that some recommendations under UPR have referred to the CBDR principle. See Chapter 3.

74 In addition, there are treaties on specialized issues such as prohibition of torture, enforced disappearance, and genocide. OHCHR webpage is a good resource for details.

75 This section draws from author's previous work, Atapattu & Schapper, supra note 2, Chapter 2.

76 As discussed in later chapters, these human rights bodies have extended impact assessments to social, health, gender issues or a combination of them.

77 *Ogoniland Case*, Atapattu & Schapper, supra note 2, p. 4.

78 *Case Concerning the Gabcikovo-Nagymaros Project (Hungary v. Slovakia)*, Judgment, 1997 I.C.J. Rep. 7 (September 25).

79 Declaration of the UN Conference on the Human Environment, held at Stockholm on June 16, 1972, https://wedocs.unep.org/bitstream/handle/20.500.11822/29567/ELGP1StockD.pdf.

80 Sands, P. & Peel, J. (2012) *Principles of International Environmental Law* (3rd edn.), Cambridge: Cambridge University Press, p. 648.

81 Ibid.

82 Convention on Access to Information, Public Participation in Decision-Making and Access to Justice in Environmental Matters (1998) http://www.unece.org/fileadmin/DAM/env/pp/documents/cep43e.pdf.

83 ECLAC, *Regional Agreement on Access to Information, Public Participation and Justice in Environmental Matters in Latin America and the Caribbean* (4 March 2018).

84 Gonzalez, C. (2013) "Environmental Justice and International Environmental Law," in *Routledge Handbook of International Environmental Law*, Alam, S., Bhuiyan, J., Chowdhury, T., & Techera, E., eds., Abingdon: Routledge, pp. 77–98; Kuehn, R. (2000) *A Taxonomy of Environmental Justice*, Environmental Law Reporter, vol. 30, p. 10681.

85 Executive Order 12898, *Federal Actions to Address Environmental Justice in Minority Populations and Low-Income Populations* (11 February 1994), https://www.archives.gov/files/federal-register/executive-orders/pdf/12898.pdf.

86 See Advocates for Environmental Rights, *International Human Rights Commission Takes Jurisdiction Over Louisiana Environmental Racism Case Residents of Mossville, LA Celebrate Landmark Decision* (21 June 2019), http://www.ehumanrights.org/news_release_mar30-10.html.

87 Petition to the IACHR Seeking Relief from Violations Resulting from Global Warming Caused by Acts and Omissions of the United States (2005) http://climatecasechart.com/wp-content/uploads/sites/16/non-us-case-documents/2005/20051208_na_petition.pdf

88 Collins, L. (2015) "The United Nations, Human Rights and the Environment," in *Research Handbook on Human Rights and the Environment*, Grear, A., & Korze, L., eds., Cheltenham, Edward Elgar, p. 219.

89 Alston, P. (1984) "Conjuring Up New Human Rights: A Proposal for Quality Control," *American Journal of International Law*, vol. 78, no. 3, p. 607, doi: 10.2307/2202599.

90 See IACtHR, *Advisory Opinion OC-23/17* (15 November 2017) ["Advisory Opinion"], https://www.corteidh.or.cr/docs/opiniones/seriea_23_ing.pdf.

91 Report of the World Commission on Environment and Development: *Our Common Future* (4 August 1987), p. 43.

92 See Copenhagen Declaration on Social Development, 1995 and Johannesburg Declaration on Sustainable Development, 2002.

93 Our Common Future, Annex 1: *Summary of Proposed Legal Principles for Environmental Protection and Sustainable Development Adopted by the WCED Experts Group on Environmental Law*, Draft principle 1 (4 August 1987), http://www.un-documents.net/ocf-a1.htm.

94 *Resolution 45/94, Need to ensure a healthy environment for the well-being of individuals*, A/RES/45/94 (14 December 1990) (emphasis added).

95 *Review of Further Developments in the Fields with which the Sub-commission Has been Concerned. Human Rights and the Environment: Final report prepared by Mrs. Fatma Zohra Ksentini, Special Rapporteur*, E/CN.4/Sub.2/1994/9 (6 July 1994).

96 *Ian Fry: Special Rapporteur on Climate Change*, https://www.ohchr.org/en/specialprocedures/sr-climate-change/ian-fry.

97 CIDCE, *Draft of the International Covenant on the Human Right to the Environment* (2017), https://cidce.org/wp-content/uploads/2016/08/Draft-of-the-International-Covenant-on-the-Human-Right-to-the-Environment_15.II_.2017_EN.pdf.

98 Draft Global Pact for the Environment (24 June 2017), https://globalpactenvironment.org/uploads/EN.pdf.

99 *Resolution 72/277*, A/RES/72/277 (14 May 2018).

100 "All peoples shall have the right to a general satisfactory environment favorable to their development," African Charter on Human and Peoples' Rights (1981), art. 24, https://www.achpr.org/legalinstruments/detail?id=49.

101 Council of Europe, European Convention on Human Rights, art. 2 and art. 8, https://www.echr.coe.int/documents/convention_eng.pdf.
102 Advisory Opinion, supra note 90, ¶ 47.
103 Ibid., ¶ 59.
104 Ibid., ¶ 62.
105 Ibid., ¶ 63. In addition, the Court discussed the issue of extraterritoriality in the context of transboundary damage and endorsed the effective control test but warned that it should be used with caution.
106 See De Schutter, O., supra note 16, at 988.
107 *Asian Human Rights Charter*, http://www.refworld.org/pdfid/452678304.pdf. The Charter contains a provision on environmental protection and sustainable development but is framed as an objective and not as a right. The AHRC, on the other hand, is an independent non-governmental body.
108 ASEAN, *Human Rights Declaration* (19 November 2012), https://asean.org/asean-human-rights-declaration.
109 Ibid., art. 28(f).
110 LAS, *Arab Charter on Human Rights* (22 May 2004), http://hrlibrary.umn.edu/instree/loas2005.html.
111 Ibid., art. 38.
112 Ibid., art. 48.
113 See Hay, K. (2009) "A Pacific Human Rights Mechanism: Specific Challenges and Requirements," *Victoria University of Wellington Law Review*, vol. 40, no. 1, http://www.victoria.ac.nz/law/research/publications/about-nzacl/publications/special-issues/hors-serie-volume-viii,-2008/Hay.pdf, doi: 10.26686/vuwlr.v40i1.5386.
114 For a discussion of the constitutional developments, see Boyd, D. (2012) *The Environmental Rights Revolution: A Global Study of Constitutions, Human Rights, and the Environment*, Vancouver: UBC Press.
115 For a discussion of cases at the national level, see Atapattu & Schapper, supra note 2, Chapter 8.
116 *Oposa v. Factoran*, G.R. No. 101083, 224 S.C.R.A. 792 (1993) (Phil.) (emphasis added).
117 OHCHR (2018) *Framework Principles on Human Rights and the Environment*, https://www.ohchr.org/sites/default/files/FrameworkPrinciplesUserFriendlyVersion.pdf.

References

ACHPR (1981) African Charter on Human and Peoples' Rights.

Advocates for Environmental Rights (2019) International Human Rights Commission Takes Jurisdiction Over Louisiana Environmental Racism Case Residents of Mossville, LA Celebrate Landmark Decision, http://www.ehumanrights.org/news_release_mar30-10.html.

Alston, P. (1984) "Conjuring up New Human Rights: A Proposal for Quality Control," *American Journal of International Law*, vol. 78, no. 3, p. 607, doi: 10.2307/2202599.

Anton, D. & Shelton, D. (2011) *Environmental Protection and Human Rights*, Cambridge: Cambridge University Press, doi: 10.1017/CBO9780511974571.

ASEAN (2012) Human Rights Declaration, https://asean.org/asean-human-rights-declaration/

Asian Human Rights Commission (AHRC) (1988) Asian Human Rights Charter, https://www.refworld.org/pdfid/452678304.pdf

Atapattu, S. & Schapper, A. (2019) *Human Rights and the Environment: Key Issues*, Abingdon: Routledge.

Badrinarayana, D. (2022) "The Hard Expectations of Soft Law" and Atapattu, S., "A Human Right Whose Time Has Come," Centerpiece and Sidebar, Environmental Law Institute's the Environmental Forum, vol. 39 no. 3, https://www.eli.org/the-environmental-forum/hard-expectations-soft-law.

Boyd, D. (2012) *The Environmental Rights Revolution: A Global Study of Constitutions, Human Rights, and the Environment*, Vancouver: UBC Press.

Buergenthal, T. (2006) "The Evolving International Human Rights System," *American Journal of International Law*, vol. 100, no. 4, p. 783.

Case Concerning Gabcikovo-Nagymaros Project (Hungary v. Slovakia) (1997) Judgment, I.C.J. Rep. 7.

CIDCE (2017) Draft of the International Covenant on the Human Right to the Environment, https://cidce.org/wp-content/uploads/2016/08/Draft-of-the-International-Covenant-on-the-Human-Right-to-the-Environment_15.II_.2017_EN.pdf.

Collins, L. (2015) "The United Nations, Human Rights and the Environment," in Grear, A. & Korze, L. (eds.), *Research Handbook on Human Rights and the Environment*, Cheltenham: Edward Elgar, doi: 10.4337/9781782544432.00019.

Complaints Procedures under the Human Rights Treaties (2022), https://www.ohchr.org/en/treaty-bodies/human-rights-bodies-complaints-procedures/complaints-procedures-under-human-rights-treaties.

Convention on the Rights of the Child (1989), https://www.ohchr.org/en/instruments-mechanisms/instruments/convention-rights-child.

Council of Europe, European Convention on Human Rights, https://www.echr.coe.int/documents/convention_eng.pdf.

Daly, E. & May, J. (2020) *Dignity Law: Global Recognition, Cases and Perspectives*, Getzville: William S. Hein & Co.

Daly, E. & May, J. (2021) "The Indivisibility of Human Dignity and Sustainability," in *The Cambridge Handbook of Environmental Justice and Sustainable Development*, Atapattu, S., Gonzalez, C., & Seck, S., eds., Cambridge: Cambridge University Press.

De Schutter, O. (2019) *International Human Rights Law: Cases, Materials, Commentary* (3rd edn.), Cambridge: Cambridge University Press.

Declaration of the UN Conference on the Human Environment, held at Stockholm (1972), https://wedocs.unep.org/bitstream/handle/20.500.11822/29567/ELGP1StockD.pdf.

ECLAC (2018) Regional Agreement on Access to Information, Public Participation and Justice in Environmental Matters in Latin America and the Caribbean, https://www.cepal.org/en/escazuagreement.

Executive Order 12898, Federal Actions to Address Environmental Justice in Minority Populations and Low-Income Populations (1994) available at: https://www.archives.gov/files/federal-register/executive-orders/pdf/12898.pdf.

Framework Principles on Human Rights and the Environment (2018) https://www.ohchr.org/sites/default/files/FrameworkPrinciplesUserFriendlyVersion.pdf.

General Comment No. 14 (2000) E/C.12/2000/4.

General Comment No. 15 (2002), E/C.12/2002/11.

General Comment No. 36 (2019), CCPR/C/GC/36.

Global Pact for the Environment (2017) Draft Global Pact for the Environment, available at: https://globalpactenvironment.org/uploads/EN.pdf.

Gonzalez, C. (2013) "Environmental Justice and International Environmental Law," in *Routledge Handbook of International Environmental Law*, Alam, S., Bhuiyan, J., Chowdhury, T., & Techera, E., eds., 77–98, Abingdon: Routledge.

Harrington, R. (2017) "By 2050, the Oceans Could Have More Plastic Than Fish," Business Insider, available at: https://www.businessinsider.com/plastic-in-ocean-outweighs-fish-evidence-report-2017-1.

Hay, K. (2009) "A Pacific Human Rights Mechanism: Specific Challenges and Requirements," *Victoria University of Wellington Law Review*, vol. 40, no.1, available at: http://www.victoria.ac.nz/law/research/publications/about-nzacl/publications/special-issues/hors-serie-volume-viii,-2008/Hay.pdf, doi: 10.26686/vuwlr.v40i1.5386.

HRC (2022) webpage: https://www.ohchr.org/en/hrbodies/hrc/home.

Hunter, D., Salzman J., & Zelke, D. (2015) *International Environmental Law and Policy* (5th edn.), St. Paul: West Academic.

IACHR (2017) Advisory Opinion OC-23/17, https://www.corteidh.or.cr/docs/opiniones/seriea_23_ing.pdf.

Ian Fry: Special Rapporteur on climate change (2022) https://www.ohchr.org/en/specialprocedures/sr-climate-change/ian-fry.

International Covenant on Civil and Political Rights (1966) https://www.ohchr.org/sites/default/files/ccpr.pdf.

International Covenant on Economic, Social and Cultural Rights (1966) https://www.ohchr.org/en/instruments-mechanisms/instruments/international-covenant-economic-social-and-cultural-rights.

Kauffman, C. & Martin, P. (2018) "Constructing Rights of Nature Norms in the US, Ecuador and New Zealand," Global Environmental Politics, vol. 18, no. 4, p. 43, doi: 10.1162/glep_a_00481.

Keller, H. & Ulfstein, G., eds. (2012) *UN Human Rights Treaty Bodies: Law and Legitimacy*, Cambridge: Cambridge University Press.

Knox, J. (2013) Report of the Independent Expert on the issue of human rights obligations relating to the enjoyment of a safe, clean, healthy and sustainable environment (Mapping Report), A/HRC/25/53, https://undocs.org/A/HRC/25/53.

Ksentini, F.Z. (1994) UN Special Rapporteur on Human Rights and Environment, Final report, E/CN.4/Sub.2/1994/9.

Kuehn, R. (2000) "A Taxonomy of Environmental Justice," *Environmental Law Reporter*, vol. 30, p. 10681.

LAS (2004) Arab Charter on Human Rights, available at: http://hrlibrary.umn.edu/instree/loas2005.html.

OAS (1988) Additional Protocol to the American Convention on Human Rights in the Area of Economic, Social and Cultural Rights (Protocol of San Salvador), OAS Treaty Series No. 69, available at: https://www.oas.org/juridico/english/treaties/a-52.html.

OHCHR, Report of the Office of the United Nations High Commissioner for Human Rights on the relationship between climate change and human rights (2009), A/HRC/10/61.

Oposa v. Factoran (1993) G.R. No. 101083, 224 S.C.R.A. 792 (the Philippines).

Petition to the IACHR Seeking Relief from Violations Resulting from Global Warming Caused by Acts and Omissions of the United States (2005) http://climatecasechart.com/wp-content/uploads/sites/16/non-us-case-documents/2005/20051208_na_petition.pdf.

Sands, P. & Peel, J. (2012) *Principles of International Environmental Law* (3rd edn.), Cambridge: Cambridge University Press.

Shaw, M. (2021) *International Law* (9th edn.), Cambridge: Cambridge University Press.

Skogly, S. (2010) "Extraterritoriality: Universal Human Rights without Universal Obligations?" in *Research Handbook on International Human Rights Law*, Joseph, S. & McBeth, A., eds., 71–96, Cheltenham: Edward Elgar.

Social and Economic Rights Action Center and the Center for Economic and Social Rights v. Nigeria, Comm. No. 155/96 (2001) (Ogoniland case)

Sommer, L. (2021) Climate Change is the Greatest Threat to Public Health, Top Medical Journals Warn, https://www.npr.org/2021/09/07/1034670549/climate-change-is-the-greatest-threat-to-public-health-top-medical-journals-warn.

Special Procedures of the Human Rights Council (2022) https://www.ohchr.org/EN/HRBodies/SP/Pages/Welcomepage.aspx.

Subcommittee on the Prevention of Torture (2022) https://www.ohchr.org/EN/HRBodies/OPCAT/Pages/OPCATIndex.aspx.

The World Counts, Hazardous Waste Statistics, https://www.theworldcounts.com/counters/waste_pollution_facts/hazardous_waste_statistics.

Treaty bodies (2022) https://www.ohchr.org/en/treaty-bodies.

What are human rights (2022) https://www.ohchr.org/en/what-are-human-rights.

What the treaty bodies do (2022) https://www.ohchr.org/en/treaty-bodies/what-treaty-bodies-do.

Working for Your Rights (2022) https://www.ohchr.org/sites/default/files/Documents/AboutUs/Leaflet_english_web.pdf.

UDHR (1948) https://www.un.org/en/about-us/universal-declaration-of-human-rights.

UN Charter (1945) https://treaties.un.org/doc/publication/ctc/uncharter.pdf.

UN Human Rights, Instruments and mechanisms (2022) https://www.ohchr.org/EN/HRBodies/Pages/HumanRightsBodies.aspx.

UNECE (1998) Convention on Access to Information, Public Participation in Decision-Making and Access to Justice in Environmental Matters, https://unece.org/DAM/env/pp/documents/cep43e.pdf.

UNGA, The Human Right to a Clean, Healthy and Sustainable Environment (2022), A/76/L/.75.

UN Women (1946) Commission on the Status of Women, https://www.unwomen.org/en/csw#multiyear.

WCED (1987) Our Common Future, https://undocs.org/A/42/427.

WCED (1987) Our Common Future, Annex 1: Summary of Proposed Legal Principles for Environmental Protection and Sustainable Development, http://www.un-documents.net/ocf-a1.htm.

WHO (2022) Diarrheal Disease, https://www.who.int/news-room/fact-sheets/detail/diarrhoeal-disease.

WHO (2022) Environmental Health, https://www.who.int/health-topics/environmental-health.

World Bank (2018) Global Waste to Grow by 70 Percent by 2050 Unless Urgent Action Is Taken: World Bank Report, https://www.worldbank.org/en/news/press-release/2018/09/20/global-waste-to-grow-by-70-percent-by-2050-unless-urgent-action-is-taken-world-bank-report.

World Conference on Human Rights (1993) Vienna Declaration and Program of Action, https://www.ohchr.org/Documents/ProfessionalInterest/vienna.pdf.

World Economic Forum (2018) 90% of Fish Stocks Are Used up – Fisheries Subsidies Must Stop Emptying the Ocean, https://www.weforum.org/agenda/2018/07/fish-stocks-are-used-up-fisheries-subsidies-must-stop.

World Summit for Social Development (1995) Copenhagen Declaration on Social Development, A/CONF.166/9, https://undocs.org/A/CONF.166/9.

World Summit on Sustainable Development (2002) Johannesburg Declaration on Sustainable Development, A/CONF.199/20, https://undocs.org/A/CONF.199/20

Part II

Charter-Based Mechanisms

3 UN Human Rights Council and the Office of the High Commissioner for Human Rights

Two charter-based institutions – the HRC and the OHCHR – have been at the forefront of articulating environmental rights, recognizing, for the first time, the right to a healthy environment as a stand-alone right.[1] The High Commissioner for Human Rights (HCHR), as the principal human rights official of the UN,[2] has played a critical role in promoting environmental rights over the years.[3] The best-known advocate for environmental rights is Mary Robinson who was the HCHR during 1997–2002 and was appointed by the UN Secretary General (UNSG) as the Special Envoy for Climate Change in 2014.[4] In this chapter we discuss how these two institutions have addressed environmental degradation, climate change, and sustainable development including SDGs within their mandate.[5]

3.1 Human Rights Council

3.1.1 Introduction and Mandate

The HRC is the principal UN body dedicated to the protection of human rights. It was established by the UN General Assembly in 2006,[6] replacing the Commission on Human Rights and became a subsidiary body of UNGA, instead of the ECOSOC,[7] thereby raising its political standing.[8] The HRC has a wide, all-encompassing mandate, including: (a) promoting universal respect for the protection of human rights for all, without any distinction; (b) addressing situations of violations of human rights and making recommendations; (c) mainstreaming human rights within the UN system; (d) promoting human rights education, technical assistance, and capacity-building, (e) serving as a forum for dialogue on thematic issues on human rights; (f) making recommendations to the UNGA for the further development of human rights law; (g) promoting the implementation of human rights obligations by States; (h) undertaking a universal periodic review of the fulfillment by States of their human rights obligations (i) contributing toward the prevention of human rights violations and responding promptly to human rights emergencies; and (j) working closely with Governments, regional organizations, national human rights institutions, and civil society.[9]

It is an intergovernmental body consisting of 47 States responsible for the promotion and protection of all human rights around the globe.[10] Unlike other

DOI: 10.4324/9781003128847-5

human rights institutions, HRC's mandate is not limited to a particular human rights issue or a specialized group or limited geographically. Simply put, it is responsible for all human rights everywhere. Since its mandate is to protect and promote *all* human rights of everyone everywhere, it has increasingly encompassed environmental issues, especially climate change, within its mandate, culminating in the recognition, for the first time at the international level, of a right to a clean, healthy, and sustainable environment as a human right in 2021.[11] On several occasions, the HRC has urged states to take human rights into consideration when developing their environmental policies.[12]

The HRC has three subsidiary bodies: Universal Periodic Review Working Group; Advisory Committee; and the Complaint Procedure.[13] In addition, it has established *subsidiary expert mechanisms* to provide thematic expertise and *forums* to provide a platform for dialog and cooperation. They report annually to the Council.[14] In addition to its own work on the environment, the HRC has created several special procedures and mandate holders to address the link between human rights and the environment, recognizing that a conducive environment is necessary for the enjoyment of many of the protected rights. Their work is discussed in chapter 4.

3.1.2 Universal Periodic Review and Working Group Recommendations

The HRC adopts several procedures to monitor and promote human rights around the world, including the Universal Periodic Review (UPR), the complaints procedure, and the adoption of resolutions on thematic issues and countries. Among these, the UPR is a unique process which involves a review of the human rights record of all UN member states. It is a state-driven process which gives the opportunity to each state to describe the human rights situation in their country and what action they have taken to improve the situation. It is designed to ensure equal treatment for every country with the ultimate aim of improving the human rights situation in all countries and addressing violations.[15]

UNGA resolution establishing the HRC elaborates on the UPR procedure:

> The review shall be a cooperative mechanism, based on an interactive dialogue, with the full involvement of the country concerned and with consideration given to its capacity-building needs; such a mechanism shall complement and not duplicate the work of treaty bodies; the Council shall develop the modalities and necessary time allocation for the universal periodic review mechanism....[16]

Thus, it is essentially a peer review process - states themselves evaluate each other's reports and make recommendations. Currently, no similar mechanism exists elsewhere. A working group is established comprising states to evaluate reports and make recommendations. Each report has three sections: a presentation by the state under review; an interactive dialogue and responses by the state under review; and conclusions and/or recommendations. The section titled "conclusions and/or recommendations" is noteworthy: it includes recommendations that the state party has expressed support for, recommendations that the state party believes are already implemented, and those that did not receive the support of the state party. Thus, for example, the following recommendation made by Spain (working group member) did not receive Bolivia's support: "Advance in the commitment to incorporate into the national human rights law the United Nations Guiding Principles on Business and Human Rights, with special attention to labour rights, rights of indigenous communities and environmental rights."[17]

The documents that are associated with the review are: the national report prepared by the state; a report by the OHCHR that compiles relevant information from treaty bodies, special procedures and UN agencies; and a summary of civil society information prepared by the OHCHR.[18] The report compiled for the Independent Expert on Human Rights and the Environment explains the procedures relating to the UPR process.[19]

Since there are literally hundreds of state reports and working group reports, only the recommendations made by the working groups during 2013–2022 are discussed in this chapter.[20] We will use the Mapping Report No 6 as background information for the section on the UPR process for the period prior to 2013. According to Mapping Report No 6, a total of 45 states discussed their constitutional right to a healthy environment during their UPR review during the period surveyed (until 2013).[21] Mongolia, Indonesia, and Solomon Islands discussed the impacts of climate change on the enjoyment of a right to a healthy environment while Peru discussed the impact of mining on the right to a healthy environment. Many countries referred to the impact of environmental harm on human health. The sources of harm included toxic waste dumping (Ivory Coast), nuclear waste (Belarus, Kyrgyzstan), mining operations (Ghana, Brazil, Mongolia, Nauru and Bolivia), natural disasters (Greece), oil exploration (Nigeria), gold mining (Mongolia), and acid rain from volcanic explosions (Vanuatu).[22]

Many of the working group reports referred to environmental issues, climate change, and sustainable development, including SDGs. While it is impossible to discuss all the reports, most salient recommendations will be highlighted here. As noted, these are not comments provided by experts in the field, unlike the treaty bodies discussed in Chapters 5–9.

3.1.2.1 Environmental Protection

On the 2014 report of Bolivia, the working group recommended continuing efforts on environmental protection and achieving "harmonious development of people and nature" and incorporating UN Guiding Principles on Business and Human Rights, with special attention to labor rights, rights of indigenous communities, and environmental rights.[23] On its 2019 report, the working group recommended ratifying the Escazú Agreement.[24] On the report of Kazakhstan, the recommendations included: ensuring that the environment is protected, informing people of the impact of pollutants on their health and the environment, and addressing environmental hazards that affect people's health, especially children living in environmentally high-risk areas.[25] On the report of Iraq, the working group recommended taking measures to protect the environment through its sustainable development plan[26] while on the report of Bosnia and Herzegovina, the recommendations included adopting a strategy to fight air pollution which affects the right to health, especially of children and older persons.[27] On the report of Guinea, it was recommended consulting the local communities before concluding contracts relating to natural resources or projects which have a social and environmental impact[28] and ratifying ILO Convention 169.[29]

On the report of Lao, the working group recommended developing guidelines to conduct studies on the environmental impacts of foreign investment projects[30] while on the report of Congo, the recommendations included ensuring that businesses comply with laws on human rights, labor, and environment and regularizing unauthorized mining areas.[31] The working group recommended integrating a rights-based approach to environmental protection on the report of Uruguay and to implement programs that have positive environmental effects.[32]

On the report of Chile, the recommendations included: guaranteeing the right to a safe environment; implementing socio-environmental recovery programs, developing environmental quality norms in accordance with WHO standards, and applying the Guiding Principles on Business and Human Rights in relation to natural resources; establishing a regulatory framework for companies, so that their activities do not affect the exercise of ESC rights; adopting measures to address the environmental impact of development projects; strengthening policies on the right to live in a non-polluted environment; and assessing the environmental and socioeconomic damage due to industrial activity, including mining operations.[33] On its previous report, it was recommended ensuring justice and reparation for victims of environmental degradation and adopting environmental standards for pollutant emissions by foundries.[34]

On the report of Viet Nam, the recommendations included: reducing the environmental impacts of industrialization and ensuring the right to safe water and sanitation in rural areas; and better treatment of wastewater, domestic waste, and industrial waste, thus protecting the environment.[35] On the report of Nauru, the working group recommended establishing a regulatory framework to reduce the physical and environmental risks associated with phosphate mining[36] while on the report of Nigeria several recommendations were made to address the damage caused by oil pollution as well as implementing policies relating to climate change and disaster risk reduction, and supporting communities affected by oil spills by providing health care and education facilities, and creating alternative livelihood options.[37] On a previous report of Nigeria, the recommendations included monitoring the environmental impact of the oil industry on human rights of people in the Niger Delta and taking suitable measures.[38]

On the report of China, it was recommended reducing, in accordance with the ICESCR and the Guiding Principles on Business and Human Rights, the environmental impact of industrialization; speeding up pollution governance in coastal waters; protecting citizens' right to know, participating and supervising in the environmental field; implementing laws on environmental protection, air pollution, and protection of wildlife; and addressing water pollution.[39] The working group recommended Costa Rica to enact a new water law embodying the right to water, as recommended by the Independent Experts on human rights and the environment, and safe drinking water and sanitation.[40] Likewise, on the report of Suriname, it was recommended taking measures to combat land and water pollution caused by dangerous industrial waste.[41]

The working group recommended Ecuador to continue its policies on rights of nature and increase efforts on its policies on a healthy environment[42] while on the report of the Maldives, the working group recommended adopting legislation to protect the environment and establishing a joint private-public partnership to reduce the environmental impact of tourism on livelihoods and the rights of people.[43] On the report of Macedonia, it was recommended monitoring the human rights obligations relating to the enjoyment of a safe, clean, healthy, and sustainable environment, particularly with regard to air quality and industrial pollution.[44] On the report of Bhutan, it was recommended strengthening environmental conservation efforts, which contribute to safeguarding the common heritage of humankind. Reference was also made to the link between massive movement of people and environmental degradation.[45]

3.1.2.2 Climate Change

Climate change was discussed in many recent reports, especially small island nations both in relation to the interactive dialog and the recommendations. Thus, on the report of Tuvalu,[46] the recommendations included: raising awareness of the population of climate impacts and involving them in decisions on mitigation and adaptation efforts; continuing to promote international cooperation to combat climate change; working with major emitters to protect the human rights of Tuvaluans by securing global reductions in GHG emissions; participating in discussions on human rights and climate change to send a strong message to the parties to the UNFCCC on the importance of reaching a global agreement on climate change.[47] It was noted that the greatest threats facing Tuvalu are climate change and sea-level rise. On its report submitted to the 24th session, it was recommended: implementing NAPA to address the impacts of climate change with assistance from the UNDP, the Global Environment Fund and other international stakeholders; and formulating, with regional and UN mechanisms, a plan to address natural disasters including potential mass displacements of the population.[48] The issue of displacement was addressed on the report of Solomon Islands.[49] It was recommended creating a permanent national protection group responsible for addressing displacements caused by climate change or natural disasters. It was also recommended to "approve the roadmap on the reduction of emissions resulting from deforestation and forest degradation, on the role of conservation and sustainable forest management and on the reinforcement of forest carbon stocks."[50]

On the report of Italy, the working group recommended engaging women, children, and persons with disabilities in the development of legislation, and programs on climate change and disaster risk reduction.[51] A more elaborate recommendation was made on the report of Kyrgyzstan:

> Adopt a comprehensive, gender-responsive and disability-inclusive approach to climate change adaptation and mitigation policies, consistent with the United Nations Framework Convention on Climate Change and the Sendai Framework for Disaster Risk Reduction 2015–2030, to address the economic, cultural and social impacts and challenges that climate change represents, for the full and effective enjoyment of human rights for all.[52]

On the 2019 report of Bolivia, the working group recommended promoting climate change resilience and adaptation measures to guarantee the right to food and standard of living, particularly for the most vulnerable; and strengthening its policies on prevention and post-disaster recovery to mitigate the effects of climate change.[53] On the report of Spain it was recommended taking measures to prevent, reduce, and combat pollution as part of the national action plan for climate adaptation,[54] while on the report of Lesotho, it was recommended to ensure that a human rights approach is reflected in the 2020 submission of its NDC.[55] It was also recommended to implement the National Climate Change Policy and establish a system to monitor its implementation.

Extensive and wide-ranging recommendations were made on the 2019 report of Fiji.[56] These included: increasing awareness-raising programs for vulnerable groups on climate change and disaster risk reduction policies; mainstreaming climate adaptation across development activities with the assistance of the international community; responding to the negative impacts with a focus on vulnerable groups, including women and children, and promoting gender equality and adopting a gender approach to policies on climate change and disaster risk reduction; incorporating a human rights perspective by creating mechanisms for the participation of citizens in decision-making, access to justice and reparations; and strengthening the legislative and policy framework for building strong and resilient communities to ensure that all Fijians enjoy basic human rights. Recommendations also included adapting to climate change within the framework of its National Adaptation Plan; involving women in forums on climate change and natural disasters and economically empowering women who are relocated due to climate change; implementing legislative and policy frameworks to mitigate the impacts of climate change and natural disasters; ensuring the enjoyment of its people's rights in the face of climate change and adopting an inclusive and participatory approach to climate policy; revising building codes to increase resilience[57] to severe weather events and protecting rights to adequate housing, sanitation, and clean water; strengthening social protection systems so that affected families and children would receive support; supporting victims of natural disasters; and conducting a national dialogue on climate change to outline priorities.[58]

On the report of Iran, the working group recommended conducting an assessment of its disaster risk reduction management and adaptation efforts taking into account the most vulnerable people[59] while on the report of Kenya, it was recommended ensuring that women, especially rural women, are engaged in the development of climate change and disaster risk reduction legislation and policies.[60]

On the report of Nicaragua, it was recommended developing legislative frameworks to address cross-sectoral environmental challenges including climate adaptation and mitigation and reinforcing existing frameworks by integrating climate adaptation in planning, budgeting, and building institutional capacity, data, and knowledge to integrate environmental and climate change considerations into national regulatory frameworks.[61] It was also recommended to cooperate with UN bodies and other international organizations on natural disasters, especially with regard to children and displaced persons and maintain active participation in climate negotiations; continue to address climate impacts and promote international cooperation particularly with regard to technology development and transfer, financial support and capacity-building.[62]

The working group recommended Costa Rica to develop, in collaboration with elderly persons, persons with disabilities, and indigenous and other marginalized groups, measures to address disaster risk reduction and climate change[63] while on the report of Democratic People's Republic of Korea, the Committee requested information on the steps taken to include those most vulnerable to climate change in measures to address its causes and effects.[64] On the report of Comoros, it was recommended to implement measures on climate change and disaster risk reduction, ensuring a rights-based approach, adopting a gender perspective and prioritizing vulnerable groups while engaging stakeholders and the public in policy discussions and programs on climate change and disaster risk reduction.[65]

Extensive comments and recommendations were made on the report of Vanuatu both in the interactive dialog and under recommendations.[66] These included: focusing climate efforts on vulnerable groups, including women and children; reviewing policies and programs on climate adaptation and mitigation to include a rights-based approach and a gender perspective, prioritizing vulnerable groups and seeking assistance from the international community; including women, especially those living in outer islands, in the National Advisory Board on Climate Change and Disaster Risk Reduction, ensuring their participation in the planning, decision-making, and implementation processes; including the impact of climate change on women in its gender policy under the UNFCCC; promoting stronger mitigation action by main emitters to protect the fundamental rights of the people[67]; implementing the National Policy on Climate Change and Disaster Risk Reduction, the Climate Change and Disaster-Induced Displacement Policy, and the Strategic Plan 2016–2020; protecting the rights of those displaced by natural disasters; and amending the emergency protocols to include support for those who are most vulnerable to natural disasters, particularly children with disabilities.[68]

On the report of Viet Nam, the recommendations included addressing the vulnerabilities of women, children, persons with disabilities, and marginalized groups when implementing programs on adaptation, green growth, and sustainable forestry development; providing information dissemination and awareness-raising on climate impacts and mitigation measures; and participating in initiatives, within the UN system, on climate change and human rights especially by vulnerable people.[69]

The recommendations made on China's report deviated somewhat from the typical pattern on most other reports. These included: promoting South-South cooperation in responding to climate change; implementing the 2030 Agenda for Sustainable Development, while remaining committed to the Paris Agreement and implementing the Paris Agreement taking into account the impact on human rights;[70] ensuring that development and infrastructure projects inside and outside its territory respect human rights and the environment and sustainability of natural resources, in line with applicable national and international law and the 2030 Agenda for Sustainable Development; providing leadership on climate change; and strengthening measures to combat pollution and climate change.[71]

On the report of US, the recommendations included: addressing climate change within a framework of shared but differentiated responsibility[72] and together with the international community; participating in climate negotiations to adopt a strong legally binding outcome; and adopting federal legislation to prohibit environmental pollution and reducing GHG emissions to control climate change.[73]

Extensive comments and recommendations were made on the report of Tuvalu. These included: implementing mitigation policies and programs ensuring protection of human rights; promoting the right to a healthy environment, and the fight against climate change; assessing the impact of climate change and natural disasters on human rights of low-income populations, and building their resilience; formulating, with regional organizations and the UN, a plan to address the situation of migrant children and the potential massive population displacement; and integrating a gender perspective into all climate change and disaster risk reduction policies.[74]

Likewise, extensive comments were made on the report of Kiribati.[75] The recommendations included: addressing the impacts of climate change on people's access to food and clean water, particularly the most vulnerable sectors; strengthening the positive impact of Kiribati's Joint Implementation Plan on Climate Change and Disaster Risk Management by training relevant stakeholders on human rights approaches to climate change and disaster risk management; continuing its leadership role and advocacy with the international community, through the Coalition of Low Lying Atoll Nations on Climate Change and the Alliance of Small Island States, on the need for ambitious and binding targets for GHG emissions to mitigate the negative effects of climate change on human rights; and securing assistance from the international community in relation to adaptation and mitigation plans.[76] On the report of India, the working group recommended: providing access to clean and modern energy and developing climate-friendly green cities, implementing its commitments to achieve its NDCs under the Paris Agreement, and promoting children's right to education, especially climate change adaptation and mitigation.[77]

3.1.2.3 Sustainable Development and SDGs

Several recommendations addressed sustainable development[78] while almost all the working group reports since 2015 referred to SDGs. Thus, on the report of Kazakhstan, the working group recommended protecting ESC rights, paying particular attention to achieving the SDGs and consolidating the progress made toward reaching the SDGs.[79] On the 2019 report of Bolivia, the recommendations referred to several SDGs and targets as well as ensuring access to drinking water[80] while on the report of Angola, the recommendations included continuing efforts to alleviate poverty and ensuring sustainable development; implementing programs on rural development, and referred to SDG 3 in the context of free-healthcare coverage.[81] On the report of Egypt, the recommendations referred to several SDGs, its 2030 Sustainable Development Strategy and the need to work closely with relevant stakeholders in its implementation; and implementing its socioeconomic measures to provide a dignified life for all citizens without discrimination and strengthening women's empowerment.[82] On the report of Lao, the working group recommended establishing a mechanism to link recommendations made under the UPR process with the 2030 Sustainable Development Agenda and Goals[83] and to incorporate targets and goals into national plans.[84] Likewise, on the report of Ethiopia, it was recommended aligning 2030 Agenda with the human rights agenda as well as integrating the human rights recommendations with the SDGs.[85] On the report of Brunei Darussalam, the working group recommended continuing efforts to implement national priorities to achieve the SDGs[86] while several references to the SDGs were made on the report of Kenya in the context of addressing poverty, universal health coverage, and rights of persons with disabilities.[87]

3.1.2.4 Indigenous Rights

On many reports, the working group recommended protecting the rights of indigenous peoples, their ancestral lands according to their uses and customs[88] and ratifying ILO Convention No 169.[89] Recommendations also referred to ensuring adequate consultations with indigenous peoples, their participation in legislative or administrative measures affecting them and with regard to any project that may affect their land or livelihoods, including establishing procedures for free, prior and informed consent[90] and guaranteeing their constitutional rights.[91] On the report of Brazil, the working group recommended speeding up the process of demarcating and protecting indigenous lands and their rights[92] while the need to ensure the best interests of indigenous peoples in carrying out large-scale national development projects was stressed on the report of Nicaragua.[93]

On the report of Norway, the working group recommended inviting for country visits by Special Rapporteurs on: racism, migrants, and indigenous peoples. Several members referred to the need to protect the rights of indigenous peoples, ensure uniform procedures for consultations with Sami people in accordance with ILO Convention 169, protect rights of indigenous peoples to develop their effective participation,[94] review mechanisms for extractive activities on Sami lands to guarantee adequate consultation with affected Sami people, provide compensation and benefit sharing, increase efforts to protect their traditional way of life including the culture in consultation with those communities; implement the recommendations of the Committee on Racial Discrimination; and take measures to improve the legal framework for Sami land, fishing and reindeer rights.[95] The need to recognize the collective rights of indigenous peoples, adopting measures to ensure their full participation in public affairs, and the need to recognize their ethnic and cultural existence and greater visibility were recommended on the report of Uruguay.[96]

On the report of Cambodia, the working group recommended: simplifying the allocation of community land concessions/titles to indigenous peoples; implementing a coherent resettlement policy and consulting indigenous groups; and settling all land disputes, evictions and relocations in a fair and transparent manner with adequate compensation. Reference was also made to victims of land grabbing, especially indigenous peoples.[97] On the report of Chile, it was recommended to expedite the constitutional recognition of indigenous peoples and enhance their participation and prior consultation on policies that affect them; guarantee their rights, as enshrined in UNDRIP; and strengthen the implementation of ILO Convention 169 especially conducting an independent evaluation of mechanisms for their consultation and participation.[98]

Extensive comments were made on the report of New Zealand, both during the interactive dialog and under recommendations. These included, in addition to ratifying ILO Convention 169, protecting the rights of indigenous peoples through law, policy and practice in conformity with human rights obligations; enhancing cooperation between the government and indigenous peoples; harmonizing national regulations with UNDRIP and including UNDRIP and the Treaty of Waitangi in the National Plan of Action for Human Rights; and developing, in consultation with the indigenous peoples and if necessary with assistance of the Expert Mechanism on the Rights of Indigenous Peoples, an action plan to harmonize legislation and existing policies with UNDRIP.[99] On the 2015 report of the US, the working group recommended respecting the rights of indigenous peoples and ethnic minorities, consulting with them on their land, environment and other issues, and correcting historical injustices and offering compensation.[100]

On the report of Suriname, the working group recommended implementing the judgments of the Inter-American Court of Human Rights regarding the Moiwana Community and Saramaka people, and taking measures to reduce the negative impact of mining on the environment and indigenous peoples' lands, in line with international standards.[101] The working group recommended Australia to implement the recommendations made by the Special Rapporteur on the Rights of Indigenous Peoples,[102] conduct consultations with indigenous peoples in the formulation of policies affecting them, and ensure that Aboriginal and Torres Strait Islander Peoples give their consent to policies and programs that affect their communities and futures.[103]

3.1.2.5 Business Enterprises and Transnational Corporations

On several reports the working groups' recommendations related to the activities of businesses and transnational corporations in relation to environmental degradation and human rights violations. Thus, on the report of Chile, it was recommended adopting measures to implement the national action plan on business and human rights, and ensuring its implementation in order to reduce negative impacts on human rights and the environment.[104] On the report of China, it was recommended establishing a legal framework to guarantee that activities carried out by industries subject to its jurisdiction do not negatively impact human rights abroad, and extending Chinese laws, regulations, and standards such as the Guiding Principles on Business and Human rights to these companies.[105]

The working group recommended India to establish regulations to ensure that the business sector complies with international and national human rights, labor, environmental, and other standards while on UK's report it was recommended enacting legislation to ensure accountability on human rights violations and environmental damages resulting from global operations of UK companies.[106] Likewise, on the report of Ecuador, the working group recommended continuing claims with transnational companies to get them to repair the serious environmental damage to the Amazon,[107] while on the report of Solomon Islands, the recommendations included ensuring that corporations, particularly those active in the forest industry, respect human and environmental rights, based on the Guiding Principles on Business and Human Rights.[108] The working group recommended Papua New Guinea to implement the accepted recommendations from the first UPR of increasing the oversight of extractive and logging industries and related businesses to mitigate the human rights and environmental impacts by establishing more rigorous and transparent measures.[109]

3.1.3 Resolutions of the HRC

3.1.3.1 Environmental Protection/Degradation

While the HRC has adopted many resolutions relating to the environment, especially climate change and sustainable development, the pride of place must surely go to Resolution 48/13 that recognized, for the first time, the right to a clean, healthy, and sustainable environment as a human right.[110] After years of lobbying by civil society groups and academics alike, it took the human rights community almost 50 years since the 1972 Stockholm Conference[111] to recognize it. This landmark resolution, which was years in the making, recognized that "environmental degradation, climate change and unsustainable development constitute some of the most pressing and serious threats to the ability of present and future generations to enjoy human rights, including the right to life"[112] and that the human rights implications of environmental damage are felt most acutely by those who are already in vulnerable situations including indigenous peoples, older persons, persons with disabilities, and women and girls.[113] Noting that over 155 countries have already recognized some form of a right to a healthy environment, the HRC recognized the right to a clean, healthy, and sustainable environment as a human right that is important for the enjoyment of human rights, and affirmed that its protection requires the full implementation of multilateral environmental agreements.[114]

It encouraged states to build capacities to protect the environment; share good practices in fulfilling human rights obligations relating to a healthy environment; adopt policies for the enjoyment of the right to a healthy environment; and take into account human rights obligations relating to the environment when implementing the SDGs. Furthermore, it invited the UN General Assembly to consider the matter.[115] Granted, the resolution is only a soft law instrument, and its wording is even softer. However, even in this diluted form, the recognition of a stand-alone right to a healthy environment has both political and legal ramifications. It sent a clear signal to the international community that environmental rights are important, they are justiciable, and other protected rights could be jeopardized if a right to a healthy environment is not respected.[116]

This resolution is the culmination of a series of resolutions on human rights and the environment adopted by the HRC. For example, its 2011 resolution reaffirmed that human rights obligations have the potential to inform and strengthen policymaking in relation to environmental protection and urged states to take human rights into consideration when developing their environmental policies. It requested the OHCHR to conduct a detailed analytical study on the relationship between human rights and the environment.[117] By resolution 19/10, the HRC created a mandate for an independent expert on human rights and environment to study the human rights obligations relating to the enjoyment of a safe, clean, healthy, and sustainable environment; identify best practices; make recommendations toward the realization of MDGs; consider the results of the 2012 UN Conference on Sustainable Development; apply a gender perspective; and work in close coordination with other special procedures.[118]

Resolution 25/21 urged states again to comply with human rights obligations when developing their environmental policies. It referred to the important role played by human rights defenders in relation to a healthy environment, and the importance of human rights obligations when developing the post-2015 development agenda. Noting that human beings are at the center of concerns for sustainable development, the resolution stated that the right to development must be fulfilled in order to meet the developmental and environmental needs of present and future generations equitably.[119] By resolution 28/11, the HRC extended the mandate of the Independent Expert as a Special Rapporteur and called upon the Special Rapporteur to convene an expert seminar on the effective implementation of human rights obligations relating to the environment.[120] Resolution 34/20 reiterated the link between human rights and the environment:

> [t]he impact of climate change, the unsustainable management and use of natural resources, the unsound management of chemicals and waste, the

> resulting loss of biodiversity and the decline in services provided by ecosystems may interfere with the enjoyment of a safe, clean, healthy and sustainable environment, and ... environmental damage can have negative implications, both direct and indirect, for the effective enjoyment of all human rights...[121]

3.1.3.2 Climate Change

The HRC's first resolution on climate change was issued in 2007 soon after the Inuit petition was dismissed by the Inter-American Commission of Human Rights.[122] Concerned that "climate change poses an immediate and far-reaching threat to people and communities around the world and has implications for the full enjoyment of human rights,"[123] HRC requested the OHCHR to conduct an analytical study on the relationship between climate change and human rights[124] and encouraged states to contribute to the study.

In resolution 10/4 the HRC noted that climate change has a range of implications for the enjoyment of human rights including the rights to life and safe drinking water and sanitation, and recognized that the effects of climate change will be felt by those who are already in vulnerable situations due to poverty, gender, age, indigenous or minority status, and disability.[125] Affirming that human rights obligations can inform and strengthen climate policy, the HRC decided to hold a panel discussion on the relationship between climate change and human rights, and requested the OHCHR to prepare a summary of the panel discussion and make it available to the conference of the parties to the UNFCCC for consideration. The HRC welcomed the decision of the Special Rapporteur on Housing to prepare a report on the impact of climate change on housing and encouraged other mandate holders to consider climate change within their mandates.[126]

The HRC, stressing the importance of holding the temperature increase to below 2 degrees Celsius above pre-industrial levels and trying to limit it to 1.5 degrees, called upon states to consider human rights within the UNFCCC and to "adopt a comprehensive, integrated and gender-responsive approach" to adaptation and mitigation policies.[127] It decided to hold a panel discussion on women's rights and climate change focusing on best practices and requested the OHCHR to prepare an analytical study on the integration of a gender-responsive approach into climate action and invited special mandate holders to contribute to the panel discussion.[128]

Resolution 41/21 referred to the importance of facilitating interaction between human rights and climate communities at national and international levels to build capacity, taking into account the Geneva Pledge for Human Rights in Climate Action[129] and similar efforts, and expressed concern that climate change is contributing to the increased frequency and intensity of both sudden-onset and slow-onset events. It urged states to implement policies to increase the participation of persons with disabilities in climate responses and decided to hold a panel discussion on rights of persons with disabilities in the context of climate change, focusing on best practices. It requested the OHCHR to submit a summary report of the panel discussion and to conduct an analytical study on the topic and invited special mandate holders to contribute to it.[130]

Resolution 47/24 reaffirmed the commitment to implement the UNFCCC and Paris Agreement and stated that responses to climate change should be integrated with social and economic development.[131] Recognizing that climate change posed an existential threat to some countries, the HRC expressed concern that adverse effects are felt more acutely by those who are already in a vulnerable situation due to geography and other factors, it called upon states to *consider* human rights within the UNFCCC.[132] Unfortunately, by calling upon states *to consider* human rights in relation to climate action, the HRC has diluted the importance of human rights obligations. States are *legally bound* to fulfill their human rights obligations in relation to climate change, including mitigation, adaptation, and extreme weather events. The HRC has thus denigrated the legal status of human rights obligations which states may feel that they can disregard if they are so inclined.

In addition, the HRC requested the Secretary General, in consultation with states, special procedures, and relevant international and intergovernmental organizations, including the IPCC, to submit a report on the adverse impact of climate change on human rights of people in vulnerable situations.[133] It encouraged states to discuss the possible creation of a new special procedure addressing the impact of climate change on human rights.[134]

3.1.3.3 Sustainable Development and SDGs

Almost all the resolutions discussed in the previous section refer to sustainable development, SDGs and/or 2030 Agenda. For example, resolution 16/11 referred to sustainable development eight times. Referring to the Millennium Declaration, the resolution noted how respect for human rights can contribute to sustainable development for which good governance is essential. It referred to the Rio Conference on Environment and Development, Agenda 21, Johannesburg Declaration on Sustainable Development and Principle 7 of the Rio Declaration on the CBDR principle.

Similarly, resolution 47/24 referred to 2030 Agenda and Goal 13 on taking urgent action on climate change. It recognized that poverty eradication is critical to the implementation of SDGs, climate resilience, and the protection of human rights including those in vulnerable situations, inhabitants of small island states, and least developed countries who are disproportionately affected by climate change. Resolution 48/13 that recognized a right to a healthy environment also refers to Agenda 2030 and SDGs:

> [s]ustainable development, in its three dimensions (social, economic and environmental), and the protection of the environment, including ecosystems, contribute to and promote human well-being and the enjoyment of human rights, including the rights to life, to the enjoyment of the highest attainable standard of physical and mental health, to an adequate standard of living, to adequate food, to housing, to safe drinking water and sanitation and to participation in cultural life, for present and future generations.[135]

Additionally, similar to other resolutions, it identified environmental degradation, climate change and *unsustainable development* as some of the most serious threats to the enjoyment of rights. It also referred to the unsustainable management of natural resources as interfering with the enjoyment of a clean, healthy, and sustainable environment.

In addition to these resolutions, the HRC has compiled a document titled "How the universal periodic review process supports sustainable development" as part of UN Good Practices.[136] It identified three interlinked pillars that shape the UN's priorities: peace and security, sustainable development, and human rights. It notes that among the multiple mechanisms available to support states in fulfilling these pillars including achieving SDGs, the UPR process stands out. The guide stresses that the human rights framework underpins the 2030 Agenda and over 90% of the targets align with human rights standards. Another trend is the UPR process which can be used as a problem-solving tool to address challenges like climate change and the COVID-19 pandemic. It noted that as of November 2021, approximately 250 recommendations on climate change have been made during the 3rd cycle.[137]

3.2 Office of the High Commissioner for Human Rights

As noted in Chapter 2, the High Commissioner for Human Rights is the principal human rights officer and heads the OHCHR which supports all treaty bodies and special procedures. In this section, we will discuss the work of the OHCHR in relation to environmental protection, climate change, and sustainable development including SDGs. Since the OHCHR has addressed environmental degradation and climate change together, this section will do the same.

3.2.1 Environmental Protection/Degradation

OHCHR believes that the human rights framework lays down the responsibilities of duty-bearers with respect to all human rights harms, *including those caused by environmental degradation.*[138] In this regard, its work on health, poverty and right to food, water and sanitation, and land and housing is relevant. With regard to the right to health which includes promoting rights-based climate action,[139] OHCHR elaborates on its key aspects as availability, accessibility, acceptability, quality, participation, and accountability.[140] Relying on GC No 14 of the ESCR Committee,[141] the OHCHR points out that the right to health is an inclusive right extending to underlying determinants of health: access to safe and potable water and adequate sanitation; an adequate supply of safe food, nutrition, and housing; healthy occupational and environmental conditions; and access to health-related education and information.[142] It is thus clear that a healthy environment is critical to enjoy the right to health.

Closely related is access to safe drinking water and sanitation which is necessary to live a life in dignity. However, 2.1 billion people lack access to safe water while 4.5 billion people lack access to sanitation[143] which disproportionately impacts the most disadvantaged and marginalized members of society.[144] Since the recognition by the UNGA that the right to safe drinking water and sanitation is a human right, OHCHR has elaborated on various aspects of this right by compiling report on eliminating discrimination and inequalities, preparing a handbook on the right to water and sanitation, and compiling good practices. The report on Eliminating Discrimination and Inequalities notes that the right to water "entitles everyone without discrimination to sufficient, safe, acceptable, physically accessible and affordable water for personal and domestic use."[145]

The report notes that:

> Non-discrimination and equality are fundamental principles of human rights law. The terms 'equity' and 'equality' are sometimes used interchangeably, yet there is an important difference: Equality is a legally defined term and a binding principle of human rights law, while equity is a moral imperative that is open to diverse interpretations.[146]

OHCHR engages with governments and local stakeholders and supports the mandate of the Special Rapporteur on Right to Water and Sanitation, strengthening partnerships, integrating human rights to water and sanitation in the UN system, and supporting the integration of human rights under SDG 6 Global Acceleration Framework.

With regard to the right to food, OHCHR identified its four key elements as availability, accessibility, adequacy, and sustainability and noted that food should be safe for human consumption and free from adverse substances. Moreover, the right to adequate food is crucial to the enjoyment of other rights including the rights to health, life, water, adequate housing, and education.[147] Similar to other rights, the OHCHR promotes the right to food by disseminating international human rights standards and principles; providing advice and assistance to states on laws, policies, and action plans; engaging intergovernmental bodies, mainstreaming human rights in UN agencies work, and supporting the work of civil society groups, including small scale farmers.

3.2.2 Climate Change

It would not be wrong to say that the bulk of the OHCHR's work relating to the environment has been on climate change. In line with Agenda 2030 and the Paris Agreement, OHCHR seeks to promote a rights-based approach to climate action. Pursuant to HRC's 2007 resolution on climate change, the OHCHR prepared an analytical report on the link between climate change and human rights.[148] In its report, the OHCHR outlined the causes and consequences of climate change and referred to IPCC reports that present scientific consensus on climate change. It noted the unequal burden of climate change on certain countries and the role of equity and the CBDR principle.[149] Noting that while human rights treaties do not embody a right to a healthy environment but the treaty bodies have recognized the link between human rights and the environment,[150] the report went on to discuss the impact of global warming on specific rights, ranging from the right to life to the right to self-determination. It also discussed the impact on specific groups especially women, indigenous peoples, and children stating that "[t]he effects of climate change will be felt most acutely by those segments of the population who are already in vulnerable situations due to factors such as poverty, gender, age, minority status, and disability"[151] and that states must address these vulnerabilities under principles of equality and non-discrimination. The report discussed various displacement scenarios[152] and noted that "climate refugees" do not have a right of entry as they have no legal basis in international law. Likewise, while the eventual submergence of small island states is a possibility, human rights law does not provide answers to the status of these populations.[153]

Moreover, the report notes that there are human rights implications of measures that states adopt to address climate change such as agro-fuel production and the REDD program.[154] While there is broad agreement that climate change has negative implications for the enjoyment of human rights, the report cautiously concluded that "it is less obvious whether, and to what extent, such effects can be qualified as human rights violations in a strict legal sense."[155] Nonetheless, human rights obligations provide protections to people whose rights are affected by climate change and requires international cooperation among states.[156]

In its document titled 'key messages on human rights climate change,' OHCHR notes that climate change affects an array of human rights, both directly and indirectly.[157] States as duty-bearers have an obligation to take effective measures to prevent and redress these impacts and to ensure that all rights-holders have the necessary adaptive capacity. It further notes that climate justice requires that climate action is consistent with existing human rights obligations as those who contributed the least suffer its harms disproportionately. The obligations of states and other duty-bearers including businesses in relation to climate action include: (a) mitigating climate change and preventing its negative human rights impacts; (b) ensuring all persons have the necessary adaptive capacity; (c) ensuring effective and informed participation, accountability and effective remedy for human rights harms; (d) mobilizing maximum available resources for sustainable, human rights-based development; (e) ensuring equity, equality and non-discrimination; and (h) protecting from violations of human rights by businesses.[158]

The 2017 report on the panel discussion on climate change, human rights, migrants, and displacement across international borders[159] noted that both slow-onset and sudden events have a devastating impact on people and the planet. Since 2008, about 221.5 million people per year have been displaced, internally and across borders, by weather or climate-related disasters. Unjustly, climate change disproportionately harmed people who had contributed the least to it[160] and affected the enjoyment of rights by millions of people. The report pointed out that migrants fled the effects of climate change not out of choice but because of the need to escape conditions that could not provide the basic fundamental rights. Throughout their migration they faced violations of other rights including xenophobia. While human rights law did provide for the protection of those who are on the move, there are gaps with regard to those fleeing the adverse consequences of climate change, especially slow-onset impacts.[161]

Moreover, the report highlighted the need to address the underlying causes of migration. A key issue is to address the protection needs of all those compelled to move by complex factors and to design long-term solutions, particularly when returning home is not an option, paying greater attention to gender aspects as 48% of migrants are women.[162] While the majority of displaced persons remained within their own country, some crossed international borders. However, in most instances, international law did not provide a right of admission and stay. The Nansen Initiative[163] showed that human rights played a role, by: first, reducing displacement risks in countries of origin by addressing poverty, population density, bad governance, and discrimination; second, ensuring access to food, water, shelter, and health for those who are migrating. Finally, a rights-based approach could help in relation to admission to other countries in disaster-affected situations. In the absence of specific obligations to admit people displaced across borders due to climate change and other disasters, harmonizing and strengthening national approaches is crucial.[164] Participants agreed that states should protect the rights of migrants regardless of their status and millions of people were on the move due to climate change with some regions and countries being particularly affected exacerbated by the protection gap in international law.[165] They called for "a human rights-based approach to climate change that put the interests of all people, including migrants, at the centre…."[166] Some called for initiatives at the regional level as migration was often regional.

The participants called for a right-based approach to climate migration based on principles of equality, non-discrimination, and the CBDR.[167] They called upon the HRC to contribute to the Global Compact on Migration by drawing the linkages between climate change, migration, and human rights.[168] Likewise, they called for the Global Compact to recognize climate change as a driver of migration and to address migration triggers such as poverty, inequality, insecurity and natural disasters.[169] They called for the urgent implementation of the Paris Agreement and for the Task Force on Displacement established under the Paris Agreement to integrate human rights into its work plan. Moreover, climate change and migration policies should meet the needs of vulnerable groups, address protection gaps, and clarify and strengthen international environmental law. The report recommended that the human rights framework should guide the work of bodies established under the UNFCCC. Likewise, human rights mechanisms, including the treaty bodies, should support states regarding human rights obligations applicable to climate change, including extreme weather events and slow-onset processes.[170]

Pursuant to HRC resolution 38/4, the OHCHR prepared an analytical study on gender-responsive climate action.[171] It noted that diverse factors, including social status, gender and poverty affects one's capacity to adapt to climate change. While human rights law prohibits gender-based discrimination, women often face systemic discrimination including limited access to financial assets and services, education, land resources, and decision-making:

> In general, women are more likely to experience the adverse effects of climate change than men, because women constitute most of the world's poor and are often directly dependent on threatened natural resources as their primary source of food and income.[172]

Because of the global perpetuation of discrimination and patriarchal structures, women face a higher risk. Since climate change perpetuates gender inequality it is essential to address the gendered impact of climate change.

The report noted that many smallholder farmers are women whose livelihoods and food sources are at risk due to climate change which contribute to increasing food insecurity among women. Likewise, male-dominated structures that govern land ownership, unequitable distribution of domestic and care work as well as discriminatory laws that restrict women's ownership of land negatively impact their adaptation to climate change. Moreover, climate change can disproportionately affect the physical and mental health of women especially during severe weather events due to scarcity of water, poor air quality and spread of vector-borne diseases. Noting that human mobility poses unique risks to women, the report points out that they are exposed to gender-based violence as are LGBTQ+ persons. Both sudden and slow-onset events can drive human mobility affecting homes, lands, and territories. The report also addressed the situation of women human rights defenders.

Noting that states have obligations to implement gender-responsive climate policies, the report examined the relevant provisions of the ICESCR and CEDAW both of which prohibit gender-based discrimination. Moreover, UNFCCC and over 60 other documents address the need to ensure gender quality. In 2017, the parties adopted the Gender Action Plan under the Lima Work Program to guide gender-responsive approaches to climate change.[173] A gender-responsive approach requires the full participation of women and ensuring that a just transition to a low-carbon economy does not perpetuate gender inequality. Climate finance is another area that requires a gender-responsive approach.[174]

Pursuant to HRC resolution 41/21, the OHCHR prepared an analytical study on the promotion and protection of rights of persons with disabilities in the context of climate change.[175] Noting that impairment due to disability results from the interaction with attitudinal and environmental barriers which hinder their full participation in society on par with others. Thus, persons with disabilities are at risk of harm due to climate change which has both direct and indirect impacts on the enjoyment of rights. An estimated 1 billion people who have a disability may experience the adverse impacts more severely than others, especially during emergencies sustaining higher rates of morbidity and mortality. Both sudden and slow-onset events can affect their access to food and nutrition, safe drinking water, and sanitation and health care services.[176] Unfortunately, the majority of persons with disabilities live in poverty. As the IPCC notes, poorest people will continue to experience the worst impacts of climate change due to lost income, displacement, hunger, and health problems. Multiple intersecting factors of discrimination related to gender, poverty, displacement, disability, age, and indigeneity increase the risks of climate change experienced by persons with disabilities. Thus, their participation will allow for tailored action that addresses their specific needs:

> A human rights-based approach … empowers persons with disabilities as agents of change to address the harmful impacts of climate change in their day-to-day lives. If persons with disabilities are left out of decision-making, that leaves them unable to contribute by identifying risk reduction and adaptation measures that could be effective for, and carried out by, persons with disabilities.[177]

The report discusses health care inequities, food insecurity, access to accessible homes, water and sanitation, livelihoods and decent work and mobility faced by persons with disabilities. While climate-induced extreme weather events was one of the main causes of movement of 28 million people in 2018, the ability to migrate depends on resources and mobility and those who move may be in need of refugee protection. However, "persons with disabilities are at risk of being left behind in a degraded environment without social and support networks when members of their family or community move because of climate change impacts."[178] Those who do move also face challenges related to mobility, accessible transportation, accommodation and services, and resettlement possibilities are often limited. The report noted that equality and non-discrimination are fundamental principles of the Disability Rights Convention and are interconnected with human dignity. The Committee on the Rights of Persons with Disabilities has recognized that climate change exacerbates inequality and vulnerability among persons with disabilities and has recommended that states mainstream disability inclusion in their policies and programs on climate change.[179] The Committee warned that the failure to take climate action constituted a breach of human rights obligations. Referring to the UNFCCC, the Doha work program and the 2018 report of the Adaptation Committee, the report encouraged states to adopt a participatory approach and seek input from stakeholders including persons with disabilities for adaptation planning. Similar to the report on gender, the report recommended fully integrating human rights and disability into climate action.[180]

Pursuant to Human Rights Council resolution 44/7, in 2021, OHCHR prepared an analytical study on the protection of rights of older persons in relation to climate change.[181] Similar to the reports discussed above, OHCHR prepared this report after circulating a questionnaire to member states and other stakeholders including national human rights institutions and civil society to solicit their views. Noting that by 2050 there will be approximately 1.5 billion people aged 65 years and older, the report pointed out that while age itself does not make individuals more vulnerable to climate change, in combination with physical, political, economic and social factors, older persons may be at more risk as we saw during the COVID-19 pandemic.

Older persons do not constitute a homogenous group and there is enormous variance among them; neither is there an agreed definition. However, they are often overlooked in research and data collection, and individuals may not self-identify within the "older persons" category. Human rights impacts of climate change are aggravated by ageism and older persons are often considered as passive and incapable. They do not have a specific treaty dedicated to them and environmental instruments often do not mention them. At the same time, climate impacts could affect older persons disproportionately. They are especially affected by heatwaves, extreme cold and during severe weather events. During the 2003 heatwave in Europe, 80% of additional deaths occurred in persons over 75 while during Hurricane Katrina in the US, 75% of those who died were over the age of 60.[182] Moreover, their mental health is affected due to loneliness and isolation.

Another area that affects older persons disproportionately is human mobility. In emergencies, they may have difficulty reaching safety while they may not be aware of warnings and evacuation orders, especially if new technologies are used to disseminate such information. If they are displaced, they face additional hurdles in accessing humanitarian aid if they are unfamiliar with new technology to access networks and information. Other rights affected include rights to food, water and sanitation, social protection, care and support, decent work and livelihoods, and cultural rights. The report also discussed multiple and intersecting forms of discrimination in relation to gender, disabilities, minorities, and indigeneity.[183]

Discussing the key human rights instruments, the report pointed out that treaties such as the ICCPR, ICESCR, CEDAW and the CRPD protect older persons' human rights even though age is not specifically identified as a prohibited ground of discrimination. Moreover, states have legal obligations to implement climate policies that empower all people by ensuring their full participation in climate action[184] and several general comments refer to older persons.

Similar to other vulnerable groups, we should consider older persons as agents of change due to their knowledge, expertise and leadership and they may have more time and resources to devote to climate action. The younger generation, who have received attention for their innovative climate action in recent years, can learn from the experience of the older generation and build intergenerational solidarity in the climate movement. Older persons who are in positions of power can amplify the voices of younger people and validate their action.

The recommendations include strengthening the international legal system by conducting empirical research on the impact of climate change on older persons and their rights; including the rights of older persons in future climate COP decisions; adopting an international legal instrument on human rights of older person; and taking urgent action to mitigate and adapt to climate change that protects human rights of all.[185]

3.2.3 *Sustainable Development and SDGs*

While many of the reports discussed above refer to sustainable development and SDGs, OHCHR has a separate topic titled "sustainable development through human rights."[186] OHCHR notes that the SDGs are the blueprint to achieve a sustainable future for all and "they address the global challenges we face, including poverty, inequality, climate change, environmental degradation, peace and justice."[187] Noting that human rights are essential to achieve sustainable development that leaves no one behind, OHCHR stressed that 2030 Agenda is firmly anchored in human rights principles and standards.

As part of its mission, OHCHR works with states, NHRIs, civil society and other stakeholders in several areas: intergovernmental engagement by strengthening the linkage between human rights and sustainable development in the High Level Political Forum and the HRC; designing policies with the UN system to support SDG implementation; providing human rights analysis and content to national sustainable development strategies and plans; advocating for human rights sensitive indicators and a rights-based approach to the collection, analysis, and dissemination of data; and the Addis Ababa Agenda for Action[188] to revitalize the global partnership for sustainable development.

3.3 Conclusion

In this chapter we discussed how the two leading human rights bodies – the HRC and OHCHR – have addressed environmental issues within their mandate. The discussion revealed that both take the impact of environmental degradation, climate change, and unsustainable development on the enjoyment of rights very seriously. Indeed, it was the HRC that took the unprecedented step of recognizing the right to a healthy environment as a distinct right.

From the working group recommendations, resolutions and reports we discussed in this chapter, it is clear that environmental rights have a firm place on the agenda of both these institutions. Likewise, sustainable development and SDGs are firmly entrenched. Almost all the reports compiled by the OHCHR and the resolutions adopted by the HRC after 2015 refer to 2030 Agenda, the SDGs, and occasionally to its targets.

The trajectory of working group recommendations from older reports to later ones is notable. Older recommendations are not very detailed and references to the environment almost seemed to be in passing. With the worsening impacts of climate change, many of the recent reports have very detailed discussions, comments, and recommendations on environmental issues, climate change and SDGs. Some of the very detailed comments and recommendations on climate change appear on the reports of small island states and major emitters, emphasizing the need to fulfill their commitments under the Paris Agreement and cooperate with other states.

Likewise, very detailed comments and recommendations have been made in relation to indigenous rights including ratifying ILO Convention 169, implementing the free, prior and informed consent principle, and consulting with them, protecting their ancestral lands and culture, and incorporating provisions of UNDRIP into national law. A few reports have referred to the need to adopt legislation to control the activities of transnational corporations abroad while many have referred to the need to adopt environmental legislation, standards and policies, ensure consultation with affected communities, and provide reparations for past wrongs.

Notes

1 See Chapter 2.
2 See OHCHR website: https://www.ohchr.org/en/about-us/high-commissioner. In accordance with UNGA resolution 48/141 (20 December 1993) which created the post of the High Commissioner for Human Rights (HCHR), HCHR is appointed by the UNSG and approved by the UNGA. Resolution 48/141 is available at: http://www.un-documents.net/a48r141.htm.
3 Michelle Bachelet, the immediate past High Commissioner was an ardent supporter of a right to a healthy environment and called on states to take bold actions to give effect to the right to a healthy environment, following UN HRC's resolution recognizing it: https://www.ohchr.org/en/2021/10/bachelet-hails-landmark-recognition-having-healthy-environment-human-right.
4 *Secretary-General Appoints Mary Robinson of Ireland Special Envoy for Climate Change* (14 July 2014), https://press.un.org/en/2014/sga1481.doc.htm.
5 It is impossible to summarize all the relevant work here but we will highlight significant resolutions, working group recommendations under the UPR process, and other relevant documents.
6 *Resolution 60/251*, A/RES/60/251 (3 April 2006), https://www2.ohchr.org/english/bodies/hrcouncil/docs/a.res.60.251_en.pdf.
7 According to Resolution 60/251, UNGA decided to establish the Human Rights Council in replacement of the Commission on Human Rights, as a subsidiary organ of the General Assembly.
8 Cantú Rivera, H. (2018) "The UN Human Rights Council: Achievements and Challenges in its First Decade," in *International Human Rights Institutions, Tribunals, and Courts*, Oberleitner, G., ed., Singapore: Springer, p. 51, doi: 10.1007/978-981-10-5206-4_3.
9 *Resolution 60/251*, A/RES/60/251 (3 April 2006), available at: https://undocs.org/A/RES/60/251. This chapter will not engage in a debate on the pros and cons of the HRC, its successes and failures or recommendations for reform. The chapter by Cantú Rivera, supra note 8, provides a good summary.
10 See HRC webpage: https://www.ohchr.org/en/hrbodies/hrc/home.
11 *Resolution 48/13*, A/HRC/RES/48/13 (8 October 2021) ["Resolution 48/13"].
12 *Resolution 16/11*, A/HRC/RES/16/11 (12 April 2011).
13 See HRC Subsidiary Bodies: https://www.ohchr.org/en/hr-bodies/hrc/other-sub-bodies.
14 Ibid.
15 See HRC Universal Periodic Review: https://www.ohchr.org/en/hr-bodies/upr/upr-main.
16 Resolution 60/251, supra note 9, ¶ 5(e).
17 *Report of the Working Group on the Universal Periodic Review, Plurinational State of Bolivia*, A/HRC/28/7 (17 December 2014), ¶ 115.3 ["Bolivia report"]. Several other recommendations did not receive Bolivia's support.
18 This resembles the practice of "shadow reports" submitted to treaty bodies although this seems a more formalized process as the summary is prepared by the OHCHR itself.
19 *Mapping Human Rights Obligations Relating to the Enjoyment of a Safe, Clean, Healthy and Sustainable Environment, Individual Report on the UN General Assembly and the Human Rights Council, including the Universal Periodic Review Process, Report No.* 6 (December 2013) ["Mapping Report No. 6"], available at: https://www.ohchr.org/en/special-procedures/sr-environment/mapping-report.

20 A keyword search of working group reports during 2014–2021 came up with the following interesting results: "climate change" was mentioned 716 times, "sustainable development" – 692 times, "SDGs" – 289 times, "environmental" – 242 times, "indigenous" – 585 times, "health" – 243 times, "disaster" – 295 times, "water" – 146 times, "food" – 81 times, "adaptation" – 63 times, and "mitigation" – 42 times.
21 Mapping Report No. 6, supra note 19.
22 Ibid.
23 Bolivia report, supra note 17.
24 *Report of the Working Group on the Universal Periodic Review, Plurinational State of Bolivia*, A/HRC/43/7 (20 December 2019) ["2019 Bolivia report"]; ECLAC, *Regional Agreement on Access to Information, Public Participation and Justice in Environmental Matters in Latin America and the Caribbean* (27 September 2018), available at: https://www.cepal.org/en/escazuagreement.
25 *Report of the Working Group on the Universal Periodic Review, Kazakhstan*, A/HRC/43/10 (20 December 2019) ["Kazakhstan report"].
26 *Report of the Working Group on the Universal Periodic Review, Iraq*, A/HRC/43/14 (20 December 2019).
27 *Report of the Working Group on the Universal Periodic Review, Bosnia and Herzegovina*, A/HRC/43/17 (19 December 2019). This is one of the few references to rights of older persons.
28 *Report of the Working Group on the Universal Periodic Review, Guinea*, A/HRC/44/5 (20 March 2020).
29 A similar recommendation was made on the reports of Lao, Kenya, Nicaragua. See also discussion in Section 3.1.2.4.
30 *Report of the Working Group on the Universal Periodic Review, Lao People's Democratic Republic*, A/HRC/44/6 (17 March 2020) ["Lao report"].
31 *Report of the Working Group on the Universal Periodic Review, Democratic Republic of the Congo*, A/HRC/42/5 (5 July 2019).
32 *Report of the Working Group on the Universal Periodic Review, Uruguay*, A/HRC/41/18 (18 April 2019) ["Uruguay report"].
33 *Report of the Working Group on the Universal Periodic Review, Chile*, A/HRC/41/6 (2 April 2019) ["Chile report"].
34 *Report of the Working Group on the Universal Periodic Review, Chile*, A/HRC/26/5 (2 April 2014).
35 *Report of the Working Group on the Universal Periodic Review, Viet Nam*, A/HRC/41/7 (28 March 2019) ["Viet Nam report"].
36 *Report of the Working Group on the Universal Periodic Review, Nauru*, A/HRC/31/7 (16 December 2015).
37 *Report of the Working Group on the Universal Periodic Review, Nigeria*, A/HRC/40/7 (26 December 2018).
38 *Report of the Working Group on the Universal Periodic Review, Nigeria*, A/HRC/25/6 (16 December 2013).
39 *Report of the Working Group on the Universal Periodic Review, China*, A/HRC/40/6 (26 December 2018) ["China report"].
40 *Report of the Working Group on the Universal Periodic Review, Costa Rica*, A/HRC/27/12 (7 July 2014) ["Costa Rica report"].
41 *Report of the Working Group on the Universal Periodic Review, Suriname*, A/HRC/33/4 (1 July 2016) ["Suriname report"].
42 *Report of the Working Group on the Universal Periodic Review, Ecuador*, A/HRC/36/4 (10 July 2017) ["Ecuador report"].
43 *Report of the Working Group on the Universal Periodic Review, Maldives*, A/HRC/30/8 (13 July 2015).
44 *Report of the Working Group on the Universal Periodic Review, Former Yugoslav Republic of Macedonia*, A/HRC/26/10 (26 March 2014).

45 *Report of the Working Group on the Universal Periodic Review, Bhutan*, A/HRC/27/8 (1 July 2014).
46 *Report of the Working Group on the Universal Periodic Review, Tuvalu*, A/HRC/10/84 (9 January 2009), ¶ 21.
47 Ibid., ¶ 15.
48 *Report of the Working Group on the Universal Periodic Review, Tuvalu*, A/HRC/24/8 (5 July 2013).
49 *Report of the Working Group on the Universal Periodic Review, Solomon Islands*, A/HRC/32/14 (13 April 2016) ["Solomon Islands report"], It is not clear what a "national protection group" is.
50 Ibid. This is the first time a reference was made to the REDD program and forest carbon stocks.
51 *Report of the Working Group on the Universal Periodic Review, Italy*, A/HRC/43/4 (27 December 2019). The exact recommendation appears on the report of Gambia, San Marino, Kazakhstan, Angola, Iran, Madagascar, Iraq, Slovenia, Egypt, Bosnia and Herzegovina, Portugal (with an addition of indigenous and marginalized communities), Bhutan (with a reference to devising holistic strategies on managing climate change and its impacts on livelihoods), Albania, Cote d'Ivoire, Mauritius, China, and Belize. On the report of El Salvador recommendation was made to ensure that these groups participate in the implementation of the National Plan on Climate Change; take steps to mitigate the effects of climate change; strengthen the national strategy to address climate change; and establish a mechanism to respond to droughts. While Uruguay welcomed efforts to ratify the Escazú Agreement, it was not included in the recommendations, *Report of the Working Group on the Universal Periodic Review, El Salvador*, A/HRC/43/5 (2 January 2020).
52 *Report of the Working Group on the Universal Periodic Review, Kyrgyzstan*, A/HRC/44/4 (19 March 2020). The exact recommendation appears in the report on Guinea, Lao, Spain and Lesotho.
53 2019 Bolivia report, supra note 24.
54 *Report of the Working Group on the Universal Periodic Review, Spain*, A/HRC/44/7 (18 March 2020).
55 *Report of the Working Group on the Universal Periodic Review, Lesotho*, A/HRC/44/8 (18 March 2020). NDCs are the commitments that states have made to reduce their GHG emissions under the Paris Agreement.
56 *Report of the Working Group on the Universal Periodic Review, Fiji*, A/HRC/43/8 (23 December 2019).
57 Similar recommendation was made on the report of Chile, see Chile report, supra note 33.
58 Ibid.
59 *Report of the Working Group on the Universal Periodic Review, Islamic Republic of Iran*, A/HRC/43/17 (27 December 2019).
60 *Report of the Working Group on the Universal Periodic Review, Kenya*, A/HRC/44/9 (20 March 2020).
61 *Report of the Working Group on the Universal Periodic Review, Nicaragua*, A/HRC/42/16 (5 July 2019) ["Nicaragua report"]. Similar recommendation was made on the report of Ethiopia and to strengthen national disaster preparedness plans to adapt to the changing climate and to reduce the risk of future disasters and address risks of gender-based violence; continue to implement national strategy countering the effects of climate change, including adaptation and mitigation and integration of disaster risk reduction, *Report of the Working Group on the Universal Periodic Review, Ethiopia*, A/HRC/42/14 (5 July 2019). Similar recommendations were made on the report of Costa Rica, Albania, Cote d'Ivoire, Equatorial Guinea, and Brunei Darussalam.
62 Nicaragua report, supra note 61. It was also recommended to protect ancestral cultural identity of indigenous peoples.

63 *Report of the Working Group on the Universal Periodic Review, Costa Rica*, A/HRC/42/122 (5 July 2019).
64 *Report of the Working Group on the Universal Periodic Review, Democratic People's Republic of Korea*, A/HRC/42/10 (25 June 2019).
65 *Report of the Working Group on the Universal Periodic Review, Comoros*, A/HRC/41/12 (25 April 2019) ["Comoros report"].
66 *Report of the Working Group on the Universal Periodic Review, Vanuatu*, A/HRC/41/10 (5 April 2019).
67 The different formulation of the recommendation by Fiji on this report is noteworthy. On many other reports, Fiji's recommendations have been very similar, using almost identical language. Barbados and Iceland also recommended including women in the decision-making process.
68 The continued reference to *natural* disasters, even in the context of climate change is disconcerting and misleading. This can be seen in the work of most human rights bodies discussed in the book.
69 Viet Nam report, supra note 35.
70 China report, supra note 39.
71 Ibid.
72 It is noteworthy that the CBDR principle is referred to here (although the wording is modified).
73 *Report of the Working Group on the Universal Periodic Review, USA*, A/HRC/30/12 (20 July 2015) ["USA report"].
74 *Report of the Working Group on the Universal Periodic Review, Tuvalu*, A/HRC/39/8 (10 July 2018).
75 *Report of the Working Group on the Universal Periodic Review, Kiribati*, A/HRC/29/5 (13 April 2015).
76 Ibid.
77 *Report of the Working Group on the Universal Periodic Review, India*, A/HRC/35/10 (17 July 2017). On Iceland's report, it was recommended moving towards increased renewable energy production, in line with their pledges on climate action.
78 The formulation referred to by China on many of its recommendations is rather curious – sometimes it referred to economic and social sustainable development and at other times it referred sustainable economic and social development and seems to give primacy to economic and social development at the expense of the environment, ignoring the environmental dimension of sustainable development.
79 Kazakhstan report, supra note 25.
80 Bolivia report, supra note 24.
81 *Report of the Working Group on the Universal Periodic Review, Angola*, A/HRC/43/11 (11 December 2019).
82 *Report of the Working Group on the Universal Periodic Review, Egypt*, A/HRC/43/16 (27 December 2019). Similar recommendation was made on the report of Afghanistan.
83 A similar recommendation was made on the report of Comoros and the need to establish a national body to coordinate reports on human rights and SDGs in a comprehensive manner and to use the 3rd cycle recommendations to generate data to support the implementation of SDGs and improve human rights, see Comoros report, supra note 65.
84 Lao report, supra note 30. On the report on Spain, it was recommended promoting *economic, social and sustainable development* to better protect rights of vulnerable groups (emphasis added). It is not clear why economic and social development was added to sustainable development as the latter encompasses both economic and social development together with environmental protection. On the report of Lao, reference was made to *economic and social sustainable development* to reduce those living in poverty (emphasis added). Reference was made to sustainable economic and

social development on the reports of Dominica, Cote d'Ivoire, Brunei Darussalam, and Dominican Republic.

85 *Report of the Working Group on the Universal Periodic Review, Ethiopia*, A/HRC/42/14 (5 July 2019). Similar recommendation was made on the report of Cote d'Ivoire, Equatorial Guinea, Uruguay and Chile (also referred to integrating good practices of implementing UPR recommendations into national reviews on SDGs), Cyprus (place human rights at the heart of the framework for achieving SDGs and continue to implement the 2030 Agenda with a focus on all goals that are directly related to human rights), Vanuatu (promoting an integrated approach to human rights and SDGs and implement the national plan for sustainable development to reduce poverty and raise living standards).

86 *Report of the Working Group on the Universal Periodic Review, Brunei Darussalam*, A/HRC/42/11 (25 June 2019).

87 *Report of the Working Group on the Universal Periodic Review, Kenya*, A/HRC/44/9 (20 March 2020).

88 Democratic Republic of the Congo, Belize, Russian Federation (including ratifying the Convention against Statelessness and the Biosafety Protocol to the Convention on Biological Diversity). Similar recommendation on the report of Philippines, *Report of the Working Group on the Universal Periodic Review, Philippines*, A/HRC/36/12, (18 July 2017).

89 This recommendation was made for 42 countries. On a few reports, it was recommended to ratify other treaties as well. Examples include: Dominican Republic, North Macedonia, Cyprus, Bangladesh, Canada (ratify human rights treaties that have not been ratified including the one on Statelessness and Convention on Migrant Workers and Statelessness), New Zealand (ratify Optional Protocol to the ICESCR), Russian Federation (formally endorse UNDRIP), and Zambia and Japan (sign the Treaty on Prohibition of Nuclear Weapons).

90 *Report of the Working Group on the Universal Periodic Review, Brazil*, A/HRC/36/11 (18 July 2017) ["Brazil report"]. Similar recommendations were made for Ecuador, see Ecuador report, supra note 42.

91 Brazil report, supra note 90.

92 Ibid. Similar recommendation was made on the report of Venezuela.

93 *Report of the Working Group on the Universal Periodic Review, Nicaragua*, A/HRC/27/16 (1 July 2014).

94 A similar recommendation was made for Venezuela, especially with regard to their ancestral land and for Papua New Guinea, *Report of the Working Group on the Universal Periodic Review, Papua New Guinea*, A/HRC/33/10, (13 July 2016) ["Papua New Guinea report"]. On Uruguay's report it was also recommended recognizing their identity, ethnic and cultural pre-existence, Uruguay report, supra note 32.

95 *Report of the Working Group on the Universal Periodic Review, Norway*, A/HRC/42/3 (28 June 2019). On Suriname's report, it was recommended respecting the rights of indigenous people related to their land, culture and resources, see Suriname report, supra note 41.

96 Uruguay report, supra note 32. Similar recommendations on the report of Malaysia.

97 *Report of the Working Group on the Universal Periodic Review, Cambodia*, A/HRC/41/17 (5 April 2019). Similar recommendation was made on Papua New Guinea's report (take measures to prevent land grabbing and illegal logging and ensure that future actions are based on respect for human rights and environmental protection).

98 Chile report, supra note 33. Similar recommendations on the report of Niger.

99 *Report of the Working Group on the Universal Periodic Review, New Zealand*, A/HRC/41/4 (1 April 2019).

100 USA report, supra note 73.

101 Suriname report, supra note 41.

102 A similar recommendation was made to Costa Rica, see Costa Rica report, supra note 40.
103 *Report of the Working Group on the Universal Periodic Review, Australia*, A/HRC/31/14 (13 January 2016).
104 Chile report, supra note 33.
105 China report, supra note 39.
106 *Report of the Working Group on the Universal Periodic Review, UK and Northern Ireland*, A/HRC/36/9, (14 July 2017).
107 Ecuador report, supra note 42.
108 Solomon Islands report, supra note 49.
109 Papua New Guinea report, supra note 94.
110 Resolution 48/13, supra note 11.
111 United Nations Conference on the Human Environment (5–16 June 1972), Stockholm, see: https://www.un.org/en/conferences/environment/stockholm1972.
112 Resolution 48/13, supra note 11, Preamble.
113 Ibid.
114 Ibid., ¶¶ 1–3.
115 The UNGA adopted a resolution almost identical to the HRC resolution declaring a right to a clean, healthy and sustainable environment as a human right. *The human right to a clean, healthy and sustainable environment*, A/76/L.75 (26 July 2022).
116 See Badrinarayana, D. (May/June 2022) "The Hard Expectations of Soft Law," and Atapattu, S., "A Human Right Whose Time Has Come," Centerpiece and Sidebar, *Environmental Law Institute's The Environmental Forum*, vol. 39, no. 3, available at: https://www.eli.org/the-environmental-forum/hard-expectations-soft-law.
117 *Resolution 16/11*, A/HRC/RES/16/11 (12 April 2011), ¶ 1.
118 *Resolution 19/10*, A/HRC/RES/19/10 (19 April 2012).
119 *Resolution 25/21*, A/HRC/RES/25/21 (15 April 2014).
120 *Resolution 28/11*, A/HRC/RES/28/11 (17 April 2015). A similar request was made by resolution 34/20 to convene an expert seminar. See also resolution 31/8.
121 *Resolution 34/20*, A/HRC/RES/34/20 (6 April 2017).
122 Referred to in Chapter 2.
123 *Resolution 7/23*, A/HRC/RES/7/23 (28 March 2008), Preamble.
124 This study is discussed in Section 3.2.2.
125 *Resolution 10/4*, A/HRC/RES/10/4 (25 March 2009).
126 Ibid. The work of special mandate holders is discussed in Chapter 4.
127 *Resolution 38/4*, A/HRC/RES/38/4 (16 July 2018).
128 Ibid. It again encouraged special mandate holders to consider climate change within their respective mandates. The OHCHR reports referred to in these resolutions are discussed in the next section.
129 See Mary Robinson Foundation – Climate Justice, *Geneva Pledge on Human Rights and Climate Action announced* (13 February 2015), https://www.mrfcj.org/resources/geneva-pledge-human-rights.
130 *Resolution 41/21*, A/HRC/RES/41/21 (23 July 2019).
131 *Resolution 47/24*, A/HRC/RES/47/24 (26 July 2021).
132 Ibid., ¶¶ 3–4 (emphasis added).
133 See Report of the Secretary General, *The impact of climate change on the human rights of people in vulnerable situations*, A/HRC/50/57 (6 May 2022). This report is not discussed here as it is not by one of the institutions discussed in this volume.
134 A special rapporteur position was created by Resolution 48/13 on climate change and human rights.
135 Resolution 48/13, supra note 11, preamble.
136 UNDP, OHCHR & UN Development Coordination Office, *UN Good Practices: How the Universal Periodic Review Process Supports Sustainable Development*, (24 February 2022), available at: https://www.undp.org/publications/un-good-practices-how-universal-periodic-review-process-supports-sustainable-development.

137 Ibid.
138 See *OHCHR and Climate Change*: https://www.ohchr.org/en/climate-change (emphasis added).
139 See *Contributions of the Right to Health Framework to the Effective Implementation and Achievement of the Health-Related Sustainable Development Goals, Report of the United Nations High Commissioner for Human Rights*, A/HRC/38/37 (20 April 2018). The report refers to specific health-related goals and targets in Agenda 2030 which include avoiding deaths and illnesses from hazardous chemicals and air, water and soil pollution and contamination (Target 3.9).
140 See *OHCHR and the Right to Health*: https://www.ohchr.org/en/health.
141 See Chapter 6.
142 *OHCHR and the Right to Health*, supra note 140.
143 See *OHCHR and the Rights to Water and Sanitation*: https://www.ohchr.org/en/water-and-sanitation.
144 See OHCHR, *Human Rights Topics: Water and Sanitation*: https://www.ohchr.org/en/topic/water-and-sanitation.
145 UN Water, *Eliminating Discrimination and Inequalities in Access to Water and Sanitation* (8 May 2015), available at: https://www.ohchr.org/sites/default/files/Documents/Issues/Water/DiscriminationPolicy.pdf.
146 Ibid.
147 See *OHCHR and the Right to Food*: https://www.ohchr.org/en/food.
148 *Report of the Office of the UN High Commissioner for Human Rights on the Relationship Between Climate Change and Human Rights*, A/HRC/10/61 (15 January 2009) ["OHCHR report"].
149 This principle forms part of the legal architecture of the climate regime and is included as one of the principles in Article 3 of the UNFCCC.
150 OHCHR report, supra note 148, ¶ 18.
151 Ibid., ¶ 42 (footnotes omitted).
152 Ibid., ¶ 56.
153 Ibid., ¶ 60.
154 REDD program refers to Reducing Emissions from Deforestation and Forest Degradation, a solution put in place to address deforestation as well as climate change. See https://redd.unfccc.int.
155 OHCHR report, supra note 148, ¶ 70 (footnotes omitted). This view seems to have changed since then. See Torres Islanders case (Chapter 5) and *Sacchi v Argentina* (Chapter 8).
156 Ibid., ¶¶ 84–85.
157 *Key Messages on Human Rights and Climate Change* (July 2016), https://www.ohchr.org/sites/default/files/KeyMessages_on_HR_CC.pdf.
158 Ibid. In addition, several publications discuss the link between human rights and environmental degradation including desertification, biodiversity loss and COVID-19. These documents are available here: https://www.ohchr.org/en/climate-change.
159 *Summary of the Panel Discussion on Human Rights, Climate Change, Migrants and Persons Displaced across International Borders Report of the Office of the United Nations High Commissioner for Human Rights*, A/HRC/37/35 (14 November 2017) ["Panel report"], https://undocs.org/A/HRC/37/35.
160 Ibid., ¶ 6.
161 Ibid., ¶ 8.
162 Ibid., ¶ 12.
163 Nansen Initiative, *Disaster-induced Cross-border Displacement* (December 2015), https://disasterdisplacement.org/wp-content/uploads/2014/08/EN_Protection_Agenda_Volume_I_-low_res.pdf.
164 It is noteworthy that a recommendation was not made to adopt international laws to protect cross border migrants affected by climate change.

165 Under the 1951 Convention Relating to the Status of Refugees, only political refugees are protected as persecution based on certain grounds must be established. A protection gap thus exists with regard to "climate refugees" which is not a legally accepted term either.
166 Panel report, supra note 159, ¶ 40.
167 It is noteworthy that the CBDR principle is mentioned here in the context of human rights principles.
168 Panel report, supra note 159, ¶ 54.
169 It should be pointed out that reference is made, yet again, to *natural disasters*.
170 Panel report, supra note 159, ¶ 61.
171 *Analytical Study on Gender-Responsive Climate Action for the Full and Effective Enjoyment of the Rights of Women*, A/HRC/41/26 (1 May 2019) ["Gender study"], https://undocs.org/A/HRC/41/26.
172 Ibid., ¶ 4 (footnotes omitted).
173 See UNFCCC, *The Enhanced Lima Work Programme on Gender*: https://unfccc.int/topics/gender/workstreams/the-enhanced-lima-work-programme-on-gender. Other relevant instruments discussed include the 2030 Agenda, Sendai Framework for Disaster Risk Reduction, Beijing Declaration and the Addis Ababa Declaration.
174 Gender study, supra note 171, ¶ 47. In addition, a panel discussion was held on women's rights and climate change, A/HRC/42/26 (24 July 2019), summary report available at: https://undocs.org/A/HRC/42/26.
175 *Analytical Study on the Promotion and Protection of the Rights of Persons with Disabilities in the Context of Climate Change Report of the Office of the United Nations High Commissioner for Human Rights*, A/HRC/44/30 (22 April 2020) ["Disability study"], available at: https://undocs.org/A/HRC/44/30.
176 Ibid., ¶ 5.
177 Ibid., ¶ 7.
178 Ibid., ¶ 21.
179 See Chapter 9.
180 Disability study, supra note 175, ¶ 39.
181 *Analytical Study on the Promotion and Protection of the Rights of Older Persons in the Context of Climate Change, Report of the Office of the United Nations High Commissioner for Human Rights*, A/HRC/47/46 (30 April 2021), available at: https://undocs.org/A/HRC/47/46.
182 Ibid., ¶ 9.
183 Ibid., ¶ 64.
184 Ibid., ¶ 41.
185 Ibid., ¶¶ 67–68.
186 See *Human Rights Topics: Sustainable Development through Human Rights*: https://www.ohchr.org/en/topic/sustainable-development-through-human-rights.
187 Ibid.
188 *Addis Ababa Action Agenda of the Third International Conference on Financing for Development* (13–16 July 2105), available at: https://sustainabledevelopment.un.org/content/documents/2051AAAA_Outcome.pdf.

References

About UN Human Rights: High Commissioner (2022), https://www.ohchr.org/en/about-us/high-commissioner.

Addis Ababa Action Agenda of the Third International Conference on Financing for Development (2015), https://sustainabledevelopment.un.org/content/documents/2051AAAA_Outcome.pdf.

Bachelet Hails Landmark Recognition That Having a Healthy Environment Is a Human Right (2021), https://www.ohchr.org/en/2021/10/bachelet-hails-landmark-recognition-having-healthy-environment-human-right.

Badrinarayana, D. (2022) "The Hard Expectations of Soft Law" and Atapattu, S., "A Human Right Whose Time Has Come," Centerpiece and Sidebar, *Environmental Law Institute's The Environmental Forum*, vol. 39, no. 3, https://www.eli.org/the-environmental-forum/hard-expectations-soft-law.

Cantú Rivera, H. (2018) "The UN Human Rights Council: Achievements and Challenges in Its First Decade," in *International Human Rights Institutions, Tribunals, and Courts*, Oberleitner, G., ed., Singapore: Springer, doi: 10.1007/978-981-10-5206-4_3.

Contributions of the right to health framework to the effective implementation and achievement of the health-related sustainable development goals, report of the United Nations High Commissioner for Human Rights (2018), A/HRC/38/37.

ECLAC (2018) *Regional Agreement on Access to Information, Public Participation and Justice in Environmental Matters in Latin America and the Caribbean*, https://www.cepal.org/en/escazuagreement.

Human Rights Council Subsidiary Bodies (2022), https://www.ohchr.org/en/hr-bodies/hrc/other-sub-bodies.

Human Rights Topics: Sustainable Development through Human Rights (2022), https://www.ohchr.org/en/topic/sustainable-development-through-human-rights.

Human Rights Topics: Water and Sanitation (2022), https://www.ohchr.org/en/topic/water-and-sanitation.

Key Messages on Human Rights and Climate Change (2016), https://www.ohchr.org/sites/default/files/KeyMessages_on_HR_CC.pdf.

Mapping Human Rights Obligations Relating to the Enjoyment of a Safe, Clean, Healthy and Sustainable Environment, Individual Report on the UN General Assembly and the Human Rights Council, including the Universal Periodic Review Process, Report No. 6 (2013), https://www.ohchr.org/en/special-procedures/sr-environment/mapping-report.

Mary Robinson Foundation – Climate Justice (2015) *Geneva Pledge on Human Rights and Climate Action Announced*, https://www.mrfcj.org/resources/geneva-pledge-human-rights.

Nansen Initiative (2015) *Disaster-induced Cross-border Displacement*, https://disasterdisplacement.org/wp-content/uploads/2014/08/EN_Protection_Agenda_Volume_I_-low_res.pdf.

OHCHR and Climate Change (2022), https://www.ohchr.org/en/climate-change.

OHCHR and the Right to Food (2022), https://www.ohchr.org/en/food.

OHCHR and the Right to Health (2022), https://www.ohchr.org/en/health.

OHCHR and the Rights to Water and Sanitation (2022), https://www.ohchr.org/en/water-and-sanitation.

OHCHR, *Analytical study on gender-responsive climate action for the full and effective enjoyment of the rights of women* (2019), A/HRC/41/26.

OHCHR, *Analytical study on the promotion and protection of the rights of persons with disabilities in the context of climate change, Report of the Office of the United Nations High Commissioner for Human Rights* (2020), A/HRC/44/30.

OHCHR, *Analytical study on the promotion and protection of the rights of older persons in the context of climate change, Report of the Office of the United Nations High Commissioner for Human Rights* (2021), A/HRC/47/46.

OHCHR, *Promotion and Protection of Human Rights around the Globe* (2022), https://www.ohchr.org/en/hrbodies/hrc/home.

Report of the Office of the UN High Commissioner for human rights on the relationship between climate change and human rights (2009), A/HRC/10/61.
Report of the Secretary General: The impact of climate change on the human rights of people in vulnerable situations (2022), A/HRC/50/57, https://undocs.org/A/HRC/50/57.
Report of the Working Group on the Universal Periodic Review, Angola (2019), A/HRC/43/11.
Report of the Working Group on the Universal Periodic Review, Australia (2016), A/HRC/31/14.
Report of the Working Group on the Universal Periodic Review, Bhutan (2014), A/HRC/27/8.
Report of the Working Group on the Universal Periodic Review, Bosnia and Herzegovina (2019), A/HRC/43/17.
Report of the Working Group on the Universal Periodic Review, Brazil (2017), A/HRC/36/11.
Report of the Working Group on the Universal Periodic Review, Brunei Darussalam (2019), A/HRC/42/11.
Report of the Working Group on the Universal Periodic Review, Cambodia (2019), A/HRC/41/17.
Report of the Working Group on the Universal Periodic Review, Chile (2014), A/HRC/26/5.
Report of the Working Group on the Universal Periodic Review, China (2018), A/HRC/40/6.
Report of the Working Group on the Universal Periodic Review, Chile (2019), A/HRC/41/6.
Report of the Working Group on the Universal Periodic Review, Comoros, (2019), A/HRC/41/12.
Report of the Working Group on the Universal Periodic Review, Costa Rica (2014), A/HRC/27/12.
Report of the Working Group on the Universal Periodic Review, Costa Rica (2019), A/HRC/42/122.
Report of the Working Group on the Universal Periodic Review, Democratic People's Republic of Korea (2019), A/HRC/42/10.
Report of the Working Group on the Universal Periodic Review, Democratic Republic of the Congo (2019), A/HRC/42/5.
Report of the Working Group on the Universal Periodic Review, Ecuador (2017), A/HRC/36/4.
Report of the Working Group on the Universal Periodic Review, Egypt (2019), A/HRC/43/16.
Report of the Working Group on the Universal Periodic Review, El Salvador (2020), A/HRC/43/5.
Report of the Working Group on the Universal Periodic Review, Ethiopia (2019), A/HRC/42/14.
Report of the Working Group on the Universal Periodic Review, Fiji (2019), A/HRC/43/8.
Report of the Working Group on the Universal Periodic Review, Former Yugoslav Republic of Macedonia (2014), A/HRC/26/10.
Report of the Working Group on the Universal Periodic Review, Guinea (2020), A/HRC/44/5.
Report of the Working Group on the Universal Periodic Review, India (2017), A/HRC/35/10.
Report of the Working Group on the Universal Periodic Review, Iraq (2019), A/HRC/43/14.
Report of the Working Group on the Universal Periodic Review, Islamic Republic of Iran (2019), A/HRC/43/17.
Report of the Working Group on the Universal Periodic Review, Italy (2019), A/HRC/43/4.
Report of the Working Group on the Universal Periodic Review, Kazakhstan (2019), A/HRC/43/10.
Report of the Working Group on the Universal Periodic Review, Kenya (2020), A/HRC/44/9.
Report of the Working Group on the Universal Periodic Review, Kiribati (2015), A/HRC/29/5.
Report of the Working Group on the Universal Periodic Review, Kyrgyzstan (2020), A/HRC/44/4.
Report of the Working Group on the Universal Periodic Review, Lao People's Democratic Republic (2020), A/HRC/44/6.

Report of the Working Group on the Universal Periodic Review, Lesotho (2020), A/HRC/44/8.
Report of the Working Group on the Universal Periodic Review, Maldives (2015), A/HRC/30/8.
Report of the Working Group on the Universal Periodic Review, Nauru (2015), A/HRC/31/7.
Report of the Working Group on the Universal Periodic Review, New Zealand (2019), A/HRC/41/4.
Report of the Working Group on the Universal Periodic Review, Nicaragua (2014), A/HRC/27/16.
Report of the Working Group on the Universal Periodic Review, Nicaragua (2019), A/HRC/42/16.
Report of the Working Group on the Universal Periodic Review, Nigeria (2013), A/HRC/25/6.
Report of the Working Group on the Universal Periodic Review, Nigeria (2018), A/HRC/40/7.
Report of the Working Group on the Universal Periodic Review, Norway (2019), A/HRC/42/3.
Report of the Working Group on the Universal Periodic Review, Papua New Guinea (2016), A/HRC/33/10.
Report of the Working Group on the Universal Periodic Review, Philippines (2017), A/HRC/36/12.
Report of the Working Group on the Universal Periodic Review, Plurinational State of Bolivia (2014), A/HRC/28/7.
Report of the Working Group on the Universal Periodic Review, Plurinational State of Bolivia (2019), A/HRC/43/7.
Report of the Working Group on the Universal Periodic Review, Solomon Islands (2016), A/HRC/32/14.
Report of the Working Group on the Universal Periodic Review, Spain (2020), A/HRC/44/7.
Report of the Working Group on the Universal Periodic Review, Suriname (2016), A/HRC/33/4.
Report of the Working Group on the Universal Periodic Review, Tuvalu, (2009), A/HRC/10/84.
Report of the Working Group on the Universal Periodic Review, Tuvalu (2013), A/HRC/24/8.
Report of the Working Group on the Universal Periodic Review, Tuvalu (2018), A/HRC/39/8.
Report of the Working Group on the Universal Periodic Review, UK and Northern Ireland (2017), A/HRC/36/9.
Report of the Working Group on the Universal Periodic Review, Uruguay (2019), A/HRC/41/18.
Report of the Working Group on the Universal Periodic Review, USA (2015), A/HRC/30/12.
Report of the Working Group on the Universal Periodic Review, Vanuatu (2019), A/HRC/41/10.
Report of the Working Group on the Universal Periodic Review, Viet Nam (2019) A/HRC/41/7.
Secretary-General Appoints Mary Robinson of Ireland Special Envoy for Climate Change (2014), SG/A/1481-BIO/4603-ENV/DEV/144, https://press.un.org/en/2014/sga1481.doc.htm.
Summary of the panel discussion on human rights, climate change, migrants and persons displaced across international borders, Report of the Office of the United Nations High Commissioner for Human Rights (2017), A/HRC/37/35.
Summary of the panel discussion on women's rights and climate change (2019), A/HRC/42/26.
The human right to a clean, healthy and sustainable environment (2022), A/76/L.75.
Universal Periodic Review (2022), https://www.ohchr.org/en/hr-bodies/upr/upr-main.
UN Conference on the Human Environment (1972), Stockholm, https://www.un.org/en/conferences/environment/stockholm1972.
UN Water (2015) *Eliminating Discrimination and Inequalities in Access to Water and Sanitation*, https://www.ohchr.org/sites/default/files/Documents/Issues/Water/DiscriminationPolicy.pdf.
UNDP, OHCHR & UN Development Coordination Office (2022) *UN Good Practices: How the universal periodic review process supports sustainable development*, https://www.

undp.org/publications/un-good-practices-how-universal-periodic-review-process-supports-sustainable-development.

UNFCCC (1992), https://unfccc.int/resource/docs/convkp/conveng.pdf.

UNFCCC (2015) *The Enhanced Lima Work Programme on Gender*, https://unfccc.int/topics/gender/workstreams/the-enhanced-lima-work-programme-on-gender.

UNFCCC (2022) *Reducing Emissions from Deforestation and Forest Degradation*, https://redd.unfccc.int.

UNHCR (1951) The 1951 Convention Relating to the Status of Refugees, https://www.unhcr.org/en-us/3b66c2aa10.

4 Special Procedures Mandate Holders

4.1 Introduction

As noted in Chapter 2, there are several special procedures mandate holders whose mandate implicates environmental issues.[1] Of the 44 thematic mandates, several mandate holders have discussed environmental issues especially climate change to the extent these issues overlap with their mandates. These include the Special Rapporteurs on: Right to Food; Cultural Rights; Right to Development; Rights of Persons with Disabilities; Right to Health; Right to Housing and Adequate Standard of Living; Rights of Indigenous Peoples; Extreme Poverty and Human Rights; Hazardous Substances and Waste; and Safe Drinking Water and Sanitation as well as the Working Group on Human Rights and Transnational Corporations and other Business Enterprises. Some Special Rapporteurs have issued special reports on the link between climate change and their mandate.[2]

As discussed in Chapter 2, a mandate was created in 1990 specifically on human rights and the environment. Although several reports were submitted including a set of draft articles, these were never adopted by the UN. With consequences of climate change becoming increasingly visible with serious ramifications for human beings, interest in the link between human rights and environmental protection resurfaced with calls for the creation of a special mandate. This call culminated in the appointment of an independent expert on human rights and the environment in 2012 which was later converted to a special rapporteur position. A separate mandate was created on climate change and human rights in 2021. This chapter will survey these developments.

Moreover, several special mandate holders are currently working on reports on climate change and/or environmental degradation. For example, the Special Rapporteur on Racial Discrimination is doing her next report on Climate Justice, Environmental Racism, and Racial Equality and has invited comments and suggestions.[3]

4.2 Appointment of a Special Rapporteur on Human Rights and the Environment

As noted, the UN Sub-Commission on Prevention of Discrimination and Protection of Minorities appointed a Special Rapporteur in 1990 to study the relationship

DOI: 10.4324/9781003128847-6

between human rights and the environment. Ms. Fatma Zohra Ksentini was appointed to the position. She submitted four reports including a note during her tenure as Special Rapporteur and a set of draft articles on human rights and the environment.

She identified several environmental principles that have become binding on states: assessing the environmental consequences of activities; informing other states of potential transboundary impacts; consulting with potential victims where there are transboundary impacts; urgently informing other states and providing mutual assistance in the event of accidents; and providing access to remedies to citizens of the victim state on the basis of non-discrimination. The report made specific reference to Principles 1 and 21 of the Stockholm Declaration and concluded that the relationship between the environment, development, satisfactory living conditions, dignity, and well-being "constitute recognition of the right to a healthy and decent environment, which is inextricably linked, both individually and collectively, to universally recognized fundamental human rights standards and principles…"[4] While the Stockholm Declaration discusses these inter-linkages, that by itself does not mean that there exists a separate right to a healthy environment. In fact, it took the UN over 25 years since the submission of this report to recognize a distinct right to a healthy environment.

The draft principles on human rights and the environment which were appended to the final report specifically recognized that "sustainable development links the right to development and the right to a secure, healthy and ecologically sound environment"[5] and that human rights violations lead to environmental degradation and vice versa. The draft principles endorsed a substantive right to a healthy environment: "All persons have the right to a secure, healthy and ecologically sound environment. This right and other human rights… are universal, interdependent and indivisible."[6]

Other substantive rights included: freedom from pollution, the right to protection and preservation of air, soil, water, sea-ice, flora, fauna, and essential processes and areas necessary to maintain biodiversity and ecosystems, right to safe and healthy food and water, right to a healthy working environment, and right to adequate housing. While some of the rights included are established rights under international law and their inclusion here only reinforces them, the other "rights" are clearly lex ferenda and clearly go beyond the parameters of traditional human rights law.[7]

The draft articles endorsed the right of indigenous peoples to control their lands, territories, and natural resources and to maintain their traditional way of life. In addition, they have the right to protection against any action that may result in the destruction of their territories. Procedural rights are also included in the draft articles including the right to environmental and human rights education.

The draft articles are noteworthy for the inclusion of duties of states and people and provide that all persons have the duty to protect and preserve the environment. Furthermore, in order to ensure the right to a healthy environment, states have to adopt administrative, legislative, and other measures to effectively implement the rights in the Declaration with the aim of preventing environmental harm. The draft articles contain an elaborate list of measures that states must take, including disseminating environmental information; conducting the environmental impact of activities; ensuring public participation in decision-making; providing effective remedies; and licensing and regulation of substances and activities that are harmful to the environment. The draft Declaration is unprecedented as it extends its reach to cover activities of transnational corporations which pose significant challenges for both international and national law and have been responsible for gross human rights violations and severe environmental damage in host countries.[8] Unfortunately, the draft Declaration or even a watered-down version was never adopted; it seems destined to gather dust in the UN archives.

4.3 Special Rapporteur on the Implications for Human Rights of the Environmentally Sound Management and Disposal of Hazardous Substances and Wastes

Although the work of the Special Rapporteur on Hazardous Substances is not usually referenced when discussing the link between human rights and the environment, exposure to hazardous substances clearly has human rights implications. When creating the mandate in 1995, the Commission on Human Rights recognized that the illicit traffic and dumping of toxic products and wastes pose a serious threat not only to the environment, but also to the enjoyment of human rights.[9]

The mandate of the Special Rapporteur was strengthened and expanded in 2011 to cover not only the illicit movement but also the whole life-cycle of hazardous materials.[10] The Special Rapporteur was tasked "to monitor the adverse effects that the generation, management, handling, distribution and final disposal of hazardous substances and wastes may have on the full enjoyment of human rights…"[11] The expanded mandate included information on: human rights issues raised by transnational corporations and other business enterprises by hazardous substances and wastes; support and assistance to victims of human rights violations; and ambiguities in international instruments that allow the movement and dumping of such substances.[12] In 2012, the mandate was further expanded to cover the protection of human rights defenders given the increasing incidents of their victimization.[13]

Five special rapporteurs have been appointed to date under this mandate. The previous Special Rapporteur, Okechukwu Ibeanu, noted in his report that developing countries accept hazardous products and toxic wastes due to poverty but lack disposal facilities placing the human rights of local populations at risk. Moreover, women, children, and youth are particularly at risk because they are often among the poorest and therefore likely to work in polluting industries and the Special Rapporteur called for greater attention to the gender and age aspects. He pointed out that states, corporations and other private entities do not share vital information about the potential effects of pollution and irreversible damage to the environment and even when an incident occurs, they are reluctant to divulge information and often give false information misleading the public. The Special Rapporteur considered that rights to information and participation are both rights in themselves and also essential for the enjoyment of other rights and play an important role in the context of illicit movement and dumping of toxic products. He pointed out that the rights to information and participation in relation to both human rights and environmental protection are well reflected in the international legal framework under both areas of law.[14]

The current Special Rapporteur, Baskut Tuncak, noted in his preliminary report that his mandate reflects current issues in both environmental and human rights laws[15] and that modern laws "increasingly embraced a life-cycle approach to chemicals management, helping to reduce the human rights implications of hazardous substances and wastes."[16] This mandate overlaps with that of the Independent Expert (now Special Rapporteur) on Human Rights and Environment, to which we now turn.

4.4 Independent Expert on Human Rights and the Environment

The HRC, by resolution 19/10, established the mandate for an independent expert on human rights and the environment in 2012. The mandate included: studying the human rights obligations relating to the enjoyment of a safe, clean, healthy and sustainable environment; identifying best practices relating to the use of human rights obligations in relation to environmental protection; and preparing a compendium of best practices.[17] John Knox was appointed as the Independent Expert.

In his first report the Independent Expert noted that environmental rights are a latecomer to the human rights field but in recent years environmental concerns have moved from the periphery to the center of human efforts. The close relationship between human rights and the environment has led to two developments: the recognition of an explicit right to a healthy environment; and increased attention paid to the environment in relation to existing rights such as the rights to life and health.[18] At the global level, more attention has been paid to "greening" existing rights, which has resulted in identifying rights affected by environmental degradation (substantive rights); and rights that support better environmental policymaking (procedural rights).[19] The obligations that human rights law imposes in relation to environmental protection are less clearly understood and need further clarity so that states can fulfill those obligations fully.[20]

In his second report ("Mapping Report"), the Independent Expert mapped the human rights obligations relating to the environment with an extensive review of global and regional sources.[21] He identified three sets of human rights obligations relating to the environment: substantive, procedural, and those applicable to vulnerable groups. Procedural obligations – which are now firmly established – include the duty to assess environmental impacts on human rights, the obligation to provide information, facilitate participation, and provide access to remedies. Substantive obligations include the duty to adopt legal and institutional frameworks "that protect against environmental harm that interferes with the enjoyment of human rights, including harm caused by private actors."[22] Moreover, states may have additional obligations to protect particularly vulnerable groups such as women, children and indigenous groups. The Independent Expert concluded: while human rights obligations relating to the environment are still evolving, *they are already clear enough* to provide guidance to states.[23] In his good practices report, he identified two thematic areas that need further clarification: human rights obligations relating to transboundary environmental harm including climate change; and those relating to protecting the rights of environmental human rights defenders.[24]

4.5 Special Rapporteur on Human Rights and Environment

As noted, in 2015 the mandate of the Independent Expert was converted to a special rapporteur position. As Special Rapporteur, John Knox submitted four reports to the HRC – climate change (2016),[25] biodiversity (2017),[26] children's rights (2018),[27] and the final report with Framework Principles on Human Rights and the Environment appended (2018).[28]

In his report on climate change, the Special Rapporteur noted that pursuant to the HRC resolution 7/23 the OHCHR prepared an analytical report on the link between climate change and human rights. While acknowledging the link between climate change and human rights, the Special Rapporteur did not readily accept that climate change violated human rights law, but stressed the importance of bringing a human rights perspective to climate change.[29] Of course, the main challenge is the structure of human rights itself – because human rights operate vertically between states as duty bearers and people as rights holders, if damage originates from a source outside the territory, are states still liable under human rights law?[30] However, as the Committee on the Rights of the Child recognized in its groundbreaking decision in *Sacchi v. Argentina*, the global nature of climate change does not absolve states of their individual responsibility to address climate change.[31] As the more recent decision of the Human Rights Committee in *Torres Islanders case* indicated, states have obligations relating to both mitigation and adaptation and can be held liable for failure to fulfill them.[32]

In his report on biodiversity and human rights, one of the first reports to discuss this link, the Special Rapporteur noted:

> The full enjoyment of human rights, including the rights to life, health, food and water, depends on the services provided by ecosystems. The provision of ecosystem services depends on the health and sustainability of ecosystems, which in turn depend on biodiversity. The full enjoyment of human rights thus depends on biodiversity, and the degradation and loss of biodiversity undermine the ability of human beings to enjoy their human rights.[33]

He was, however, careful to note that biodiversity has intrinsic value that may not be captured within a human rights framework. He pointed out that while the importance of a healthy environment is widely recognized, the relationship between human rights and biodiversity is not well understood. Human beings depend on the environment for their very survival. However, biodiversity and ecosystems are coming under increasing threat which, in turn, has an impact on human beings and their rights. The rights that are at risk of being infringed include the rights to life, health, food, water, culture, an adequate standard of living, and non-discrimination. In addition, the report points out that loss of biodiversity often results from and reinforces existing patterns of discrimination, and some communities are disproportionately affected.

The report also discussed procedural and substantive human rights obligations relating to the conservation and sustainable use of biodiversity.[34] While states have the discretion to strike a balance between environmental protection and other legitimate societal needs, the balance must be reasonable and should not result in unjustified, foreseeable infringements of human rights. States' duty to safeguard biodiversity to protect rights from infringement extends to environmental harm from private actors, and businesses have a responsibility to respect the rights relating to biodiversity as well. The report highlighted the importance of the principle of cooperation in protecting biodiversity and ecosystem services.

4.5.1 *Framework Principles on Human Rights and Environment*

Sixteen principles were included in the Framework Principles as well as commentary to each principle. Before these principles were finalized, draft guidelines were published and circulated among various stakeholders requesting written comments. An expert seminar and a public consultation were also held before the principles were finalized. Soliciting broad participation by many stakeholders in this endeavor allowed many voices to be heard and many concerns to be addressed. The Framework Principles were appended to the final report of the Special Rapporteur and was submitted to the HRC in 2018.

The report was careful to point out that the principles included there do not create new obligations. Rather, they reflect the application of existing human rights obligations in relation to environmental issues. States should thus accept the framework principles as a reflection of actual or emerging international human rights law. While not all states may have accepted all the obligations included there, they reflect broad consensus or, at a minimum, best practices.[35] When he submitted the report to the HRC, the Special Rapporteur stated that it is high time that the international community recognized a human right to a healthy environment[36] and requested the HRC to support its recognition.[37]

While the Framework Principles are a major step forward (even though they have not yet been formally adopted), it is regrettable that many of these principles are framed in soft language. If, as the Special Rapporteur points out, these principles are already part of international law and in fact they are, then there is no reason why they should be so framed even though soft language may be politically more palatable. This could have the unfortunate effect of diluting existing obligations with states treating them as being voluntary, not obligatory.

4.5.2 David Boyd as Special Rapporteur

David Boyd succeeded John Knox as the new Special Rapporteur and began his term in August 2018. Since then, he has submitted several thematic reports: clean air;[38] safe climate;[39] good practices;[40] healthy biosphere;[41] global water crisis;[42] healthy and sustainable food;[43] and non-toxic environment to live, work, and play.[44] In his first report to the General Assembly, he summarized the developments relating to environmental rights and stressed the importance of implementing and enforcing environmental standards in a non-discriminatory and non-retrogressive manner.[45] With regard to the transboundary and/or global dimension of many of the environmental challenges, the Special Rapporteur noted that the obligation of states to cooperate requires states to work together to address transboundary and global environmental threats to human rights.[46]

Special Rapporteur discussed the vulnerability of certain groups that are most at risk from environmental degradation and the need to respect, protect, and fulfill their rights. These include indigenous peoples, children, women and girls, older persons, persons living in poverty, minorities, and displaced populations and refugees. He noted that "these vulnerabilities often overlap… compounding the risks of environmental harm and the concomitant violation of their human rights."[47] The report also highlighted the need to protect environmental human rights defenders who are targeted because of their work; on average, four defenders are killed every week and many more are subjected to violence.[48] The Special Rapporteur called upon the UN to formally recognize the human right to a healthy environment, pointing out that "it is beyond debate that human beings are wholly dependent on a healthy environment in order to lead dignified, healthy and fulfilling lives."[49] He believed that if the UDHR were to be drafted today, it is hard to imagine that it will not include a right to a healthy environment and enumerated the tangible benefits of such recognition. He identified several ways to recognize this right: adopt a new international treaty such as the Global Pact for the Environment; develop an additional protocol to an existing human rights treaty, perhaps the ICESCR; and the most expeditious approach is to adopt a UNGA resolution. He believed that the right to a healthy environment meets each of the substantive and procedural requirements to recognize a new right as acknowledged by the GA in its resolution 41/120 of 1986.[50] He noted that this right is an umbrella term encompassing both procedural and substantive components. Substantive components are: access to clean air; access to clean water and adequate sanitation; healthy and sustainably produced food; a safe climate; healthy biodiversity and ecosystems; and non-toxic environments in which to live, work and play.[51] The procedural components are access to information, public participation, and access to justice.[52] Each of the subsequent reports submitted by him elaborated on the substantive components.

Boyd's first thematic report focused on clean air as a vital element of the right to a healthy environment.[53] The report noted that air pollution causes a wide range of health effects including respiratory illness and infections,[54] and that more than 90% of the world's population lives in regions that exceed WHO guidelines for ambient air quality which means over 6 billion people including 2 billion children are breathing polluted air. Moreover, similar to most environmental issues, both indoor and outdoor pollution disproportionately affects poor people and poor communities. Just as there are environmental injustices within nations, there is a widening disparity in air quality between nations. While more affluent countries have seen improvements in air quality, it has worsened in poorer countries. Moreover, there is a relationship between air pollution and climate change and many of the same activities that harm air quality also contribute to climate change.

Noting that poor air quality has implications for a wide range of human rights from rights to life to an adequate standard of living, the Special Rapporteur expressed concern that while the GA has adopted numerous resolutions on the right to water, it has never adopted one on the right to clean air.[55] He pointed out that both are essential to life, health, dignity, and well-being and that the UN HCHR had noted that "there can be no doubt that all human beings are entitled to breathe clean air."[56] Moreover, special procedures as well as treaty bodies have called on states to address air pollution. They are also a vital element of the SDGs and improving air quality is essential to achieving several of its targets. In addition to evaluating state obligations relating to addressing air pollution and improving air quality and standards, the report examined business obligations. The Special Rapporteur noted that they should comply with the Guiding Principles on Business and Human Rights as well as Children's Rights and Business Principles but noted that there are many examples from all over the world of businesses violating the right to breathe clean air.[57] The report pointed out that the failure to respect, protect and fulfill the right is inflicting a terrible toll on people:

> Given the devastating impacts of poor air quality on people's lives, health and human rights, actions must be taken rapidly and systematically, with a priority focus on ameliorating conditions for the most vulnerable. Fulfilling the right to breathe clean air goes hand in hand with achieving the Sustainable Development Goals, including healthy lives for all, sustainable cities, universal access to clean energy and effective action to address climate change.[58]

In his thematic report on climate change, the Special Rapporteur discussed the causes of the global climate crisis as well as its consequences.[59] Noting that "society is addicted to fossil fuels," the Special Rapporteur observed that despite 27 years of commitments dating back to the UNFCCC (now 30 years), "the world is neither headed in the right direction, nor addressing the crisis at an adequate pace."[60] Since 1990, global energy consumption has increased by 57% while GHG emissions have *increased* by 60 per cent. Instead of phasing out fossil fuels, states provided subsidies amounting to $5.2 trillion in 2017.

The report notes that climate change is having a major impact on a wide range of human rights ranging from the right to life to the right of self-determination. Moreover, climate change is disproportionately affecting those living in poverty. Referring to the report of the Special Rapporteur on Extreme Poverty who warned about a future of climate apartheid, Boyd pointed out that these vulnerable groups contributed little to the problem and lack resources to protect themselves. Other such communities include those whose vulnerabilities are caused by gender, age, disability, geography, and cultural or ethnic background. Although at risk, they also have the potential to contribute to climate solutions and should be considered as agents of change, not just victims. With regard to the human rights obligations relating to climate change, the SR pointed out that states agreed, by signing on to the UNFCCC to ensure a safe climate "which is vital to the enjoyment of a broad range of human rights."[61] Both the Cancun decision and the Paris Agreement refer to the need to protect rights in the context of climate change and the link between climate change and human rights has been explored by the Human Rights Council, treaty bodies, judiciaries and many international bodies with the Inuit Petition playing a catalyst role. The SR stressed that states have obligations to protect human rights from environmental harm and the potentially catastrophic effects of climate change require states to take immediate action to prevent those harms:

> To comply with their international human rights obligations, States should apply a rights-based approach to all aspects of climate change and climate action. Applying a rights-based approach clarifies the obligations of States and businesses; catalyses ambitious action; highlights the plight of the poorest and most vulnerable; and empowers people to become involved in designing and implementing solutions.[62]

Stressing that climate change is already harming billions of people and their rights, and exacerbate inequality and injustice, the SR called upon states to address the ambition gap and submit ambitious NDCs by 2020 that will put the world on track to reducing GHG emission by 45% by 2030 as recommended by the IPCC. They should prepare "rights-based deep decarbonization plans" to achieve net zero emissions by 2050.[63] He stressed that the "failure to fulfill international climate change commitments is a prima facie violation of the state's obligations to protect the human rights of its citizens."[64] He called upon states to immediately terminate all fossil fuel subsidies and to ramp up climate finance to fulfill the commitment made by developed countries to mobilize at least US $100 billion per year to finance mitigation and adaptation needs of developing countries. Furthermore, states should provide financing for loss and damage perhaps through an air travel levy, a levy on fuels used by aviation or shipping industries, or a climate damages levy on the revenue of fossil fuel companies.[65]

Continuing his exploration of proposed substantive elements of a right to a healthy environment, Boyd submitted his fifth report on healthy and sustainable food.[66] He noted that while food is essential for life, today's food systems are major drivers of climate change, biodiversity loss, soil pollution, water depletion, and zoonotic diseases. Although the world produces enough food to feed everybody, about 2 billion people lack access to safe, nutritious food and about 800 million people suffer from daily hunger which increased to 130 million due to the COVID-19 pandemic.[67] The report discussed the environmental impact of food systems and pointed out that they are responsible for 21%–37% of global GHG emissions while agriculture accounts for around 70% of freshwater use globally of which nearly one third is used for livestock.

With regard to the impact of unsustainable food systems on human rights, the report pointed out that the environmental impact is primarily caused by industrial food systems and interfere with the enjoyment of a wide range of rights from the right to life to cultural rights. The report discussed the rights of children and other vulnerable and marginalized groups whose rights may be jeopardized by the environmental impacts of food systems: "These groups often have fewer resources, are disproportionately impacted and have less access to health care services, increasing the risk of illness or death."[68] A critical issue is lack of formal land and tenure rights that affects the right to food of millions of indigenous peoples, peasants, Afro-descendants, women, and the poor.

Referring to General Comment No 12 of the Committee on ESC Rights,[69] the report pointed out that sustainability requires food to be accessible for present and future generations. According to the Framework Principles, there are three types of state obligations: substantive, procedural, and special obligations toward vulnerable populations. With regard to substantive obligations, the report noted that states must not violate the right to food, the right to a healthy and sustainable environment, or other human rights affected by the environmental consequences of food systems.[70] Moreover, states must take steps to protect rights from being violated by third parties, in particular businesses, take positive action to fulfill these rights, and avoid discrimination and retrogressive measures. Furthermore, states should integrate the right to food and the right to a healthy environment into national laws, and policies. They have additional obligations toward indigenous peoples, local communities, Afro-descendants, and peasants. Given the important role played by agribusinesses, they must adopt human rights policies, conduct human rights due diligence, and establish grievance mechanisms. The Special Rapporteur concluded:

> A rights-based approach, focused on the right to food and the right to a healthy environment, is an essential catalyst for accelerating the transformation from today's unsustainable food systems to a future where everyone enjoys healthy and sustainable food, workers are treated fairly and degraded ecosystems are restored. *This is an obligation for States, not an option.*[71]

The Special Rapporteur continued his examination of the substantive components of the right to a healthy environment with his report on human rights and the global water crisis.[72] Over 2 billion people lack access to safe drinking water, 785 million lack even basic water services, and over 4 billion people lack access to safely managed sanitation. Waterborne diseases cause nearly 2 million deaths annually with the majority being children under the age of five years. Water pollution, water scarcity, and water-related disasters affect a range of human rights from right to life, health, food, education, culture, and a healthy environment. With regard to water, states have human rights obligations which are legally enforceable.

States must apply a rights-based approach to allocating, using, conservation, and protecting water. Referring again to the Framework Principles, the Special Rapporteur noted that there are three categories of state obligations: substantive, procedural, and special obligations toward those in vulnerable situations.[73] Moreover, fulfilling the rights to water, sanitation, and a healthy environment is essential for achieving SDG 6 and to address climate change. Thus, in order to implement the rights-based approach to ensuring healthy aquatic systems, states must take: international action; maintain and improve water quality; prevent or alleviate water scarcity; improve water governance; and achieve water and climate co-benefits.[74]

In his 2022 report, the first report to be submitted after the HRC recognized the right to a healthy environment, the Special Rapporteur focused on the right to a non-toxic environment in which he noted that communities are exposed to extreme levels of pollution and toxic contamination including "sacrifice zones."[75] The report pointed out that pollution and toxic substances cause at least 9 million premature deaths, 15 times more than from all wars, murders, and other forms of violence, with air pollution being the largest environmental contributor to premature deaths. Noting that "toxification of planet Earth is intensifying," the report noted that the production of chemicals doubled between 2000 and 2017 and will double again by 2030. Extraction and processing of fossil fuels as well as petrochemicals and plastic industries produce a massive amount of pollution and toxic chemicals. While all humans are exposed to pollution and toxic chemicals, a disproportionate burden is borne by individuals and communities that are already facing poverty, discrimination, and marginalization:

> The disturbing phenomenon of poor and marginalized communities being more heavily affected by pollution is a form of environmental injustice. Environmental injustices related to pollution and the production, export, use and disposal of toxic substances are rooted in racism, discrimination, colonialism, patriarchy, impunity and political systems that systematically ignore human rights.[76]

The Special Rapporteur defined sacrifice zones as places where residents suffer "devastating physical and mental health consequences and human rights violations as a result of living in pollution hotspots and heavily contaminated areas"[77] and most hazardous facilities tend to be located near poor and marginalized communities. Stressing that sacrifice zones is "a stain upon the collective conscience of humanity," the Special Rapporteur pointed out that they are the antithesis of sustainable development, harming present and future generations.[78] Such areas exist in both the Global North and South and examples abound from every part of the world.

With regard to the human rights obligations relating to pollution and toxic substances ranging from the right to life to cultural rights, the report pointed to the work of UN treaty bodies, special mandate holders, regional and national courts, and national human rights institutions. He pointed out that the recent recognition of the right to a healthy environment marks a turning point and that achieving a non-toxic environment is a legal obligation, not a policy option.[79] Regarding substantive obligations, the Special Rapporteur noted that states must not cause pollution or expose people to toxic substances that violate the right to a healthy environment; protect the right from being violated by third parties, especially businesses, and take positive action to fulfill that right by strengthening legislation and developing plans to prevent pollution, eliminating toxic substances, and rehabilitating contaminated sites. In this regard, prevention, precaution, non-discrimination, non-regression, and special duties toward vulnerable populations are important. To fulfill their obligation to ensure a non-toxic environment, states should urgently detoxify sacrifice zones and eliminate environmental injustices; strengthen national efforts and fulfill procedural rights.[80]

Furthermore, the Special Rapporteur pointed out that all human rights ultimately depend on a healthy biosphere: "Without healthy, functioning ecosystems, which depend on healthy biodiversity, there would be no clean air to breathe, safe water to drink or nutritious food to eat."[81] Moreover, healthy ecosystems regulate the Earth's climate, filter air and water, and recycle nutrients. Most of terrestrial biodiversity is found in forests – not only are they home to different species, billions of people rely on natural medicines for their healthcare and half of prescription drugs are derived from nature. The COVID-19 pandemic which killed millions of people worldwide and made even a bigger number sick, not to mention the socioeconomic impact, is due to the human interference with the biosphere – the latest infectious disease to jump from animals to humans. Noting that the damage to the biosphere is having a major impact on a wide range of human rights such as the rights to a healthy environment, life, health, food, water, sanitation, an adequate standard of living, development, and culture, the Special Rapporteur urged humanity to re-evaluate its relationship with nature pointing to the call made by scientists in 2019 for urgent and transformative change.[82] The report further elaborated on the obligations of states, responsibilities of businesses, and those of conservation organizations and stressed that while it is not too late to respond to the global nature emergency, time is running out.[83]

4.6 Special Rapporteur on Cultural Rights

The Special Rapporteur on cultural rights, Karima Bennoune, devoted one of her reports to climate change.[84] Noting that the climate emergency remains one of the greatest threats to humanity, she pointed out that conditions that allow all people to access, participate in, and contribute to cultural life is jeopardized by the climate emergency. She stressed, referring to one of her previous reports: "The universality of human rights, including cultural rights, has no meaning today without a livable environment in which they can be enjoyed."[85] While most human rights are affected by climate change, cultural rights risk being wiped out in many instances which has not been addressed adequately in climate initiatives. She recognized the disproportionate impact of climate consequences:

> The Intergovernmental Panel on Climate Change observed that "people who are socially, economically, culturally, politically, institutionally or otherwise marginalized are especially vulnerable to climate change." Those with pronounced cultural connections to land, sea, natural resources and ecosystems, including indigenous, rural and fisher peoples, face disproportionate devastation of their individual and collective cultural lives.[86]

Noting that while culture is imperiled, it remains an important key to successful climate adaptation, the report recognized that traditional knowledge and indigenous understanding will be crucial going forward. Moreover, climate change and cultural rights are intrinsically linked – culture is closely connected to ecosystems, especially for indigenous peoples and rural communities, and culture influences our understanding of the environment. Despite this close relationship, many environmental policies do not address culture while many cultural policies do not refer to the environment.[87]

Noting that we cannot be passive observers of cultural extinction, the Special Rapporteur stressed that addressing the climate-cultural rights nexus requires a transnational approach committed to climate culture justice as those who contributed least are affected most and could result in "a terrible climate culture apartheid," where the history and cultural traces of the biggest victims of climate change are allowed to disappear while the traces of those most responsible are better protected.[88]

Climate change will have a grave impact on cultural heritage of all humankind and for related human rights of millions of people. Losses will be physical, economic, social, and cultural. Cultural heritage is a human rights issue which is connected to other rights including the right to education and rights of future generations whose rights will be stolen from them by choices made today. Moreover, tangible heritage sites face irreversible damage due to, inter alia, rising temperatures and sea levels, soil erosion, and severe weather events.[89] While climatic conditions have always affected cultural heritage, climate change has fast tracked damage and in some instances triggered disappearance. Small island states and low-lying areas are especially vulnerable facing the destruction of their natural and cultural heritage including cultural identities with entire nations facing cultural extinction. The threat was created transnationally, thus require a transnational response: "Those facing such levels of damage to their cultural lives are entitled to robust international solidarity, support, cooperation and compensation."[90] Furthermore, movement away from homelands could also affect cultural practices that may be linked to certain cultural sites or natural resources. Unfortunately, some heritage losses are inevitable.[91]

In addition to the impact on cultural heritage, climate change could affect cultural diversity and socio-cultural interactions by forcing communities to change their ways of life or to migrate. As the UNESCO Universal Declaration on Cultural Diversity notes, "cultural diversity is as necessary for humankind as biodiversity is for nature."[92] Climate displacement threatens cultural survival and affects traditional livelihoods. It is hard to distinguish between forced and voluntary migration while some are unable to move. This leaves people with a terrible choice – whether to remain with cultures that sustain them or move to protect their lives and livelihoods: "Cultural losses related to migration will be especially serious for those living in entirely unique landscapes. As one expert asked, where can the Inuit find another Arctic environment?"[93] Cultural rights are a primordial component of "migration with dignity".

Another aspect negatively affected by climate change is traditional knowledge, including the very knowledge needed to respond to such change. Women are especially affected including the nutritional needs of breastfeeding or pregnant women, when traditional agriculture and fishing are no longer feasible due to climate change. The gendered impact of climate change, resource scarcity and disasters may leave women with less time to participate in cultural life. Moreover, nomadic pastoralism as a way of life may be entirely at risk.

Another group whose cultural rights and cultural heritage are disproportionately affected comprise indigenous peoples for whom connections to place, land and relationships with culturally important animals, plants, and ecosystems play an important role. Lack of respect for land rights exacerbate their vulnerabilities, especially for indigenous women who care for the land and face resource scarcity of traditional foods and medicines.[94] Not just climate impacts but also adaptation and mitigation measures taken in response to climate consequences without free, prior and informed consent of indigenous peoples or without their participation could further undermine cultural rights: "Taken together, the results of the climate emergency are significant changes to the social and cultural fabric of entire groups, and put their very cultural survival at risk."[95]

The Special Rapporteur, however, recognized that cultural rights may be subjected to limitations in certain instances but as stressed by the ESCR Committee, they should be the last resort and should be in accordance with international human rights law. Social and cultural values can contribute to climate change and certain objections to climate action on the basis of culture may have to be overridden in accordance with human rights norms. Real tensions could arise between environmental goals and lived cultures and traditions which require a dialog with the relevant stakeholders and their participation.[96]

Stressing that we must take a holistic approach to culture, cultural rights, and climate change, bringing together natural, tangible, and intangible cultural heritage and all forms of cultural expressions, the Special Rapporteur made several recommendations: adopt a human rights-based global action plan to save cultures and protect cultural rights; prioritize global efforts to prevent the cultural extinction of populations facing threats from the climate emergency, such as those in polar and coastal regions, indigenous peoples, and inhabitants of small island states; take cultural rights and cultural impacts into consideration in climate action; include harm to culture, cultural heritage and cultural rights in inventories and environmental impact and climate vulnerability assessments; monitor the impacts of climate change on cultural heritage; ensure an integrated approach to climate change, culture, and cultural rights; promote information sharing; develop remedies, compensation, and accountability mechanisms for damage to culture, cultural rights, and cultural heritage; ensure that the groups most affected by climate change are involved in climate-related policy processes; ensure gender mainstreaming throughout climate action; and incorporate cultural rights and cultural heritage into adaption plans.[97] She called upon UN treaty bodies to adopt a joint general comment on the climate emergency and human rights and to draw attention to the threat to all rights including cultural rights.[98]

4.7 Special Rapporteur on the Right to Food

The former Special Rapporteur on the Right to Food, Hilal Elver, discussed environmental issues, especially climate change in the context of the right to food on several occasions. Another former Special Rapporteurs, Jean Ziegler, discussed the issue of biofuels in his 2007 report.[99] Noting that global levels of hunger continue to rise, the former Special Rapporteur pointed out that more than 6 million children die every year from hunger and related issues before their fifth birthday and drew attention to the potentially negative impact of biofuels on the right to food:

> The sudden, ill-conceived, rush to convert food – such as maize, wheat, sugar and palm oil – into fuels is a recipe for disaster. There are serious risks of creating a battle between food and fuel that will leave the poor and hungry in developing countries at the mercy of rapidly rising prices for food, land and water. If agro-industrial methods are pursued to turn food into fuel, then there are risks that unemployment and violations of the right to food may result, unless specific measures are put in place to ensure that biofuels contribute to the development of small-scale peasant and family farming. Instead of using food crops, biofuels should be made from non-food plants and agricultural wastes, reducing competition for food, land and water.[100]

The World Food Program has identified climate change as one of the obstacles to the right to food and noted that rising demand for biofuels is pushing up global grain prices.[101] NGOs have called for a global moratorium on the expansion of agrofuels until their social, environmental, and human rights impact can be examined. The increased interest in agrofuels has led to massive investment and setting of ambitious renewable fuel targets in Western countries. The EU thus requires that agrofuels provide 10% of transport power by 2020 while the US target is to increase use of agrofuel for energy to 35 billion gallons per year. As oil prices rise, it becomes more viable to invest in alternative sources of energy but if no conscious effort is made to ensure that "producing biofuels does not bring greater hunger in its wake, then the poor and hungry will be the victims of these new fuels."[102]

The report noted that the two main types of agrofuel – bioethanol and biodiesel – are produced from a variety of food crops such as sugar cane, maize, sugar beet, potatoes, wheat, manioc, soya, palm or rapeseed, peanuts, coconuts, and other oil-rich plants. Most of the plants used for agrofuels are food products and staple foods of millions of people in poorest regions of the world who are already suffering from high food prices.[103] While global consumption of agrofuels is low, it will likely rise under the targets set by the EU, US, and Latin America. However, these targets cannot be met by agricultural production in industrialized countries alone; thus, they are very interested in the production is southern countries to meet these needs.

While biofuels could bring positive benefits for climate change, the former Special Rapporteur stressed that we need to examine its negative aspects including greater hunger: "The greatest risk is that dependence on the agro-industrial model of production will fail to benefit poor peasant farmers and will generate violations of the right to food."[104] The report highlighted a number of key concerns: increasing food prices; increasing competition over land and forests, and forced evictions; deteriorating conditions of work; and increasing prices and scarcity of water. Rather than encouraging people to use less energy, agrofuels suggest that we can continue by changing fuels and according to studies agrofuels may not even be carbon neutral. The Special Rapporteur recommended a five-year moratorium on biofuel production; moving to second generation technologies which would reduce competition between food and fuel; adopting technologies that use non-food crops; and ensuring that biofuel production is based on family agriculture, not industrial models of agriculture.[105]

Noting that climate change is already having a significant impact on approximately 1 billion of the world's poor, the 2015 report of former Special Rapporteur Hilal Elver noted that, without mitigation action to combat climate change, the number of hungry people could increase by 20% by 2050[106]:

> Climate change has negative impacts on agriculture while current agricultural practices and food systems are responsible for harming the environment, affecting social and environmental determinants of health and accelerating human-induced climate change. Moreover, climate change is undermining the right to food, with disproportionate impacts on those who have contributed least to global warming and are most vulnerable to its harmful effects. Urgent action must be taken to prevent climate change from intensifying, to mitigate greenhouse gas emissions and to adapt to its unavoidable effects.[107]

Relying on GC No 12 of the ESCR Committee, the Special Rapporteur identified the necessary elements of the right to food as availability, accessibility, and adequacy and added a fourth element – sustainability – which implies that food will be accessible for both present and future generations and places emphasis on principles of participation, non-discrimination, transparency and empowerment.[108] Moreover, food sustainability and security are dependent on an adequate diet, clean water, sanitation, and health care.[109] The report identified Sub-Saharan and North Africa, and the Middle East as being most affected while the largest population of hungry people live in Asia. In addition, there are many vulnerable groups spread across the world who are disproportionately affected by climate change: those living in poverty, especially in rural areas, smallholder farmers, women, and indigenous peoples.

Just as climate change has negative consequences for the right to food, agriculture and food systems negatively affect climate change, also jeopardizing the right to food. Food systems is a significant contributor of GHG emissions with crops and livestock accounting for about 15% of global emissions. In addition to the direct emissions, agriculture and food production are responsible for indirect emissions (including industry, transport, and energy) as well as fertilizers, herbicides and pesticides, energy for tillage, irrigation, harvesting, and transportation.

Despite the inclusion of food security in the UNFCCC, it has received scant attention in policies and programs. A human rights framework requires states to reduce GHG emissions with a view to reducing their negative effect on the enjoyment of rights. In the run up to the Paris Agreement several special mandate holders issued an open letter calling on state parties to ensure full coherence between human rights obligations and climate action – and to include human rights language in the outcome document.

The report also discussed the adverse impact of mitigation policies on the right to food and examined biofuel production, bioenergy, water-diversion for energy production, and emission reduction strategies such as the REDD+ program. These have resulted in forced relocations of smallholder farmers and indigenous peoples. First-generation biofuels have resulted in food v. fuel conflicts while hydropower projects have created conflicts between water for energy and water for agriculture. The report stressed that adaptation policies should focus on helping farmers reduce their vulnerability and strengthen their resilience. While several funds have been established to help developing countries, they have failed to secure adequate funds. Moreover, there is lack of participation in allocating funding.

Given the enormous challenges in relation to meeting food security in the face of climate change, the former Special Rapporteur proposed agroecology as an alternative to industrial agriculture:

> Agroecology is an ecological approach that integrates agricultural development with relevant ecosystems. It focuses on maintaining productive agriculture that sustains yields and optimizes the use of local resources while minimizing the negative environmental and socioeconomic impacts of modern technologies. Recycling nutrients and energy…, integrating crops and livestock and improving interactions and productivity throughout the agricultural system …are also important components of agroecology. It is a system that foregoes the use of synthetic inputs, such as synthetic fertilizers and pesticides, veterinary drugs, genetically modified seeds and breeds, preservatives, additives and irradiation.[110]

Benefits of agroecology include improving soil quality, plant health and biodiversity, increased resilience of crops and farms and benefits to poor rural communities all of which contribute to the enjoyment of the right to food. The Special Rapporteur's recommendations included: protecting human rights in all climate-related actions; ensuring policy coherence between the UNFCCC and other international treaties relevant to climate change and food security to promote climate justice and the right to food; reviewing policies that promote subsidies and production targets for their negative impact on the right to food; a separate category of "climate refugees" be recognized under international law to avoid human catastrophe; a human rights impact assessment be carried out including public participation before mitigation and adaptation projects are authorized; alternative energy and mitigation policies be scaled back; the pivotal role in food production of smallholder farmers, women, and indigenous and local communities be recognized; evaluating agricultural and trade policies to avoid price volatility and financial vulnerabilities; prioritizing social protection measures to eliminate hunger and avoid food insecurity; and reducing excessive consumption and eliminate food waste.[111]

The 2017 report addressed the impact of (natural) disasters on the right to food.[112] It noted that the main causes of reversing progress toward eliminating hunger include armed conflict, natural disasters, *climate change-induced extreme weather events*, economic slowdown and lack of effective social protection.[113] The report noted that global warming-triggered natural disasters, and droughts as well as floods affect food security, infrastructure, crops, livestock, food supplies, and food production systems, in addition to the significant loss of lives and property. The report notes: "According to FAO, worldwide economic losses from natural disasters have reached a staggering average of $250 billion to $300 billion a year. Yet, we know comparatively little about the full impact of such disasters on the agricultural sectors."[114] The report noted that climate change not only affects crops and livestock but also give rise to extreme weather events which are expected to become more frequent in the future: "In times of disaster, while the impacts on food availability are often of immediate concern, the gradual effects of natural hazards on accessibility, adequacy and sustainability are of equal importance – they may be less visible and yet more enduring."[115] The report pointed out that natural disasters can have a long term impact on the right to food by threatening environmental resources and entire ecosystems vital for sustainable food production. Contamination of soil and water, salinization of water bodies, reducing nature's defense capacity, landslides, and biodiversity loss[116] all negatively affect the right to food.

The report also discussed the impact of disasters on people with special needs as disasters affect the most vulnerable disproportionately; more than 75% of the world's poor depend on natural resources for their livelihoods[117] and have little savings or safety nets. In particular, local and traditional communities such as nomadic, indigenous peoples, and peasants are especially affected by loss of access to traditional foods, loss of traditional knowledge and loss of biodiversity. Other groups disproportionately affected include women, subsistence farmers, those with disabilities, the elderly, people with HIV/AIDS, children, and internally displaced persons, migrants, and refugees.

Disasters created massive human suffering; while a plethora of frameworks and treaties ranging from human rights to climate change are applicable, there is no comprehensive multilateral treaty with general application governing disasters.[118] Despite the fragmented landscape, there has been a positive move toward implementing a human rights approach to disasters, including providing food aid and assistance:

> A human rights-based approach has crucial advantages in disaster settings, such as avoiding discrimination, prioritizing vulnerable communities, and providing measurable and enforceable obligations through accountability mechanisms. Yet, accountability remains challenging at the national and international levels due to the complexity of the humanitarian system.[119]

The report, relying on the work of the ESCR Committee, noted that states have a joint and individual responsibility to cooperate in providing disaster relief and humanitarian assistance in time of emergencies including assistance to refugees and IDPs and noted that:

> [t]he right to food in emergencies needs positive action by States not only to respect and protect, but also to fulfil the normative content of international human rights principles… Article 11 of the Covenant does not make any differentiation about the causes of difficult times in relation to State obligation.[120]

The report concluded by noting that climate-related disasters are increasing in frequency and intensity and their impacts are devastating. For example, Hurricane Harvey in the US in 2017 cost $85 billion. Many of the negative impacts can be averted with planning and investment in infrastructure and addressing the root causes. Moreover, impacts on food insecurity could lead to conflicts in countries that have limited coping capacity, which could "become trapped in a vicious cycle of conflict, disaster and food insecurity."[121]

The recommendations included: developing national laws and monitoring systems to incorporate a rights-based approach and get the participation of the private sector in emergency situations in line with the Guiding Principles on Business and Human Rights; adopt legislative and budgetary measures to focus on prevention and disaster risk reduction measures to avoid environmental degradation, biodiversity loss and increase forest conservation and watershed management; consider establishing "crop and disaster insurance" for victims and take measures to protect seeds; prioritize the needs of the most vulnerable, especially children; set policies to address the role of women in disaster and post-disaster situations acknowledging their leadership capacity; enhance coordination among humanitarian agencies; and consider the negotiation of a multilateral treaty on disasters. The Special Rapporteur called upon the ESCR Committee to adopt a general comment on human rights in times of disasters and to clarify obligations of states and the international community.[122]

The final report of the former Special Rapporteur addressed the issue of food systems and food crises.[123] In this context, she again addressed climate change and disasters and made extensive references to the SDGs. Noting that the current industrial model has serious disadvantages, the report pointed out:

> It generates food loss and waste, mistreats animals, emits greenhouse gases, pollutes ecosystems, displaces and abuses agricultural and fishery workers, and disrupts traditional farming communities. Put simply, the human rights of food system actors, including agricultural workers, smallholder farmers and consumers, are often ignored or their rights violated.[124]

The Special Rapporteur discussed several obstacles to the fulfillment of right to food such as trade agreements, subsidies, and neoliberalism; marginalization of small holder farmers and peasants; land grabs; worker exploitation and exposure to dangerous pesticides; "supermarketization" of food and increasing malnutrition; and loss of biodiversity and environmental degradation. While globalization of food systems can increase access to food and increase its diversity, in practice it has perpetuated global inequity and undermined access to food for the most vulnerable populations. Fiscal policies that promote farm subsidies mostly benefit large corporations and landowners. Most heavily subsidized areas experience the worse pollution and an increase in GHG emissions from agriculture.[125]

Similarly, globalization has enabled big agricultural corporations to dominate the market which has marginalized smallholder farmers and peasants who produce 70% of locally consumed food. Land grabs – the acquisition of large areas of land by companies for investment purposes often without any consultations with local communities – is another area of concern. Commodification of land has destructive impacts on local livelihoods, especially those without formal land grants. Pointing out that biodiversity is vital to sustainable agriculture and food production but is declining at unprecedented rates, the report quoted the FAO:

> [f]ewer than 200 plant species make major contributions to food production and just 3 crops – wheat, maize and rice – account for more than half the world's plant-based calories. Nearly one third of fish stocks are overfished and nearly 26 per cent of the 7,745 local livestock breeds are at risk of extinction.[126]

Although indigenous peoples are custodians of 80% of remaining biodiversity, they are facing severe food insecurity, extreme poverty, and other human rights violations. Moreover, hydroelectric projects, mining projects, and designation of protected areas have jeopardized their rights, despite the protections in UNDRIP.

Stressing that climate crisis is an existential threat to human survival, the report noted that it is a key driver of hunger and malnutrition leading to violations of human rights. It pointed out that the three post-2015 frameworks – 2030 Agenda for Sustainable Development, the Paris Agreement, and the Sendai Framework for Disaster Risk Reduction – should provide the foundation for sustainable and resilient development in a changing climate. Extreme weather events forcibly displaced more than 20 million people in 2017, in addition to impacting livelihoods. Massive fires have destroyed the environment and threatened the lives and food security of people. The Special Rapporteur made several recommendations including: formally recognizing ESC rights in times of peace and conflict; implementing a rights-based approach to governance; monitoring rights and guaranteeing access to justice for extraterritorial violations; financing human rights institutions and removing the silos separating international organizations;[127] adopting economic reform to address poverty and inequality of marginalized populations; empowering women and girls and promoting gender diversity; investing responsibly in technology, agroecology and traditional knowledge, and regulating innovation;[128] and enhancing the role of civil society and protecting human rights defenders. Pointing out that the right to food is not unattainable, only unrealized, the Special Rapporteur stressed that states must implement human rights obligations ensuring that all players are included in the decision-making process, not just the powerful ones.[129]

4.8 Special Rapporteur on Extreme Poverty and Human Rights

The Special Rapporteur on Extreme Poverty, Philip Alston, devoted his 2019 report to climate change[130] noting that climate change will have devastating consequences for people in poverty:

> Even under the best-case scenario, hundreds of millions will face food insecurity, forced migration, disease and death. Climate change threatens the future of human rights and risks undoing the last 50 years of progress in development, global health and poverty reduction.[131]

The report noted that climate change threatens the enjoyment of a wide range of rights; while adaptation can mitigate its impact, people in poverty must be protected from the worst impacts. Quoting World Bank data, the report stressed that with warming of 2 degrees Celsius, 100–400 million more people could face hunger, and climate change will exacerbate health shocks as many poor people are uninsured.

As noted, climate change will exacerbate existing poverty and inequality with developing countries bearing an estimated 75%–80% of the cost of climate change. Worst still, it threatens to undo the last 50 years of progress. According to the World Bank, climate change could push 120 million more people into poverty by 2030 unless states take immediate action. In South Asia alone, 800 million people live in climate hotspots. The report highlights the inequality (and the injustice) of the situation, both in relation to the impact of climate change and the ability to cope with them:

> Perversely, the richest people, who have the greatest capacity to adapt and are responsible for and have benefited from the vast majority of greenhouse gas emissions, will be the best placed to cope with climate change, while the poorest, who have contributed the least to emissions and have the least

> capacity to react, will be the most harmed. The poorest half of the world's population – 3.5 billion people – is responsible for just 10 per cent of carbon emissions, while the richest 10 per cent are responsible for a full half. A person in the wealthiest 1 per cent uses 175 times more carbon than one in the bottom 10 per cent.[132]

Noting the failures to act on the part of governments and corporate actors, the report point out that fossil fuel companies are the main drivers of climate change with that industry accounting for 91% of global industrial GHG emissions and 70% of all human-made emissions. Although it had known for decades about rising CO_2 levels, not only did it not take action to change its business model, it also embarked on a campaign to thwart meaningful action to reduce emissions. States have been complicit with corporate actors by subsidizing the fossil fuel industry amounting to US $5.2 trillion *per year.*[133]

In addition to transforming the economy and society, the report also advocated for transforming the international human rights regime. Stating that "an extraordinary challenge demands an extraordinary response" the report called upon states to acknowledge the urgency of transformational change with human rights being part of the solution. The report urged states to acknowledge the risk that climate change poses to democracy, rule of law and civil and political rights in addition to ESC rights. Additionally, large-scale movement of people both internally and across borders will pose unprecedented challenges to governance and could give rise to nationalist, xenophobic, and racist responses.

Calling on states to revitalize ESC rights, the report stressed the need to adopt policies that ensure respect for ESC rights. In addition, the report pointed out that corporate actors will not engage in the economic and social transformation that climate mitigation demands. In a harsh critique of governments and human rights bodies, the Special Rapporteur stated:

> The human rights community, with a few notable exceptions, has been every bit as complacent as most Governments in the face of the ultimate challenge to mankind represented by climate change. The steps taken by most United Nations human rights bodies have been patently inadequate and premised on forms of incremental managerialism and proceduralism that are entirely disproportionate to the urgency and magnitude of the threat. Ticking boxes will not save humanity or the planet from impending disaster… It has also sought to highlight the fact that the group that will be most negatively affected across the globe are those living in poverty. *Climate change is, among other things, an unconscionable assault on the poor.*[134]

In his interim report, the current Special Rapporteur on Extreme Poverty, Olivier de Schutter, discussed "just transition" in relation to economic recovery and eradicating poverty within planetary boundaries[135] which was presented in the face of the economic crisis due to COVID-19 pandemic as a result of which 176 million more people will face extreme poverty. The report pointed out that while adopting pro-poor policies and strengthening social protection are necessary, business as usual is not the answer – while the impact of the pandemic is immediate, we are facing a deeper emergency – the environmental crisis: "We therefore need to "build back better," relying on the integrated approach at the heart of the 2030 Agenda for Sustainable Development, and to reconcile poverty eradication with planetary boundaries."[136] While the measures proposed in the report can help solve the triple challenges of environmental sustainability, employment opportunities, and poverty eradication, social transformation requires moving from unsustainable consumption-driven growth toward wealth redistribution and fighting against obsolescence of consumer goods.[137]

Pointing out that the pandemic provided us with a once-in-a-generation opportunity to redefine development trajectories in accordance with the SDGs, the report stated that economic recovery plans can help transition to a low-carbon economy while creating employment opportunities and ensuring access to goods and services essential to the enjoyment of human rights[138]:

> This requires, inter alia, protecting workers and communities from the impacts on their livelihoods; investing in areas such as energy, buildings, food and mobility, to capture the "triple dividend" of a cleaner environment, decent jobs and affordable goods and services; moving away from unsustainable growth and the extractive and waste economy; reducing inequalities and unsustainable quest for economic growth; and adopting national action plans on social dialogue with the participation of people living in poverty.[139]

4.9 The Working Group on Human Rights and Transnational Corporations and Other Business Enterprises

A few reports issued by the Working Group on Transnational Corporations address environmental issues. One of the reports examines how the business and human rights agenda and anti-corruption efforts are interconnected.[140] Noting that corruption is a complex issue affecting all states, the report noted that it takes many forms including bribery, nepotism, extortion, cronyism, embezzlement, and fraud. It adopts the definition of corruption in the UN Convention against Corruption.[141]

The connection between human rights and corruption has long been acknowledged and corruption has devastating impacts on human rights. In the 2030 Agenda and other instruments, states have expressed their commitment to fight corruption. The report notes that one area where corruption thrives is in relation to land and natural resources where there is weak land governance and where rights to land and natural resources are not properly documented or enforced. Thus, "governments need to strengthen policy coherence regarding environmental impact assessments, large-scale development projects, land management and forest conservation."[142]

Another area where significant corrupt activities take place is the extractive sector. While the Extractive Industries Transparency Initiative,[143] and good practices regarding supply chains and human rights due diligence have been endorsed by many stakeholders, corruption remains rampant. This has led to significant human rights impacts especially in indigenous communities where extensive mining and oil and gas extraction have also resulted in impacting the rights to health and a safe environment. Granting concessions without proper environmental or social risk assessments has led to human rights abuses affecting the local population. Complaints include lack of consultations with local communities, especially indigenous peoples, and non-compliance with the free, prior and informed consent requirement.[144]

The report adopted the "three pillars" approach in the Guiding Principles on Business and Human Rights. Under the state's duty to protect against abuse by business enterprises the report recommended policy coherence, linked incentives, and policy reform which, in turn, requires procurement reform, expanding the scope of governance mechanisms and adopting mandatory due diligence regulation. In this regard, the report referred to the commitment by the European Commission to introduce rules for mandatory corporate environmental and human rights due diligence.[145]

With regard to the second pillar – corporate responsibility to respect human rights – the report pointed out that key actors, including OECD and the UN Global Compact, have called for a holistic, integrated approach to responsible business conduct.[146] The report noted that companies have adopted good practices to reduce human rights and corruption risk by creating a corporate culture of integrity, and ensuring that lawyers and managers are aware of the Guiding Principles; adopting a code of conduct or ethics by the board; building capacity and training; and integrating corruption risk into human rights due diligence. The third pillar – ensuring an effective remedy – requires that the procedure for the provision of remedy should be impartial, protected from corruption, and free from political and other influences.[147]

4.10 Special Rapporteur on Contemporary Forms of Racism, Racial Discrimination, Xenophobia, and Related Intolerance

The Special Rapporteur on Racism, Tendayi Achiume, discussed global extractivism and racial equality in her 2019 report.[148] She noted that the inequalities that characterize the global political economy are present in the extractivism economy:

> Powerful States and their transnational corporations, and the political elites of weaker States that are territories of extraction, emerge as the clear winners. The populations of these territories of extraction bear the brunt of the extractivism economy,[149] too often paying with their very lives.[150]

She noted that a defining feature of extractivism is that it involves the removal of raw materials from territories that were previously colonized, while the sale and consumption of those materials benefits nations, transnational corporations, and consumers in the global North. Territories rich in natural resources have experienced severe forms of underdevelopment referred to as "structural exploitation."[151] The socioeconomic and political devastation in the global South is the result of global extractivism economy rooted in structural inequality.

The Special Rapporteur stated that her report focuses on dominant modes of global extractivism which are environmentally unsustainable and are a human rights concern. While the report focuses on equality and non-discrimination, the reality is that global extractivism economy will make our planet unlivable for humans. Discussing the colonial racial past of extractivism economy, the report notes that it led to environmental destruction and the exploitation of non-white labor. Continuing sovereign inequality and constraints on self-determination are at the core of the extractivism economy and despite the politicization of the ecological crisis and climate change, the extractivist form of appropriation of nature continues.[152]

Referring to a report by the Inter-American Commission of Human Rights which highlighted the marginalized status of indigenous peoples and people of African descent and the human rights impact of the extractivism economy, the Special Rapporteur pointed out that host governments and private corporations oversee the destruction of ecosystems through, *inter alia*, water pollution, deforestation, destruction of biodiversity, and soil pollution:

> Extractivist projects can threaten the very physical and cultural existence of these groups as peoples and, on account of their devastating environmental impact, also result in gross violations of the rights to health and life, by causing illness and death. The recent collapse of a dam owned by an iron ore mining corporation, Vale S.A., in Brazil, in addition to killing hundreds and releasing almost 12 million cubic metres of mining waste, also threatens the very existence of indigenous groups in the area.[153]

Noting that the environmental and health consequences of extractivist processes are well documented, the Special Rapporteur pointed out that using mercury and other chemicals in gold extraction is illustrative. Examples include environmental destruction in French Guiana caused by mercury poisoning resulting in indigenous children born with development disabilities; and exposing non-whites to dangerous jobs during the colonial era and apartheid in South Africa where these gold miners contracted silicosis.[154]

Likewise, hydraulic fracking by its very nature alters the environment and there is evidence of its harmful effects on the environment and the health of local inhabitants. Air pollution, groundwater contamination, and surface water pollution are some consequences that lead to health problems. In 2011, a malfunctioning fracking well in the US spewed thousands of gallons of contaminated water into the environment and a study found that drinking water near fracking wells had dangerous levels of methane. Another example is contamination by oil spills – in 2008, the Royal Dutch Shell oil spill poured tens of thousands of barrels of crude oil into the water in Nigeria, leading to many environmental and human rights violations. While Shell reached a settlement that paid about $3,000 to the claimants, this "cannot come close to remedying the devastating effects of the spill on the affected communities."[155]

Moreover, the vulnerability of these marginalized communities in the extractivism economy is exacerbated by land grabs which entail a change in land use and ownership from local populations' food production to corporate and industrial purposes excluding local population from access to productive agricultural land. Sub-Saharan Africa is highly susceptible with more than 10 million hectares being subjected to land grabs.[156]

Additionally, according to submissions received, within African countries, women peasant farmers often experience worst forms of land dispossession, pollution, violence, and health consequences associated with extractivist processes. They are excluded from negotiations concerning these activities by male community leaders, state agents, and corporations. Moreover, gender is a "salient axis of subordination and exclusion where labour rights are concerned." For example, women are responsible for about 80% of food crop production in Uganda. Women in some regions of Uganda have indicated that access to agricultural land had been affected by oil exploration.[157]

The Special Rapporteur recommended that states, multilateral actors, and transnational corporations must respect principles of sovereign equality, the right to self-determination and right to development of all peoples. She stressed that businesses must implement the "respect, protect and remedy" framework of the Guiding Principles on Business and Human Rights, carry out corporate due diligence, and ensure prior, informed consent of affected communities. She stressed that procedural mechanisms should not be "untethered from the substantive norms these mechanisms are designed to serve." States in the global South must ensure the permanent sovereignty over natural resources of their peoples which should be understood to include the right of peoples, especially those negatively affected by the extractivism economy, and to say no to extractivism. Finally, the Special Rapporteur recommended that states should "reject color-blindness and gender blindness."[158]

4.11 Special Rapporteur on Indigenous Peoples

In his report on the situation of indigenous peoples in the US, the Special Rapporteur on Rights of Indigenous Peoples, James Anaya, discussed the evolution of federal policy and legislation as well as the contemporary legal and policy regime.[159] He noted that Native Americans, especially those on reservations have disproportionately high poverty rates, nearly double the national average plus poor health conditions, high rates of disease and illness, alcoholism, and suicide.[160]

Noting that the disadvantage of indigenous peoples is not mere happenstance, the Special Rapporteur pointed out that they stem from a history of taking vast areas of indigenous lands and resources along with the suppression of their culture, entrenched patterns of discrimination, forced removal from ancestral territories and even brutality.[161] In addition to millions of acres of lands lost, extractive and other activities[162] on or near indigenous lands, including nuclear weapons testing and uranium mining, have resulted in widespread environmental harm, and serious health problems among Native Americans. During his visit to the US, the Special Rapporteur heard concerns about several projects that could cause environmental harm to indigenous habitats, including the Keystone XL pipeline and the Pebble Mine project in Alaska which could seriously threaten the sockeye salmon fisheries in the area:[163]

> In many places, including in Alaska and the Pacific Northwest in particular, indigenous peoples continue to depend upon hunting and fishing, and the maintenance of these subsistence activities is essential for both their physical and their cultural survival, especially in isolated areas. However, indigenous peoples face ever-greater threats to their subsistence activities due to a growing surge of competing activities, restrictive state and federal regulatory regimes, and environmental harm.[164]

The Special Rapporteur acknowledged the attention to indigenous peoples' concerns by acts of Congress and federal programs in recent years and new initiatives to develop consultation policies and spaces of dialog with tribes including cleaning up environmental pollution caused by natural resource extraction. While consultation procedures have been improved, many indigenous leaders complained that they have yet to see significant change in the decision-making by government agents, particularly in relation to lands outside their controlled areas but affect their access to natural or cultural resources or environmental well-being. He heard concerns about lack of sufficient funding for, inter alia, environmental remediation[165] and pointed to the need to address the severed connections with culturally significant places and sacred sites from environmental pollution.

The recommendations included addressing environmental degradation and improving existing measures to address the concerns of indigenous peoples in the US and for developing new measures to advance toward reconciliation. UNDRIP represents an international standard accepted by the US and is a benchmark for all relevant decision-making by the federal bodies, Congress, and the judiciary as well as by the states.[166]

The 2016 report of the former Special Rapporteur on Indigenous Peoples, Victoria Tauli-Corpuz, addressed the violation of rights of indigenous peoples in the context of ever-expanding conservation measures which her predecessors had also noted.[167] The focus of the report was terrestrial protected areas[168] including World Heritage sites. While protected areas contribute to the protection of biodiversity, they have also resulted in human rights violations of indigenous peoples:

> For over a century, conservation was carried out with the aim of vacating protected areas of all human presence, leading to cultural destruction and large-scale displacements of indigenous peoples from their ancestral lands in the name of conservation. Past conservation measures caused complex and

> multiple violations of the collective and individual human rights of indigenous peoples.[169]

The report noted that protected areas nearly doubled over a period of two decades, from 8.7 million square kilometers in 1980 to 16.1 million square kilometers in 2000 with a significant spatial overlap between indigenous ancestral land and areas with high biological diversity: about 50% of protected areas worldwide has been established on traditional indigenous lands and this proportion is highest in the Americas, where it may be as high as 90% in Central America. Bolivia, Brazil, Chile, and Colombia, as well as Canada and the US, have a high percentage of protected areas on indigenous territory.[170] Indigenous peoples have strong spiritual ties with trees, animals, and their lands and conserving them is their sacred duty. However, they are not seen as conservationists as they do not call themselves that. Most indigenous territories hold intact ecosystems and they are the most sustainable form of conservation. Yet, to date, the important role played by indigenous peoples in conservation has not been acknowledged. According to UNEP, in 2014 less than 5% of world's protected areas were governed by indigenous peoples and local communities.[171]

In most countries, conservation efforts have been led by states, expropriating lands, and driving indigenous peoples out of their traditional lands, depriving them of their subsistence way of life. However, government authorities lacked the will or the capacity to protect them which has exposed these lands to illegal logging, extractive activities, illegal settlements, agribusiness expansion, and infrastructure projects, sometimes with the authorization of state agencies. While mobilization by indigenous groups has resulted in the recognition of their collective right to their traditional territories under international law, they continue to struggle to get legal recognition of their rights over ancestral lands, especially if they have been declared as protected areas.

The Special Rapporteur identified several human rights obligations that states have in this regard: the right to self-determination and land rights; participation and free, prior and informed consent; and protection against forced displacement and the right to reparation including restitution. These are principles that are found in international and regional human rights law and case law.

> From the perspective of indigenous peoples, the creation of protected areas was perceived as colonialist, as the consequences for indigenous peoples who experienced them spelled subjugation and the loss of lands, autonomy and self-governance, livelihood resources as well as the rupture of cultural and spiritual links. Protected areas under State control imposed new laws and forms of control by Government institutions. In this sense, protected areas were seen as a vehicle for coercive assimilation by indigenous peoples.[172]

New approaches to conservation have emerged and at the World Parks Congress held in Durban the international community announced a new paradigm for conservation areas which would respect the rights of indigenous peoples and local communities. To implement this vision, the Durban Accord and Action Plan were adopted. It called upon the Conference of the Parties to the Convention on Biological Diversity to ensure that indigenous peoples and local communities participate in the establishment and management of protected areas and that they share the benefits from these areas.[173] However, the Durban Accord has not been satisfactorily implemented. The report identified several challenges and opportunities: forced displacement and the failure to recognize collective rights to lands, territories and natural resources; inconsistent national legislation and their poor application; World Heritage sites and tourism mainly because these sites have been declared without prior consultation with indigenous groups; and providing for indigenous management of protected areas.[174]

4.12 Special Rapporteur on Climate Change and Human Rights

As noted, the HRC created a mandate for a Special Rapporteur on Climate Change and Human Rights by resolution 48/14[175] and Ian Fry was appointed to the position in March 2022. The mandate includes: studying how the adverse effects of climate change, including sudden and slow onset disasters, affect the full enjoyment of human rights; make recommendations to prevent adverse effects and to integrate human rights into policymaking, legislation, and plans to address climate change; identifying challenges, including financial challenges to protecting human rights and make recommendations regarding protecting human rights in the context of mitigation and adaptation policies, investments, and projects; synthesizing knowledge, including traditional knowledge and identifying good practices; exchanging views on lessons learned and best practices related to the adoption of "human rights-based, gender-responsive, age-sensitive, disability-inclusive and risk-informed approaches to climate change adaptation and mitigation policies," to contribute to the achievement of the Paris Agreement, the UNFCCC, and SDGs 13 and 14; to raise awareness of rights affected by climate change; to seek views from states and other stakeholders; to conduct country visits; closely coordinate work with the Special Rapporteurs on human rights and the environment, toxics, and safe drinking water and sanitation; and report annually to the HRC and the UNGA.[176]

While the Special Rapporteur has not yet submitted a report, he has identified priority areas as: climate change displacement; mitigation, adaptation, and finance; and loss and damage. His first country visit will be to Bangladesh. He is currently seeking input for the report on human rights implications of climate change displacement including legal protection of people displaced across international borders.[177]

4.13 Complaint Submitted to Special Mandate Holders on Rights of Indigenous Peoples in Addressing Climate-Forced Displacement

In January 2020, a complaint was submitted to ten special mandate holders whose mandates intersect with environmental rights[178] by the Alaska Institute for Justice on behalf of five tribes located in Louisiana and Alaska faced with climate-forced displacement as a result of human rights violations by the US government.[179] Despite their geographic differences the tribes are facing similar human rights violations because of the US government's failure to protect each tribe's right to self-determination to protect them from climate impacts. The complaint notes that these violations cut across several thematic mandates of the UN Special Rapporteurs "because climate-forced displacement threatens the full enjoyment of a wide range of human rights. These rights include the rights to life, health, housing, water, sanitation, a healthy environment, and food, among others."[180] The tribes[181] urged the Special Rapporteurs to find that climate-forced displacement is a human rights crisis and requested that they make recommendations to the US federal government and the state governments of Louisiana and Alaska including: recognizing the self-determination and sovereignty of all tribes; recognizing their collective rights to land, subsistence, and cultural identities and the right to return to and maintain access to their ancestral homelands; assisting the tribes in protecting and restoring their homelands to the extent possible; obtaining their free, prior and informed consent for all infrastructure developments, and coastal resiliency master plans; protecting their cultural heritage and access to sacred sites, cultural sites, fishing and hunting sites; allocating funding for adaptation measures as well as for tribal-led relocation processes and respect their sovereignty of resettlement decisions. After highlighting the impacts of climate change on these tribes and the human rights provisions that the US government has violated, the complaint concluded:

> The United States government's failure to protect the Tribal Nations named in this complaint from both the human-made and natural effects of the climate crisis has resulted in significant human rights violations that affects these tribal nation's ability to secure basic human rights and continue to lead to individual and community displacement from their land. In accordance with international law and universally held human rights norms, the U.S. government must take immediate action to redress the human rights violations enumerated in this complaint.[182]

In September 2020, the Special Rapporteurs submitted a communication to the US government[183] informing that they have received a complaint on alleged failure to protect indigenous peoples along the coastal regions of Louisiana and Alaska from the impacts of climate change, development projects and oil and gas exploration "affecting their rights to life, health, food, water, housing, a safe, clean, healthy and sustainable environment, self-determination, cultural and religious rights, and leading to the displacement of indigenous peoples from their traditional lands."[184] They expressed their concern about the impacts of climate crisis on the human rights of these indigenous communities and that measures have been taken without the inclusion and full participation of the affected communities and failed to provide them with sufficient protection. They requested a response within 60 days on the measures the government has taken to address the alleged violations and to ensure the full participation of the tribes in decision-making affecting them and to obtain their free, prior and informed consent.[185] Interestingly, the communication included an annex that summarized the international human rights law norms and standards applicable to this situation.[186]

4.14 Conclusion

This chapter discussed the work of special procedures mandate holders to see how they have addressed the issues under discussion in this volume. Of special relevance is the work of the Special Rapporteurs on Human Rights and the Environment but other special mandate holders have also discussed environmental pollution, climate change, and sustainable development including SDGs to the extent that those issues intersect with their respective mandates. There is no doubt that they have contributed to the development and consolidation of law in various areas. Although not binding, their reports carry significant weight as many of the special mandate holders are leading experts in the field.[187] After the HRC endorsed the right to a healthy environment in October 2021, the special procedures mandate holders issued a press release encouraging states to call upon the UNGA to recognize the right to a clean, healthy, and sustainable environment,[188] pointing out that "A General Assembly resolution on the right to a healthy environment would catalyse urgent and accelerated action to achieve environmental justice, by addressing the climate crisis, protect and restore nature and end toxic pollution."[189] Thanks to the tireless efforts of these mandate holders (and scholars and civil society groups) the UNGA did adopt a resolution endorsing a stand-alone right to a healthy environment, even though it was half a century after the landmark Stockholm Conference on the Human Environment.

Notes

1 Parts of this chapter are drawn from author's previous work: Atapattu, S. & Schapper, A. (2019) *Human Rights and the Environment: Key Issues*, Abingdon: Routledge, Chapter 2. Updated to include developments since the publication of the book and reproduced with publisher's permission.
2 See discussion later in this chapter.
3 Personal invitation to attend the workshop, July 2022. The Special Rapporteur is also soliciting comments by academics in the field.
4 Ksentini, F.Z. (1994): *Review of further developments in the fields with which the sub-commission has been concerned, Human Rights and the Environment: Final report prepared by Mrs. Fatma Zohra Ksentini, Special Rapporteur*, E/CN.4/Sub.2/1994/9, UN Economic and Social Council, ¶ 31, http://www1.umn.edu/humanrts/demo/HRandEnvironment_Ksentini.pdf
5 Ibid., Annex I: Draft Principles on Human Rights and the Environment.
6 Ibid., part I, art. 2.
7 The Special Rapporteur may have been far ahead of her time as the emergence of rights of nature seems to be pointing towards their direct protection. See IACHR, *Advisory Opinion OC-23/17* (15 November 2017) ["Advisory Opinion"], referred to in Chapter 2, https://www.corteidh.or.cr/docs/opiniones/seriea_23_ing.pdf.

8 In this regard, the UN Guiding Principles on Business and Human Rights are important. Although soft law, they impose an explicit obligation on business enterprises to respect human rights which now include a right to a healthy environment. See https://www.ohchr.org/sites/default/files/documents/publications/guidingprinciplesbusinesshr_en.pdf
9 Commission on Human Rights Resolution 1995/81 (1995). Adverse effects of the illicit movement and dumping of toxic and dangerous products and wastes on the enjoyment of human rights.
10 Resolution 18/11, A/HRC/RES/18/11 (13 October 2011).
11 Pallemaerts, M., Report of the Special Rapporteur on the implications for human rights of the environmentally sound management and disposal of hazardous substances and wastes, A/HRC/24/39 (13 August 2013).
12 Tuncak, B., *Report of the Special Rapporteur on the implications for human rights of the environmentally sound management and disposal of hazardous substances and wastes*, A/HRC/27/54, (27 August 2014), https://undocs.org/A/HRC/27/54.
13 Ibid.
14 Ibeanu, O., *Report of the Special Rapporteur on the adverse effects of the illicit movement and dumping of toxic and dangerous products and wastes on the enjoyment of human rights*, A/HRC/7/21 (18 February 2008), https://undocs.org/A/HRC/7/21. During the brief period that Professor Marc Pallemaerts was the Special Rapporteur before his untimely death in 2014, he submitted one report in which he identified the regulatory gaps in international law.
15 Tuncak report, supra note 12.
16 Ibid.
17 Resolution 19/10, *Human rights and the environment*, A/HRC/RES/19/10 (19 April 2012), https://undocs.org/A/HRC/RES/19/10.
18 Knox, J., *Report of the Independent Expert on the issue of human rights obligations* relating *to the enjoyment of a safe, clean, healthy and sustainable environment*, A/HRC/22/43 (24 December 2012), https://undocs.org/A/HRC/22/43.
19 Ibid., ¶ 36.
20 Ibid.
21 Knox, J., *Report of the Independent Expert on the issue of human rights obligations* relating *to the enjoyment of a safe, clean, healthy and sustainable environment (Mapping Report)*, A/HRC/25/53 (30 December 2013) ["Mapping Report"], https://undocs.org/A/HRC/25/53.
22 Ibid., ¶ 84.
23 Ibid. (emphasis added).
24 *Report of the Independent Expert on the issue of human rights obligations relating to the enjoyment of a safe, clean, healthy and sustainable environment*, A/HRC/28/61 (3 February 2015), ¶ 106, https://undocs.org/A/HRC/28/61.
25 *Report of the Special Rapporteur on the Issue of human rights obligations relating to the enjoyment of a safe, clean, healthy and sustainable environment*, A/HRC/31/52 (1 February 2016), https://undocs.org/A/HRC/31/52.
26 *Report of the Special Rapporteur on the Issue of human rights obligations relating to the enjoyment of a safe, clean, healthy and sustainable environment*, A/HRC/34/49 (19 January 2017), https://undocs.org/A/HRC/34/49. This is possibly the first time that an explicit link was made between biodiversity and human rights.
27 *Report of the Special Rapporteur on the Issue of human rights obligations relating to the enjoyment of a safe, clean, healthy and sustainable environment*, A/HRC/37/58 (24 January 2018), https://undocs.org/A/HRC/37/58.
28 *Report of the Special Rapporteur on the Issue of human rights obligations relating to the enjoyment of a safe, clean, healthy and sustainable environment*, A/HRC/37/59 (24 January 2018), https://undocs.org/A/HRC/37/59.

29 Knox, J., *Report of the Special Rapporteur on the issue of human rights obligations relating to the enjoyment of a safe, clean, healthy and sustainable environment*, A/HRC/31/52 (1 February 2016) ¶¶ 85–89, https://undocs.org/A/HRC/31/52.
30 The question is whether a national in one state can claim against another state for infringing his/her rights by that state when there is no link of nationality with the wrongdoing state. This was one of the issues in the Inuit petition before the Inter-American Commission of Human Rights in 2005 where Canadian Inuit claimed against the US. The petition was dismissed by the Commission. The petition, decision, reply and testimony at hearing available at: http://climatecasechart.com/non-us-case/petition-to-the-inter-american-commission-on-human-rights-seeking-relief-from-violations-resulting-from-global-warming-caused-by-acts-and-omissions-of-the-united-states.
31 See Chapter 8 in this volume.
32 See Chapter 5
33 Knox, J., *Report of the Special Rapporteur on the issue of human rights obligations relating to the enjoyment of a safe, clean, healthy and sustainable environment*, A/HRC/34/49 (19 January 2017), ¶ 5, https://undocs.org/A/HRC/34/49.
34 These obligations are the same as those identified in the Mapping Report, supra note 21.
35 Knox, J., *Human rights obligations relating to the enjoyment of a safe, clean, healthy and sustainable environment*, A/73/188 (19 July 2018), ¶ 8, https://undocs.org/A/73/188.
36 OHCHR press release (2018) "It's time we all recognise the human right to a healthy environment" https://www.ohchr.org/en/press-releases/2018/06/world-environment-day-5-june-2018
37 Quoting Victor Hugo, he said that the right to a healthy environment is an idea whose time has come: http://srenvironment.org/2018/07/30/newsletter-no-28. As discussed in Chapter 3, the HRC recognized this right three years later.
38 *Issue of human rights obligations relating to the enjoyment of a safe, clean, healthy and sustainable environment, report of the Special Rapporteur*, A/HRC/40/55 (8 January 2019), https://undocs.org/A/HRC/40/55.
39 *Human Rights Obligations relating to the enjoyment of a safe, clean, healthy and sustainable environment*, A/74/161, (15 July 2019), https://undocs.org/A/74/161.
40 *Right to a healthy environment: good practices, Report of the Special Rapporteur*, A/HRC/43/53 (30 December 2019), https://undocs.org/A/HRC/43/53. Each thematic report includes an annex containing good practices relating to that theme. These are available at: https://www.ohchr.org/en/special-procedures/sr-environment/annual-thematic-reports.
41 *Human Rights Obligations relating to the enjoyment of a safe, clean, healthy and sustainable environment*, A/75/161 (15 July 2020), https://undocs.org/A/75/161.
42 *Human Rights and the global water crisis: water pollution, water scarcity and water-related disasters, report of the special rapporteur*, A/HRC/46/28 (19 January 2021), https://undocs.org/A/HRC/46/28.
43 *Human Rights obligations relating to the enjoyment of a safe, clean, healthy and sustainable environment on the theme healthy and sustainable food: reducing the environmental impacts of food systems on human rights*, A/76/179 (19 July 2021), https://undocs.org/A/76/179.
44 *The right to a clean, healthy and sustainable environment: non-toxic environment, report of the special rapporteur*, A/HRC/49/53 (12 January 2022), https://undocs.org/A/HRC/49/53.
45 *Human Rights Obligations relating to the enjoyment of a safe, clean, healthy and sustainable environment*, A/73/188 (19 July 2018), https://undocs.org/A/73/188.
46 Ibid., ¶ 19.
47 Ibid., ¶ 24.

48 Ibid., ¶¶ 26–27.
49 Ibid., ¶ 37.
50 These are: "(a) Be consistent with the existing body of international human rights law; (b) Be of fundamental character and derive from the inherent dignity and worth of the human person; (c) Be sufficiently precise to give rise to identifiable and practicable rights and obligations; (d) Provide, where appropriate, realistic and effective implementation machinery, including reporting systems; (e) Attract broad international support," 2018 Report, supra note 45, ¶ 50.
51 However, there is no consensus on these elements and all elements may not be necessary in a given situation. For example, in a case involving a polluted river, the important element of a right to a healthy environment would be "access to safe water." Moreover, articulating the right to a healthy environment as encompassing all these elements could make it harder to achieve consensus and get support from states. Procedural components on the other hand, are well-established under human rights law.
52 These also form part of the procedural components of sustainable development. See Birnie, P., Boyle, A. & Redgwell, C. (2009) *International Law & the Environment* (3rd edn.), Oxford: Oxford University Press, p. 123.
53 2019 Report, supra note 44. See also Good Practices Report, supra note 39, and *Additional good practices in the implementation of the right to a safe, clean, healthy and sustainable environment, Supplementary information on the report of the Special Rapporteur on the issue of human rights obligations relating to the enjoyment of a safe, clean, healthy and sustainable environment*, A/HRC/43/53/Annex III (13 December 2019), https://www.ohchr.org/en/special-procedures/sr-environment/annual-thematic-reports; and *Good Practices of States at the national and regional levels with regard to human rights obligations relating to the environment*, A/HRC/43/54 (23 January 2020), https://undocs.org/A/HRC/43/54; which summarizes key points from an expert seminar he convened in June 2019 which focused on good practices, challenges, barriers and opportunities relating to the right to a healthy environment.
54 2019 Report, supra note 38, ¶ 2.
55 Ibid., ¶ 44.
56 Ibid.
57 Ibid., ¶¶ 79–80.
58 Ibid., ¶ 109. In order to do so, states must implement the seven steps outlined in ¶¶ 63–78 of the report (from monitoring to evaluation) and review existing laws, policies and standards to ensure that they are consistent with their human rights obligations relating to air quality.
59 July 2019 Report, supra note 39.
60 Ibid., ¶ 16.
61 Ibid., ¶ 32.
62 Ibid., ¶ 62.
63 Ibid., ¶ 75.
64 Ibid., ¶ 74.
65 Ibid., ¶¶ 91–92.
66 2021 Report, supra note 43.
67 Ibid., ¶ 4.
68 Ibid., ¶ 60.
69 See Chapter 6 in this volume.
70 2021 Report, supra note 43, ¶ 72.
71 Ibid., ¶ 89 (emphasis added).
72 Water Report, supra note 42.
73 Ibid., ¶ 54.
74 Ibid., ¶ 89
75 Non-toxic Report, supra note 44.

76 Ibid., ¶ 22.
77 Ibid., ¶ 27.
78 Ibid., ¶ 29.
79 Ibid., ¶ 47.
80 Ibid., ¶ 89.
81 Boyd, 2022 report, ¶ 3.
82 Ibid., ¶ 31.
83 Ibid., ¶ 81.
84 Bennoune, K., *Field of cultural rights*, A/75/298 (10 August 2020) ["Bennoune Report"], https://undocs.org/A/75/298.
85 Bennoune, K., *Universality, cultural diversity and cultural rights*, A/73/227 (25 July 2018), ¶ 38, https://undocs.org/A/73/227.
86 Bennoune Report, supra note 84, ¶ 7 (footnotes omitted).
87 Ibid., ¶ 21.
88 Ibid.
89 Ibid., ¶ 30. A 2014 study identified 130 World Heritage Cultural sites that were at risk due to sea level rise.
90 Ibid., ¶ 37.
91 Ibid., ¶ 39.
92 Ibid., ¶ 42. UNESCO, Universal Declaration on Cultural Diversity (2002), https://www.un.org/en/events/culturaldiversityday/pdf/127160m.pdf.
93 Ibid., ¶ 43.
94 Ibid., ¶ 53. Intersectionality is a good lens to discuss how these interact with one another to create even bigger vulnerabilities and injustices. Compound injustice is another lens. See Crenshaw, K. (1989) "Demarginalizing the Intersections of Race and Sex: A Black Feminist Critique of Antidiscrimination Doctrine, Feminist Theory, and Antiracist Politics," *University of Chicago Legal Forum* pp. 139–167.
95 Ibid., ¶ 54.
96 Ibid., ¶ 62.
97 Ibid., ¶¶ 81–84.
98 Ibid., ¶ 83(c).
99 *The right to food*, A/62/289 (22 August 2007), https://undocs.org/A/62/289.
100 Ibid., summary, p. 2.
101 Ibid., ¶¶ 21–22.
102 Ibid., ¶ 26.
103 Ibid., ¶ 27.
104 Ibid., ¶ 33.
105 Ibid., ¶ 44. In addition, the report discussed the protection gaps for those fleeing from hunger. Millions of people especially in sub-Saharan Africa regularly face hunger, famine and other violations of the right to food. While the vast majority of such people remain within their own countries, about 2 million try to enter the EU illegally every year – some of whom die in the process. The majority of these immigrants are detained and forcibly repatriated. The report noted that it is a tragedy that people fleeing from hunger are not granted adequate protection. While existing legal frameworks provide protection, they need to be "seriously strengthened" to cover "refugees from hunger." (¶ 54).
106 *Right to Food*, A/70/287 (5 August 2015), ¶ 2, https://undocs.org/A/70/287.
107 Ibid., ¶ 3.
108 Ibid., ¶¶ 19–20.
109 Ibid., ¶ 20.
110 Ibid., ¶ 73 (footnotes omitted).
111 Ibid., ¶ 89.
112 *Report of the Special Rapporteur on the right to food*, A/HRC/37/61 (25 January 2018), https://undocs.org/A/HRC/37/61.

113 Ibid., ¶ 2 (emphasis added). It is noteworthy that the report distinguishes between natural disasters and extreme weather events caused by climate change (at least their relationship) and does not refer to all disasters as *natural* disasters. There is nothing "natural" about the extreme weather events triggered by climate change. The report defines emergencies and disasters and notes that there is a tendency to distinguish between natural disasters and man-made disasters. While the two are linked, the interaction between nature and human activity needs to be better understood conceptually. In this context, the report discussed the definitions adopted by the HRC Advisory Committee and International Law Commission. The latter adopts a rights-based approach to disasters endorsing human dignity and recognizes the responsibility of the international community to provide assistance based on international co-operation, not voluntarism which is still the dominant view. See ¶¶ 62–63.
114 Ibid., ¶ 7.
115 Ibid., ¶ 15.
116 Here the Special Rapporteur referred to the report of the Special Rapporteur on Human Rights and Environment on biodiversity and human rights.
117 Ibid., ¶ 34.
118 Ibid., ¶ 57.
119 Ibid., ¶ 58.
120 Ibid., ¶ 69.
121 Ibid., ¶ 107.
122 Ibid., ¶ 112. See in this regard Chapter 7 for the General Recommendation adopted by the CEDAW Committee on Gender and Disasters. See also Commission on Human Rights, *Report submitted by the Special Rapporteur on the right to food, Jean Ziegler, in accordance with Commission on Human Rights resolution 2002/25*, E/CN.4/2003/54, (10 January 2003) which discusses the link between the right to food and the right to water; https://undocs.org/E/CN.4/2003/54.
123 *Critical perspective on food systems, food crises and the future of the right to food, Report of the Special Rapporteur on the right to food*, A/HRC/43/44 (21 January 2020), https://undocs.org/A/HRC/43/44.
124 Ibid., ¶ 6.
125 Ibid., ¶ 9.
126 Ibid., ¶ 38.
127 The report points out that institutional fragmentation and silos within and between Rome-based organizations and Geneva-based human rights mechanisms have weakened efforts to mainstream human rights into food policy agendas: "More cohesive, coordinated action between New York, Rome and Geneva that builds upon the findings of the Human Rights Council, the special procedures, the treaty bodies and the universal periodic review is vital for the effective promotion of the right to food," ¶ 72.
128 The report notes that corporate agribusinesses have increased investment in biotech and plant-based meat alternatives which shows that there are ways to make profit without contributing to climate change or environmental destruction, ¶ 82.
129 Ibid., ¶ 91.
130 *Report of the Special Rapporteur on extreme poverty and human rights*, A/HRC/41/39 (17 July 2019), https://undocs.org/A/HRC/41/39.
131 Ibid., summary, p. 1.
132 Ibid., ¶ 14 (footnotes omitted).
133 Ibid., ¶ 37.
134 Ibid., ¶ 88 (emphasis added).
135 *Extreme poverty and human rights*, A/75/181/Rev.1 (7 October 2020), https://undocs.org/A/75/181/Rev.1.
136 Ibid., ¶ 4 (footnotes omitted).
137 Ibid., ¶ 43.

138 Ibid., ¶ 55.
139 Ibid., ¶ 56.
140 *Connecting the business and human rights and the anti-corruption agendas: Report of the working group on the issue of human rights and transnational corporations and other business enterprises*, A/HRC/44/43 (17 June 2020), https://undocs.org/A/HRC/44/43.
141 UN Convention against Corruption, adopted by the UNGA by resolution 58/4 9 (2003), entered into force 14 December 2005, text available here: https://www.unodc.org/documents/treaties/UNCAC/Publications/Convention/08–50026_E.pdf.
142 Ibid., ¶ 15.
143 Extractive Industries Transparency Initiative (EITI), is the global standard to promote the open and accountable management of oil, gas and mineral resources. It requires "the disclosure of information along the extractive industry value chain from the point of extraction, to how revenues make their way through the government, and how they benefit the public." https://www.state.gov/extractive-industries-transparency-initiative-eiti.
144 Ibid., ¶ 21. The report also addressed the gender dimension of corruption and noted that corruption enables intersecting forms of discrimination, exclusion, prejudice and inequality. As women experience corruption differently than men, a gender lens should be applied to all anti-corruption efforts.
145 Ibid., ¶ 44.
146 Ibid., ¶ 50.
147 Ibid., ¶ 57. In addition, the report noted that civil society actors, including whistle-blowers, play a critical role in monitoring and reporting on corruption and multi-stakeholder engagement is key to addressing corruption and business-related human rights abuses, ¶ 70.
148 *Report of the Special Rapporteur on contemporary forms of racism, racial discrimination, xenophobia and related intolerance*, A/HRC/41/54 (14 May 2019), available at: https://undocs.org/A/HRC/41/54.
149 Ibid., ¶ 6. The Special Rapporteur defines "extractivism economy" as "industries, actors and financial flows, as well as to the economic, material and social processes and outputs, associated with the globalized extraction of natural resources." It includes mineral and fossil fuel extraction, and large-scale agricultural, forestry and fishery operations.
150 Ibid., ¶ 5.
151 Ibid., ¶ 7.
152 Ibid., ¶ 35.
153 Ibid., ¶ 50 (footnotes omitted). This quotation clearly shows the close link between extractivism, environmental degradation, human rights violations and marginalized communities.
154 Ibid., ¶ 52.
155 Ibid., ¶ 53. The Special Rapporteur referred to a 2012 report by the Special Rapporteur on the Rights of indigenous Peoples, that in addition to millions of acres of lands lost, extractive and other activities near or within indigenous lands, including nuclear weapon testing and uranium mining in the US led to widespread environmental harm and caused serious health problems among Native Americans (¶ 55).
156 Ibid., ¶ 56. The report also highlighted the killings and abuse of human rights defenders fighting on behalf of indigenous and afrodescendant communities and referred to report of Special Rapporteur on human rights defenders who identified the most dangerous countries for environmental human rights defenders as Brazil, Cambodia, Colombia, Guatemala, Honduras, India,v Mexico, Peru, the Philippines and Thailand and the extractivist industry was the sector with most violations. He pointed out that conflicts over the environment stemmed from resource exploitation which did not address the concerns of local communities.

157 Ibid., ¶ 64. The Special Rapporteur referred to a report of the Special Rapporteur on violence against women which reported that extractive industries had led to an increase in sexual violence against women.
158 Ibid., ¶ 67.
159 *Report of the Special Rapporteur on the Rights of Indigenous Peoples, James Anaya, Addendum*, A/HRC/21/47/Add.1 (30 August 2012), https://undocs.org/A/HRC/21/47/Add.1.
160 Ibid., ¶ 34.
161 Ibid., ¶¶ 37–38.
162 See the discussion on extractivism economy and racial discrimination in this chapter.
163 Anaya Report, supra note 159, ¶ 41.
164 Ibid., ¶ 42.
165 Ibid., ¶¶ 69–70.
166 Ibid., ¶ 93. Appendix I of the report contains a summary of federal programs, policies and legislation relating to indigenous peoples. This includes the Environmental Protection Agency and its offices and grants and those relating to climate change:

> North American Tribal/First Nations/Indigenous Climate Change Adaptation Project: American Indian Environmental Office is a lead partner with other federal agencies, the Canadian government, and a Canadian indigenous not-for-profit organization in an effort to design a workshop scheduled for September 2012 to focus on climate change adaptation needs of North American indigenous communities in the area of food security and traditional plant use.

It refers to the need for sustainable development and communication between indigenous peoples and the government regarding environmental issues in the context of Yamasi People.
167 *Report of the Special Rapporteur of the Human Rights Council on the rights of indigenous peoples, Victoria Tauli-Corpuz*, A/71/229 (29 July 2016), https://undocs.org/A/71/229.
168 Ibid., ¶ 12. Defined as "a geographically defined area which is designated or regulated and managed to achieve specific conservation objectives."
169 Ibid., ¶ 13.
170 Ibid., ¶ 14.
171 Ibid., ¶ 15.
172 Ibid., ¶ 36.
173 Ibid., ¶ 41.
174 Ibid., Section VII, p. 18.
175 *Resolution 48/14 Mandate of the Special Rapporteur on the promotion and protection of human rights in the context of climate change*, A/HRC/RES/48/14 (13 October 2021), https://undocs.org/A/HRC/RES/48/14.
176 Ibid.
177 https://www.ohchr.org/en/specialprocedures/sr-climate-change#:~:text=about%20the%20mandate-,Current%20mandate%20holder,Mr
178 The Special Rapporteurs are those on: internally displaced persons; Indigenous peoples; human rights and environment; right to food, cultural rights; health; adequate housing; extreme poverty; racial discrimination, and safe drinking water and sanitation.
179 http://climatecasechart.com/wp-content/uploads/sites/16/non-us-case-documents/2020/20200116_USA-162020_complaint.pdf. Summary available here: http://climatecasechart.com/non-us-case/rights-of-indigenous-people-in-addressing-climate-forced-displacement.
180 Complaint, p. 3.
181 Four Tribes in Louisiana – Isle de Jean Charles Band of Biloxi-Chitimacha-Choctaw Indians of Louisiana; Pointe-au-Chien Indian Tribe; Grand Caillou/Dulac Band of

Biloxi-Chitimacha Choctaw Tribe; the Atakapa-Ishak Chawasha Tribe of the Grand Bayou Indian Village and the Native Village of Kivalina in Alaska are the five Tribes named in the complaint.

182 p. 48.

183 Letter from Special Rapporteurs in Reference to AL USA 16/2020 (15 September 2020), http://climatecasechart.com/wp-content/uploads/sites/16/non-us-case-documents/2020/20200915_USA-162020_na.pdf.

184 Ibid., p. 1.

185 Ibid., p. 8.

186 Ibid., annex. This complaint is currently pending.

187 https://www.ohchr.org/en/special-procedures-human-rights-council/special-procedures-human-rights-council. See also, Ramcharan, B. (2009) *The Protection Roles of UN Human Rights Special Procedures*, Leiden: Nijhoff, doi: 10.1163/ej.9789004171473.i-214.

188 Press Release, Special Procedures, *UNGA must Affirm a Right to Healthy Environment: UN Experts* (6 July 2022), https://www.ohchr.org/en/press-releases/2022/07/un-general-assembly-must-affirm-right-healthy-environment-un-experts.

189 Ibid.

References

Additional good practices in the implementation of the right to a safe, clean, healthy and sustainable environment, Supplementary information on the report of the Special Rapporteur on the issue of human rights obligations relating to the enjoyment of a safe, clean, healthy and sustainable environment (2019), A/HRC/43/53/Annex III, https://www.ohchr.org/en/special-procedures/sr-environment/annual-thematic-reports.

Alaska Institute for Justice (2020) *Complaint, Rights of Indigenous People in Addressing Climate-Forced Displacement*, http://climatecasechart.com/wp-content/uploads/sites/16/non-us-case-documents/2020/20200116_USA-162020_complaint.pdf.

Annual thematic reports: Special Rapporteur on Human Rights and the Environment (2022), https://www.ohchr.org/en/special-procedures/sr-environment/annual-thematic-reports.

Atapattu, S. & Schapper, A. (2019) *Human Rights and the Environment: Key Issues*, Abingdon: Routledge.

Bennoune, K. (2018) *Universality, cultural diversity and cultural rights*, A/73/227, https://undocs.org/A/73/227.

Bennoune, K. (2020) *Field of cultural rights*, A/75/298, https://undocs.org/A/75/298.

Birnie, P., Boyle, A. & Redgwell, C. (2009) *International Law & the Environment* (3rd edn.), Oxford: Oxford University Press.

Commission on Human Rights (1995) *Resolution 1995/81*, https://digitallibrary.un.org/record/227158?ln=en.

Commission on Human Rights (2003) *Report submitted by the Special Rapporteur on the right to food, Jean Ziegler, in accordance with Commission on Human Rights resolution 2002/25*, E/CN.4/2003/54, https://undocs.org/E/CN.4/2003/54.

Crenshaw, K. (1989) "Demarginalizing the Intersections of Race and Sex: A Black Feminist Critique of Antidiscrimination Doctrine, Feminist Theory, and Antiracist Politics," *University of Chicago Legal Forum*, vol. 1989, no. 1, art. 8.

Critical perspective on food systems, food crises and the future of the right to food, Report of the Special Rapporteur on the right to food (2020), A/HRC/43/44.

Extractive Industries Transparency Initiative (EITI) (2003), https://www.state.gov/extractive-industries-transparency-initiative-eiti.

Extreme poverty and human rights (2020), A/75/181/Rev.1.

Good Practices of States at the national and regional levels with regard to human rights obligations relating to the environment (2020), A/HRC/43/54.

Human Rights and the global water crisis: water pollution, water scarcity and water-related disasters, report of the special rapporteur (2021), A/HRC/46/28.

Human Rights Obligations relating to the enjoyment of a safe, clean, healthy and sustainable environment (2018), A/73/188.

Human Rights Obligations relating to the enjoyment of a safe, clean, healthy and sustainable environment (2019), A/74/161.

Human Rights Obligations relating to the enjoyment of a safe, clean, healthy and sustainable environment (2020), A/75/161.

Human Rights obligations relating to the enjoyment of a safe, clean, healthy and sustainable environment on the theme healthy and sustainable food: reducing the environmental impacts of food systems on human rights (2021), A/76/179.

IACHR (2017) *Advisory Opinion OC-23/17*, https://www.corteidh.or.cr/docs/opiniones/seriea_23_ing.pdf.

Ibeanu, O. (2008) *Report of the Special Rapporteur on the adverse effects of the illicit movement and dumping of toxic and dangerous products and wastes on the enjoyment of human rights*, A/HRC/7/21.

Issue of human rights obligations relating to the enjoyment of a safe, clean, healthy and sustainable environment, report of the Special Rapporteur (2019) A/HRC/40/55.

Knox, J. (2012) *Report of the Independent Expert on the issue of human rights obligations* relating *to the enjoyment of a safe, clean, healthy and sustainable environment*, A/HRC/22/43.

Knox, J. (2013) *Report of the Independent Expert on the issue of human rights obligations* relating *to the enjoyment of a safe, clean, healthy and sustainable environment (Mapping Report)*, A/HRC/25/53.

Knox, J. (2016) *Report of the Special Rapporteur on the issue of human rights obligations relating to the enjoyment of a safe, clean, healthy and sustainable environment*, A/HRC/31/52.

Knox, J. (2017) *Report of the Special Rapporteur on the issue of human rights obligations relating to the enjoyment of a safe, clean, healthy and sustainable environment*, A/HRC/34/49.

Knox, J. (2018) *Human rights obligations relating to the enjoyment of a safe, clean, healthy and sustainable environment*, A/73/188.

Ksentini, F.Z. (1994) *Review of Further Developments in the Fields with which the Sub*-commission *Has been Concerned Human Rights and the Environment: Final report prepared by Mrs. Fatma Zohra Ksentini, Special Rapporteur*, E/CN.4/Sub.2/1994/9, http://www1.umn.edu/humanrts/demo/HRandEnvironment_Ksentini.pdf.

OHCHR (2011) *UN Guiding Principles on Business and Human Rights*, https://www.ohchr.org/sites/default/files/documents/publications/guidingprinciplesbusinesshr_en.pdf.

OHCHR press release (2018) *It's Time We All Recognise the Human Right to a Healthy Environment*, https://www.ohchr.org/en/press-releases/2018/06/world-environment-day-5-june-2018

Pallemaerts, M. (2013) *Report of the Special Rapporteur on the Implications for Human Rights of the Environmentally Sound Management and Disposal of Hazardous Substances and Wastes*, A/HRC/24/39, https://www.ohchr.org/en/documents/reports/report-special-rapporteur-implications-human-rights-environmentally-sound.

Petition to the IACHR seeking relief from violations resulting from global warming caused by acts and omissions of the US (2005), http://climatecasechart.com/non-us-case/petition-to-the-inter-american-commission-on-human-rights-seeking-relief-from-violations-resulting-from-global-warming-caused-by-acts-and-omissions-of-the-united-states.

Press Release, Special Procedures: UNGA Must Affirm a Right to Healthy Environment: UN Experts (2022), https://www.ohchr.org/en/press-releases/2022/07/un-general-assembly-must-affirm-right-healthy-environment-un-experts.

Ramcharan, B. (2009) *The Protection Roles of UN Human Rights Special Procedures*, Leiden: Nijhoff, doi: 10.1163/ej.9789004171473.i-214.

Report of the Independent Expert on the issue of human rights obligations relating to the enjoyment of a safe, clean, healthy and sustainable environment (2015), A/HRC/28/61.

Report of the Special Rapporteur on contemporary forms of racism, racial discrimination, xenophobia and related intolerance (2019), A/HRC/41/54.

Report of the Special Rapporteur on extreme poverty and human rights (2019), A/HRC/41/39.

Report of the Special Rapporteur of the Human Rights Council on the rights of indigenous peoples, Victoria Tauli-Corpuz (2016), A/71/229.

Report of the Special Rapporteur on the Issue of human rights obligations relating to the enjoyment of a safe, clean, healthy and sustainable environment (2016), A/HRC/31/52.

Report of the Special Rapporteur on the Issue of human rights obligations relating to the enjoyment of a safe, clean, healthy and sustainable environment (2017), A/HRC/34/49.

Report of the Special Rapporteur on the Issue of human rights obligations relating to the enjoyment of a safe, clean, healthy and sustainable environment (2018), A/HRC/37/58.

Report of the Special Rapporteur on the Issue of human rights obligations relating to the enjoyment of a safe, clean, healthy and sustainable environment (2018), A/HRC/37/59.

Report of the Special Rapporteur on the right to food (2018), A/HRC/37/61.

Report of the Special Rapporteur on the Rights of Indigenous Peoples, James Anaya, Addendum (2012), A/HRC/21/47/Add.1.

Report of the working group on the issue of human rights and transnational corporations and other business enterprises (2020), A/HRC/44/43.

Right to a healthy environment: Good practices, report of the Special Rapporteur (2019), A/HRC/43/53.

Right to food (2015), A/70/287.

Special Procedures of the HRC (2022), https://www.ohchr.org/en/special-procedures-human-rights-council/special-procedures-human-rights-council.

Special Rapporteur on climate change (2022), https://www.ohchr.org/en/specialprocedures/sr-climate-change#:~:text=about%20the%20mandate-,Current%20mandate%20

Special Rapporteur (2018) *Newsletter no. 28*, http://srenvironment.org/2018/07/30/newsletter-no-28.

Special Rapporteurs (2020) *Letter in Reference to AL USA 16/2020*, http://climatecasechart.com/wp-content/uploads/sites/16/non-us-case-documents/2020/20200915_USA-162020_na.pdf.

Special Rapporteurs (2020) *Summary, Rights of Indigenous People in Addressing Climate-Forced Displacement*, http://climatecasechart.com/non-us-case/rights-of-indigenous-people-in-addressing-climate-forced-displacement.

The right to a clean, healthy and sustainable environment: Non-toxic environment, report of the special rapporteur (2022), A/HRC/49/53.

The right to food (2007), A/62/289.

Tuncak, B. (2014) *Report of the Special Rapporteur on the implications for human rights of the environmentally sound management and disposal of hazardous substances and wastes*, A/HRC/27/54.

UNESCO (2002) *Universal Declaration on Cultural Diversity*, https://www.un.org/en/events/culturaldiversityday/pdf/127160m.pdf.

Part III

Treaty-Based Mechanisms

5 UN Human Rights Committee

5.1 Introduction and Mandate

The International Covenant on Civil and Political Rights (ICCPR) was adopted by the international community in 1966 together with its twin, the International Covenant on Economic Social and Cultural Rights (ICESCR). Both instruments have their foundation in the historic Universal Declaration of Human Rights (UDHR), adopted by the international community in 1948 in the aftermath of World War II, all of which have their origins in the UN Charter itself.

The Human Rights Committee is established under Article 28 of the ICCPR which provides that "there shall be established a Human Rights Committee" consisting of 18 members of high moral character with recognized competence in human rights who are elected to serve in their personal capacity. Articles 29–39 proceed to cover procedural issues such as, *inter alia*, elections of Committee members, how vacancies are to be filled, and meetings of the Committee.

Articles 40–45 embody provisions of a more substantive nature detailing the obligations of state parties, duties of the Committee, how decisions are to be conveyed, and reporting to the UNGA. Under Article 40, state parties have undertaken to submit "reports on the measures they have adopted which give effect to the rights recognized herein and on the progress made in the enjoyment of those rights"[1] within one year of the Covenant entering into force for the state party and whenever the Committee requests thereafter. After studying the reports submitted by state parties (referred to as "country reports"), the Committee is required to submit its report and comments to the state party (called "concluding observations"). The state party may submit its observations to the Committee on any comments made by the Committee.

Under Article 41, a state party may declare that it recognizes the competence of the Committee to receive communications that another state party is not fulfilling its obligations under the Covenant. If a matter referred to under this provision is not resolved to the satisfaction of the parties concerned, the Committee may, with their prior consent, appoint an *ad hoc* Conciliation Commission whose good offices will be available to the parties to arrive at an amicable solution.[2]

The Committee is required to submit, through the Economic and Social Council, an annual report on its activities to the UNGA. State parties have undertaken

DOI: 10.4324/9781003128847-8

to respect and to ensure to *all individuals* within its territory[3] and subject to its jurisdiction the rights recognized in the Covenant without any distinction of any kind. They have undertaken to take necessary steps to adopt legislative and other measures to give effect to the rights in the Covenant; ensure that any person whose rights are violated has an effective remedy and their rights determined by a competent judicial, administrative or legislative authority and that they will enforce such remedies when granted.[4]

Under the First Optional Protocol to the ICCPR, the Committee can receive communications from individuals claiming to be victims[5] of violations of any rights in the Covenant and state parties have recognized the competence of the Committee to receive such communications:

> A State Party to the Covenant that becomes a Party to the present Protocol recognizes the competence of the Committee to receive and consider communications from individuals subject to its jurisdiction who claim to be victims of a violation by that State Party of any of the rights set forth in the Covenant. No communication shall be received by the Committee if it concerns a State Party to the Covenant which is not a Party to the present Protocol.[6]

Individuals who claim that their rights under the Covenant have been violated may submit a written communication to the Committee, provided they have exhausted all available domestic remedies.[7] However, any communication that is anonymous or appears to be an abuse of the rights of submission, the Committee shall consider them inadmissible. The Committee must inform the State party concerned any communications received who must submit, within six months, a written explanation to the Committee clarifying the matter and the remedy it has taken. After considering the communication in light of the written information, the Committee will forward its view to the State party and the individual.[8]

As with most human rights treaties, the Covenant does not contain a provision on environmental protection or sustainable development. It has not, however, stopped the Committee from addressing these issues. This chapter examines how the Committee has addressed environmental pollution and degradation, climate change, and sustainable development, including the SDGs in its activities, especially General Comments, Concluding Observations, and decisions in individual communications. This chapter uses as its foundation the individual report on the ICCPR prepared for the mapping exercise undertaken by the former UN Special Rapporteur on Human Rights and the Environment, John Knox.[9]

5.2 General Comments

While there is no specific general comment addressing environmental issues *per se*, GC No 36 on the right to life[10] which replaced GC Nos 6 and 14[11] refers to environmental degradation including climate change. Noting that the right to life is the supreme right from which no derogation is permitted, even in situations of armed conflict and other public emergencies that threaten the life of the nation, the GC provides that it is most precious in its own right and a prerequisite for the enjoyment of all other rights. Moreover, the right to life should not be interpreted narrowly. It includes the right to enjoy a life of dignity and applies to all human beings without distinction of any kind.

The GC elaborates on the duty to protect life which includes a duty to take positive measures to protect life. States are under a due diligence obligation to take reasonable, positive measures that do not impose a disproportionate burden on them in response to *reasonably foreseeable threats to life* originating from private persons and entities whose conduct is not attributable to the state.[12] The duty to protect the right to life requires states to take special measures to protect persons in vulnerable situations whose lives have been placed at particular risk. These include human rights defenders, victims of gender-based violence, children, indigenous peoples, minorities, displaced persons, asylum seekers, refugees, people with disabilities, and stateless persons.[13]

Moreover, the duty to protect life implies that states should take appropriate measures to address general conditions that may give rise to direct threats to life or prevent individuals from enjoying their right to life with dignity. These conditions may include, inter alia, industrial accidents, *degradation of the environment*, and deprivation of indigenous peoples' land, territories, and resources.[14] Moreover, states should develop contingency plans and disaster management plans to address natural and man-made disasters that may affect the enjoyment of the right to life, such as hurricanes, tsunamis, earthquakes, radioactive accidents, and massive cyber attacks.

The final section of the GC discusses the relationship between Article 6 and other provisions of the Covenant and other legal regimes. In this context, the GC discusses the principle of *non-refoulement* and the rights of human rights defenders and notes that states must take necessary measures to respond to death threats and provide protection to human rights defenders including creating a safe and enabling environment for defending human rights. It notes that returning individuals to countries where there are substantial grounds for believing that they face a real risk to their lives violates both articles 6 and 7 of the Covenant.[15]

The GC considers environmental degradation, climate change, and unsustainable development as constituting serious threats to the right to life:

> Environmental degradation, climate change and unsustainable development constitute some of the most pressing and serious threats to the ability of present and future generations to enjoy the right to life. *The obligations of States parties under international environmental law should thus inform the content of article 6 of the Covenant, and the obligation of States parties to respect and ensure the right to life should also inform their relevant obligations under international environmental law.* Implementation of the obligation to respect and ensure the right to life, and in particular life with dignity, depends, inter alia, on measures taken by States parties to preserve the environment and protect it against harm, pollution and climate change caused by public and private actors.[16]

This paragraph clearly recognizes the link between environmental conditions and the right to life as well as the link between human rights obligations and environmental obligations. It calls upon states to address environmental degradation, climate change and unsustainable development in order to protect the right to life of present and future generations and notes that international environmental law principles should inform the content of Article 6. It then elaborates on this obligation:

> States parties should therefore ensure sustainable use of natural resources, develop and implement substantive environmental standards, conduct environmental impact assessments and consult with relevant States about activities likely to have a significant impact on the environment, provide notification to other States concerned about natural disasters and emergencies and cooperate with them, provide appropriate access to information on environmental hazards and pay due regard to the precautionary approach.[17]

Referring to the threat or use of weapons of mass destruction, especially nuclear weapons, the GC notes that weapons that cause destruction of such a catastrophic scale are incompatible with the respect for the right to life and *may* amount to a crime under international law.[18] It calls upon states to take all necessary measures to stop the proliferation of weapons of mass destruction and to provide adequate reparation to victims whose right to life has been affected by the testing or use of such weapons in accordance with principles of international law.[19]

GC No 23 elaborates on Article 27 of the ICCPR which embodies the rights of minorities.[20] It notes that the Covenant draws a distinction between the right to self-determination and the rights protected under Article 27. The former is a right belonging to peoples and is not cognizable whereas the latter is an individual right and is cognizable under the Optional Protocol. It further notes that the right to enjoy one's culture may consist in a way of life which is closely associated with territory and use of its resources which is especially true of indigenous communities.[21] With regard to the issue of who is entitled to rights under Article 27, GC No 23 provides that those who belong to a group and share a common culture, religion, and/or language are entitled to the rights and they need not be citizens of the state party. Thus, the obligations in Article 2, except those reserved for citizens (such as those rights under Article 25) cannot be restricted to citizens alone.

The most relevant provision for the present discussion provides that:

> With regard to the exercise of the cultural rights protected under article 27, the Committee observes that culture manifests itself in many forms, including a particular way of life associated with the use of land resources, especially in the case of indigenous peoples. That right may include such traditional activities as fishing or hunting and the right to live in reserves protected by law. The enjoyment of those rights may require positive legal measures of protection and measures to ensure the effective participation of members of minority communities in decisions which affect them.[22]

5.3 Concluding Observations

Most recommendations relating to environmental protection, climate change and sustainable development have been made under Article 2 on the right of self-determination (in relation to indigenous peoples) Article 6 on the right to life, Article 25 on the right to participation, and Article 27 on rights of minorities (mostly in relation to indigenous peoples). However, compared to other treaty bodies that are discussed in this volume, the references to environmental issues and climate change are fewer and less extensive.

In several Concluding Observations the Committee has noted that environmental degradation can affect the enjoyment of protected rights. Thus, for example, in its Concluding Observations on a report submitted by Colombia, the Committee pointed out that it wished to receive information on measures taken to address the ecological deterioration of the Amazon region and whether Colombia "had encountered any problems in reconciling development of its oil reserves with the maintenance of a balanced ecosystem."[23] Similar observations were made with respect to Ecuador that it wished to receive information on how ecological deterioration of the Amazon region was affecting indigenous communities and the measures taken to address that situation.[24]

Similar observations have been made on the reports of Mexico,[25] Guyana,[26] Argentina,[27] Belize,[28] and Sweden.[29] These observations provide guidance as to how to balance economic development (i.e. exploiting oil reserves) with environmental protection. They also recognize the importance of providing affected communities with relevant information and giving them an opportunity to participate in the decision-making process. The fact that a human rights institution is referring to the need to maintain a balanced ecosystem is itself a major step forward.

5.3.1 Environmental Pollution, Environmental Management and Natural Resource Management

The Committee noted, on the report of Guinea, that public participation in decision-making relating to investment projects with a social and environmental impact and in natural resource management is extremely limited.[30] It recommended that the state party hold consultations with local communities before concluding contracts on natural resource management and with regard to projects that have a social and environmental impact in order to obtain their free, prior, and informed consent.[31]

On the report of Israel, the Committee expressed concern over water shortages disproportionately affecting the Palestinian population of the West Bank, because construction of water and sanitation infrastructure is prevented, and construction of wells prohibited.[32] The Committee also expressed concern at allegations of pollution by sewage water of Palestinian land, invoking Articles 6 and 26 of the ICCPR. The Committee recommended that the state party ensure equal access to water by all residents, in accordance with the World Health Organization quality and quantity standards.[33] Further, the State party should allow the construction of water and sanitation infrastructure and wells, and address sewage and wastewater emanating from Israel.

The Committee made extensive comments on the report of USA in 2006 and noted discriminatory practices during and in the aftermath of Hurricane Katrina involving poor, especially African-Americans regarding evacuation plans and reconstruction plans.[34] The Committee recommended reviewing practices and policies to ensure the implementation of its obligation to protect life and of the prohibition of discrimination, as well as of the United Nations Guiding Principles on Internal Displacement,[35] in relation to disaster prevention and preparedness, emergency assistance, and relief measures.

On the report of Austria, the Committee questioned whether any steps had been taken to provide the population with a healthy environment by curbing pollution.[36] In his reply the state representative stated that the Code of Criminal Procedure had been amended in 1987 to impose greater penalties for pollution-related offenses, but discussions were still continuing.[37] In responding to the questions on environmental problems on the report of Poland, its representative pointed out that 11% of the territory was threatened by pollution because of the industrialization policy of the former regime.[38] A law to monitor the ecological conditions was adopted in 1990 under which industrial polluters had been closed down or penalized. Other recommendations made in a policy statement still needed to be implemented. However, the economic situation posed serious constraints although measures such as the conversion of foreign debt into ecological investments, were steps in the right direction. On the report of Peru, the Committee noted that a chapter on crimes perpetrated against the environment and public health had been included in the new Penal Code.[39] On the report of Belgium, the Committed questioned whether the government was addressing the affirmative aspects of the right to life, such as the right to health, the elimination of epidemics and *pollution-related issues*.[40] On the report of Ukraine, the Committee requested information on the measures that had been taken to protect the right to life against the risk of nuclear disaster and environmental pollution, especially after the Chernobyl accident.[41] In response the representative explained that after the Chernobyl accident, a "high-risk" area within a radius of 300 kilometers had been designated and the inhabitants had been moved into new housing. The high-risk area had been extended and a decree concerning liability in cases of non-compliance with environmental provisions had been adopted. A moratorium on the construction of new nuclear plants and power stations had also been imposed.[42]

5.3.2 Climate Change

The Committee referred climate change on several occasions. For example, it welcomed the measures taken by Dominica to protect people from climate impacts, including the adoption of the 2018 Climate Resilience Act, and mitigation, and adaptation programs.[43] It noted the vulnerability of the state party as a small island state to climate change, and the significant challenges it faced. While commending the State party for the measures taken, the Committee noted that there is no information on the steps taken to get the informed participation of the population in projects that affect sustainable development and resilience to climate change. The Committee recommended continuing efforts to develop resilience to climate change through adaptation and mitigation measures. Referring to paragraph 62 of GC No 36, the Committee stated that all such projects should be developed with the meaningful and informed participation of the population.[44]

The Committee discussed climate change and the right to life on the report of Cabo Verde[45] and expressed concerns similar to those on the report of Dominica and made similar recommendations referring to paragraph 62 of the GC No 36. In addition, the Committee stated that the State party should develop mechanisms:

> [t]o ensure the sustainable use of natural resources, develop and implement environmental standards, conduct environmental impact assessments, provide appropriate access to information on environmental hazards and adopt a precautionary approach to protect persons in the State party, including the most vulnerable, from the negative impacts of climate change and natural disasters.[46]

5.3.3 *Indigenous Peoples*

While welcoming the efforts made to promote the rights of indigenous peoples in Dominica, the Committee noted the absence of information about legal and policy frameworks regarding indigenous land and their right to free, informed, and prior consultation in relation to programs impacting them.[47] In this regard, the Committee stated that the state party *should consider* ensuring that meaningful consultations are held with the indigenous peoples concerned with a view to obtaining their *free, prior, and informed consent* with regard to any measure that could have a substantial impact on their way of life and culture.[48]

On the report of New Zealand, the Committee expressed concern that the replacement of the Foreshore and Seabed Act 2004 by the Marine and Coastal Area (Takutai Moana) Act 2011 had a discriminatory effect on Māori claims to their customary land and their right to cultural development.[49] The Committee recommended revising the Act to ensure respect of the customary rights of Māori on their lands and resources, and their cultural development. With regard to the Treaty of Waitangi, the Committee recommended strengthening its role in the Constitution, guaranteeing the informed participation of indigenous communities in all relevant consultations and decision-making processes, and developing programs to implement SDG 5 with particular focus on Māori and Pasifika women and girls, and those with disabilities.[50]

While welcoming the decree on the sustainable development of indigenous peoples in the North Siberia and the Far East, and the action plan for 2009–2011 in the Russian Federation, the Committee expressed concern about the adverse impacts on indigenous peoples due to the 2004 amendment and the process of consolidation of territories of the Russian Federation.[51] The Committee also voiced concern over the exploitation of lands, fishing grounds and natural resources of indigenous peoples through granting of licenses to private companies for development projects such as the construction of pipelines and hydroelectric dams. The Committee requested the state party to provide, in its next periodic report, information on the impact of these measures on the way of life and economic activities of indigenous peoples and on their enjoyment of rights under article 27 of the Covenant.[52]

On the report of Sweden, the Committee noted that while the state party had delegated some responsibilities for reindeer husbandry to the Sami Parliament, it expressed concern at the Sami Parliament's limited participation in the decision-making process on issues affecting land and traditional activities of the Sami people.[53] It also expressed concern about the limited progress on the passage of legislation on Sami land and resource rights. The Committee thus recommended that steps should be taken to involve the Sami in the decisions concerning the natural environment and necessary means of subsistence.

Similar concerns were expressed on the 2002 report of Sweden where the Committee noted the limited extent of Sami Parliament in decision-making regarding Sami land and their economic activities such as hydroelectricity, mining, and forestry projects.[54] The Committee recommended that the "State party should take steps to involve the Sami by giving them greater influence in decision-making affecting their natural environment and their means of subsistence."[55] In response to questions, the representative of Finland explained that the difference in legislation between Sweden and Finland regarding reindeer breeding was due to differences in economic and ecological problems of the regions. When discussing an even earlier report, Sweden pointed out that because about 2500 Sami were engaged in their traditional livelihood and reindeer-herding involving one third of Sweden's land area, it was felt necessary to limit the number of reindeer-herding groups. The state party stated that its policy was the result of careful balancing of interests of society and respect for Sami culture.[56]

On the report of the USA, the Committee noted that no action had been taken by the state party to address its previous recommendations relating to the termination of indigenous rights.[57] It noted with concern the situations in which tribal property rights could be extinguished without due process and fair compensation and recommended that the state party review its policy toward indigenous peoples as regards the termination of aboriginal rights by the plenary power of Congress regarding Indian affairs and grant them the same degree of judicial protection that is available to others. The Committee recommended taking steps to secure the rights of all indigenous peoples under articles 1 and 27 of the Covenant, and give them greater influence in decision-making affecting their natural environment, their means of subsistence, and their culture.[58]

In response to questions on Article 27 of the Covenant, the representative of Colombia noted that Article 7 of the Constitution guaranteed ethnic and cultural diversity of the nation and all cultures had equal status and dignity before the law.[59] The representative also pointed out that the Constitution recognized the inalienable right of indigenous peoples to certain lands, which had been accorded the status of self-governing territorial entities. The State was required to invest in those entities to improve the living conditions of the people, who had full control of such funds and natural resources could be developed only with the consent and participation of the community. Moreover, minorities are entitled to establish special jurisdictions within their territories and provision had been made for the protection of biodiversity and fauna and flora in the Amazon and other regions. During the past five years the state had recognized the collective ownership by indigenous peoples of approximately 15 million hectares of land in the Amazon region which had curbed the influx of businesses seeking to acquire property for development.[60]

5.4 Individual Communications

As noted, under the first Optional Protocol individuals can submit a communication to the Human Rights Committee alleging the violation of their rights in the Covenant. Although the ICCPR does not include environmental rights, several communications filed before the Committee have involved environmental issues. Two cases specifically deal with climate change: the Torres Strait Islands dispute decided in September 2022 and the Teitiota case decided in 2020. Given their significance they will be discussed in some detail in this section.[61]

In one of the early cases – the case of *EHP et al v. Canada* – Port Hope Environmental Group submitted a communication against Canada on behalf of present and future generations of Port Hope, Ontario, Canada, including 129 Port Hope residents who had specifically authorized the author to act on their behalf.[62] The communication alleged the violation of the right to life of the residents including future generations due to the radioactive waste remains in Port Hope. The author claimed that it is a threat to the life of present and future generations of Port Hope, because exposure to radioactivity is known to cause cancer and genetic defects.[63] Although the petition was deemed inadmissible for non-exhaustion of local remedies, the Committee recognized that the right to life could be impacted by environmental harm[64] and that it raised questions about the obligation of states to protect human life under Article 6. With regard to the reference to "future generations" the Committee treated it as "an expression of concern purporting to put into due perspective the importance of the matter raised in the communication."[65]

The case of *Vaihere Bordes and John Temeharo v. France*[66] involved the announcement by France that it intended to conduct a series of underground nuclear tests on two atolls in the South Pacific. The authors of the communication were all French citizens living in French Polynesia. They claimed that the tests represented a threat to their right to life and the right not to be subjected to arbitrary interference with their privacy and family life and relied on GC Nos 6[67] and 14.[68] The latter provides that "designing, testing, manufacture, possession and deployment of nuclear weapons are among the greatest threat to the right to life which confront mankind today"[69] and urged the international community to prohibit the production of nuclear weapons and to recognize it as a crime against humanity.[70] They further alleged that the French authorities had failed to take sufficient measures to protect their life and security and failed to show that underground nuclear tests do not constitute a danger to the health of people and the environment. The authors relied on studies that showed the dangers on the health of people and the indirect risk through contamination of the food chain.

France argued, inter alia, that the authors did not qualify as "victims" under the Optional Protocol because there had been neither a violation nor a real threat. The Committee pointed out that the issue here is whether the conduct of underground nuclear tests amounted to a violation of rights to life and family life and noted, based on the information provided, that the authors had not proved that the conduct of nuclear tests entitled them to be considered "victims" under the ICCPR.[71] Referring to the authors' assertion that nuclear tests would deteriorate the geological structure of the atolls and increase the likelihood of an accident, the Committee noted that it was highly controversial even in scientific circles, and it was not possible for the Committee to ascertain its validity.[72] The Committee stressed, however, relying on GC No 14, that although the authors did not show that they were victims, nuclear weapons presented a great threat to the right to life.[73]

The decision here is perplexing given the Committee's own position that activities involving nuclear weapons *including testing* pose a grave threat to the right to life of people. It was no secret that France was engaged in nuclear testing and that the complainants were living in the vicinity of these tests. The reasoning suggests that the Committee considered the threat of harm to be rather remote and hence the authors did not fall within the definition of a victim. Nonetheless, this case is important because it recognized the link between nuclear tests and the right to life. However, while it recognized that the threat of harm should not be too remote, it was unclear why it was held to be so in this case because clearly the complainants were living in close proximity to the nuclear test sites. This case also raises questions about how the Committee would evaluate scientific evidence. Had there been scientific consensus about the danger nuclear testing posed to the environment of these atolls, would the Committee have taken a different position?

In the submission in *Susila Malani Dahanayake* [74] the authors invoked Article 6 claiming that

> the right to life guaranteed by Article 6 of the Covenant has been interpreted by other treaty bodies, also by the Human Rights Committee, in a broad manner, and consequently claim a violation of their right to life, which includes a right to a healthy environment.[75]

The subject matter of this communication related to the proposed Southern expressway by the Road Development Authority of Sri Lanka. Two possible routes were examined by the proponent in the environmental impact assessment (EIA) but a third route – the "final trace" – which was not examined in any EIA report was ultimately selected for implementation. This route affected the authors' property and no new approval from the Central Environmental Authority (the approving agency) was sought. When the authors complained about this in a fundamental rights case before the Supreme Court of Sri Lanka, the Court, while recognizing that the authors' fundamental right to equality under Article 12(1) of the Constitution and principles of natural justice were infringed, did not order the project to be halted. The authors then brought this submission before the UN Human Rights Committee claiming that no new EIA was prepared for the third route and no opportunity was given to the authors to be heard as required by law. The State party responded that it did not intend to interfere with the authors' right to live in a healthy environment,[76] indicating that it accepted that the right to life included a right to live in a healthy environment. While the Committee concluded that the authors had not sufficiently substantiated their claim that they were victims of a violation of their right to life under Article 6 because they were deprived of a healthy environment, both parties seemed to have accepted the link between the right to life and the right to a healthy environment. Because they received compensation and costs for the violation of a fundamental right by the Supreme Court of Sri Lanka, the Committee was of the view that they were no longer "victims" within the meaning of the Optional Protocol.

Interestingly, many of the communications under the ICCPR that relate to the environment have arisen under Article 27 on minority rights, mostly involving indigenous groups and their right to culture. In *EP et al v. Colombia*,[77] the authors invoked Article 27 on the basis that they had been dispossessed of their lands; that the Colombian government had encouraged massive development of the islands that caused severe environmental damage that affected the water table making farming impossible and destroyed their traditional livelihoods; and the government had permitted destruction of mangrove swamps by allowing electric power plants to freely dump hot, polluted water and granted fishing rights to Honduras without regard to native interests depriving them of their traditional means of survival. They alleged that government policies and development activities resulted in damage to the environment where the authors lived making farming impossible, depleting other traditional sources of sustenance, and depriving them of fishing rights."[78] The Committee found that the communication was inadmissible due to non-exhaustion of local remedies.

The communication of *Lubicon Lake Band v. Canada*[79] involved an indigenous group whose rights had been violated due to oil and gas development on their traditional lands leading to environmental destruction and violated a number of rights in the ICCPR. The author claimed that oil development had destroyed the subsistence livelihood of its people and the delay in litigation would cause it irreparable harm and a permanent injunction would not constitute an effective remedy:

> The recognition of aboriginal rights or even treaty rights by a final determination of the courts will not undo the irreparable damage to the society of the Lubicon Lake Bank, will not bring back the animals, will not restore

> the environment, will not restore the Band's traditional economy, will not replace the destruction of their traditional way of life and will not repair the damages to the spiritual and cultural ties to the land.[80]

Moreover, the author alleged that the state party's actions had led to the deaths of 21 persons and the ability of the community to survive was in serious doubt as the number of miscarriages and stillbirths had skyrocketed constituting a violation of Article 6. Furthermore, it was claimed that "the appropriation of the Band's traditional lands, the destruction of its way of life and livelihood and the devastation wrought to the community constituted cruel, inhuman and degrading treatment within the meaning of Article 7..."[81] Canada contested many of the claims arguing that even in areas that were covered by timber leases the forest remained intact and sustained sufficient animal population to satisfy those who wished to engage in traditional activities. In holding that a violation of Article 27 had occurred, the Committee was mindful of the historic inequities that had contributed to the violation of rights enshrined in Article 27 and the violations here were serious enough as to question the ability of the community to survive: "Historical inequities, to which the State party refers, and certain more recent developments threaten the way of life and culture of the Lubicon Lake Band, and constitute a violation of article 27 so long as they continue."[82]

The case of *Andre Brun v. France*[83] involved the alleged contamination of the author's crops by the dissemination of genetically modified products by Biogemma, a private company, and the consequent destruction of transgenic crops by the author to protect his crops. The author, relying on the jurisprudence of the European Court of Human Rights, argued that the Committee should proceed by analogy and adopt an extensive interpretation of article 17 of the ICCPR[84] to encompass the right to live in a healthy environment under the ambit of private and family life.[85] He also invoked the precautionary principle and contended that the medium and long-term risk of genetically modified organisms (GMOs) on health and the environment should be taken into account.[86] He further argued that Article 6 was violated and that the promotion of a healthy environment contributed to the protection of the right to life. He cited the *Port Hope case*[87] to support the link between the right to life and the right to a healthy environment. In conclusion, the author argued that planting of transgenic crops in open fields resulted in the destruction of conventional crops and that the minimum distance between genetically modified fields and non-GM fields was ineffective. Therefore, the destruction of transgenic crops became necessary and that he had acted out of the necessity to protect the environment.

The complainant further contended that the interference with privacy and family life protected under Article 17 "must be justified and proportionate in the light of the provisions, aims and objectives of the Covenant."[88] The interference in this case consisted of the government's failure to take the necessary measures to prevent the threats to the author's health and environment associated with the dissemination of GMOs in open fields. The Committee decided that there was no actual violation or an imminent threat of violation of his rights to life and privacy, family and home and, therefore, the author could not claim to be a victim and the communication was inadmissible. The Committee pointed out that Article 2(3)(b) of the Covenant guarantees protection to alleged victims if their complaints are "sufficiently well-founded to be arguable" under the Covenant. It is noteworthy the author relied on the jurisprudence of the ECHR to substantiate his claim under the ICCPR, pointing to the potential for cross-fertilization of jurisprudence.

In *Ángela Poma Poma v. Peru*[89] the author brought a claim under Articles 1(2),[90] 2(3)(a),[91] 14(1)[92] and 17[93] of the ICCPR claiming that the diversion of the course of the river Uchusuma by the government and drilling of wells reduced the water supply for human and animal consumption. This caused wetlands to dry up where llama-raising is practiced by descendants of the Aymara people which has been part of their way of life for thousands of years.[94] The author claimed that the state party violated Article 1(2) because the diversion of groundwater from her land had destroyed the ecosystem of the Altiplano and caused the degradation of the land and the drying up of wetlands. As a result, thousands of livestock had died and the community's only means of survival had collapsed, leaving them in poverty and depriving it of its livelihood.[95] It further constituted an interference in the life and activities of her family, in violation of Article 17 as the lack of water had seriously affected their only means of subsistence.

The Committee, recalling GC No 23,[96] noted that the right to culture manifests itself in many forms, including ways of life associated with the use of land and resources, especially in the case of indigenous peoples and may include traditional activities as fishing or hunting and the right to live in reserves protected by law.[97] Regarding the question whether the water diversion had a substantive negative impact on the author's right to culture, the Committee noted that thousands of livestock had died because of the degradation of 10,000 hectares of Aymara pastureland as a direct result of a project implemented by the government:

> The admissibility of measures which substantially compromise or interfere with the culturally significant economic activities of a minority or indigenous community depends on whether the members of the community in question have had the opportunity to participate in the decision-making process in relation to these measures and whether they will continue to benefit from the traditional economy. The Committee considers that participation in the decision-making process must be effective, which requires *not mere consultation* but the free, prior and informed consent of the members of the community. In addition, the measures must respect the principle of proportionality so as not to endanger the very survival of the community and its members.[98]

The Committee observed that neither the author nor the community was consulted at any time by the State regarding the construction of the wells. Nor has the state requested studies to be undertaken by a competent body. It further observed that the author could not benefit from her traditional economic activity and, therefore, concluded that the State's action had substantively compromised the author's way of life and culture as a member of her community.[99] Here the principle of proportionality was used as the yardstick to measure whether the activities endangered the very survival of the community. This is too high a threshold to establish, as most measures could still have a considerable impact on the lives of the people while not necessarily endangering the survival of the community.

A high threshold was also established in *Teitiota v. New Zealand*,[100] the first time that the UN Human Rights Committee was confronted with the issue of "climate refugees."[101] The petitioner, Mr Teitiota, is a citizen of the Republic of Kiribati and his application for refugee status in New Zealand was rejected and he and his family were deported to Kiribati. The communication to the UN Human Rights Committee was the culmination of a series of failed applications by the petitioner for refugee status in New Zealand.[102] The 1951 Geneva Convention on Refugees[103] protects only political refugees, and climate refugees is not a legal category recognized by international law.

Kiribati, a small island state in the Pacific Ocean and approximately 50 other small island states, are facing the prospect of total inundation due to sea level rise and extreme weather events associated with climate change. These islands could become uninhabitable within the next 10–15 years. Some people have already left the island.[104] In evaluating the application for asylum, the Immigration and Protection Tribunal of New Zealand ("Tribunal") paid particular attention to the National Adaptation Programme of Action (NAPA) prepared by the Republic of Kiribati in 2007 which described many concerns including coastal erosion and accretion affecting housing, land, and property. Despite having 60 sea walls in place by 2005 in South Tarawa, storm surges and high tides had caused flooding forcing some people to relocate. Additionally, food insecurity had increased contributing to the deterioration of the health of the population. Expert testimony stated that the country is in crisis because of climate change and population pressure, and that conditions in Kiribati are dire.

The Tribunal rejected Teitiota's application for asylum but accepted that environmental degradation could "create pathways into the Refugee Convention or protected person jurisdiction."[105] Relying on the Committee's previous decisions, the Tribunal stated that the risk of a violation of the Covenant must be "imminent." The Supreme Court of New Zealand agreed, and Mr. Teitiota and his family were deported to Kiribati. He then submitted a complaint to the UN Human Rights Committee under the ICCPR. In his communication, Mr. Teitiota claimed that New Zealand had violated his right to life by rejecting his application for refugee status in New Zealand and removing him to Kiribati. He claimed that the consequences of climate change and sea level rise forced him to migrate from his home island of Tarawa to New Zealand:

> The situation in Tarawa has become increasingly unstable and precarious due to sea level rise caused by global warming. Fresh water has become scarce because of saltwater contamination and overcrowding on Tarawa. Attempts to combat sea level rise have largely been ineffective. Inhabitable land on Tarawa has eroded, resulting in a housing crisis and land disputes that have

caused numerous fatalities. Kiribati has thus become an untenable and violent environment for the author and his family.[106]

The Committee believed that the question is whether the author has substantiated his claim that deportation would cause a real risk of irreparable harm to his right to life[107] and is a 'victim' within the meaning of Article 1 of the Optional Protocol. The Committee noted that the author's claims are not hypothetical future harm, but "a real predicament caused by lack of potable water and employment possibilities, and a threat of serious violence caused by land disputes."[108] Based on the information provided by the author, the Committee decided that he had sufficiently demonstrated, for the purposes of admissibility, that due to the adverse consequences of climate change and associated sea level rise, he faced a real risk of impairment to his right to life under Article 6 of the Covenant and Articles 1 and 2 of the Optional Protocol did not constitute an obstacle to admissibility.

With regard to the question whether the author had substantiated his claim that deportation would cause a real risk of irreparable harm to his right to life[109] the Committee examined the author's claims that: increasing scarcity of habitable land on Tarawa had led to violent land disputes leading to fatalities; saltwater contamination due to sea level rise had resulted in lack of access to potable water; and his crops had been destroyed due to salt deposits on the ground leading to a violation of his means of subsistence.

The author claimed that by removing him to Kiribati, the State party subjected him to a risk to his life and did not properly assess the risk associated with his removal. The Committee noted, relaying on GC No 31,[110] that states are under an obligation not to remove a person from their territory when there are grounds to believe that there is *a real risk of irreparable harm* under Articles 6 and 7 of the Covenant. Such risk must be personal and cannot derive from general conditions in the receiving state, *except in most extreme cases*. The Committee noted that the scope of the *non-refoulement* principle under Article 6 is wider than its scope under international refugee law, since it also requires the protection of aliens not entitled to refugee status. Stating that the right to life cannot be interpreted in a restrictive manner and requires states to adopt positive measures, the Committee noted that the obligation to protect life "extends to reasonably foreseeable threats and life-threatening situations that can result in a loss of life."[111] Relying on GC No 36, the Committee pointed out that "environmental degradation, climate change and unsustainable development constitute some of the most pressing and serious threats to the ability of present and future generations to enjoy the right to life."[112] The Committee referred to the jurisprudence of regional human rights tribunals which has established that environmental degradation can affect the enjoyment of the right to life and adversely affect an individual's well-being leading to a violation of the right to life.[113]

The Committee accepted that both sudden and slow onset events can trigger the movement of people seeking protection from climate-related harm and that sea level rise could likely render the Republic of Kiribati uninhabitable in 10–15 years which "could allow for intervening acts by the republic of Kiribati, with the assistance of the international community, to take affirmative measures to protect and, where necessary, relocate its population."[114] Although the Committee did not hold that the author's rights under Article 6 of the Covenant were violated, its observation below indicates that in the future the situation in these islands will become so extreme that inhabitants may likely be deprived of a life of dignity:

> The Committee is of the view that without robust national and international efforts, the effects of climate change in receiving states may expose individuals to a violation of their rights under articles 6 or 7 of the Covenant, thereby triggering the *non-refoulement* obligations of sending states. *Furthermore, given that the risk of an entire country becoming submerged under water is such an extreme risk, the conditions of life in such a country may become incompatible with the right to life with dignity before the risk is realized.*[115]

The decision of the Committee rested on three main grounds: first, the threat to the author's right to life was not imminent; second, the Republic of Kiribati was actively pursuing adaptation measures, and finally, the author failed to establish personal injury. Given, however, the pervasive nature of climate change, it may be difficult to establish personal injury above and beyond that experienced by everyone else.

Despite the fact that the author's communication was unsuccessful and the threshold laid to establish a violation was very high, this decision breaks new ground for several reasons: first, the Committee recognized that both sudden events and slow onset events associated with climate change can trigger the movement of people, essentially accepting the notion of "climate refugees;" second, it accepted that climate change could, in extreme situations, affect the right to life, triggering the application of the *non-refoulement* principle; third, sea level rise could render the Republic of Kiribati (and, by implication, other small island states) uninhabitable; and finally, it indirectly referred to the obligation of the international community to assist these countries to take affirmative action, including relocating the populations.[116]

In another groundbreaking decision, the Committee decided in favor of a group of eight Torres Strait Islanders against Australia[117] who alleged that Australia's contribution to GHG emissions and its failure to adopt adaptation measures violated their human rights[118] under Article 2 on non-discrimination, read alone and in conjunction with Articles 6,17 and 27, each read alone. They also claimed violations of the rights of named children under Article 24(1)[119] read alone and in conjunction with Articles 6, 17 and 27.[120] The authors claimed that the state party had failed to adopt both adaptation measures (infrastructure against the impact of climate change especially sea level rise) and mitigation measures to reduce GHG emissions and to stop the promotion of fossil fuel extraction and use. They relied on GC No 36 to show that climate change is a matter of human rights.

The complainants inhabit a group of islands off the northern tip of Queensland, Australia which are highly vulnerable to climate change impacts including sea level rise, storm surge, coral bleaching, and ocean acidification.[121] The islanders want Australia to do more to protect the Torres Strait from climate change and requested Australia to: (a) provide $20 million for emergency measures such as seawalls, and investment in long-term adaptation measures to ensure that the islands can continue to be inhabited; (b) reduce its emissions by 65% below 2005 levels by 2030 and going net zero before 2050; and (c) phase out thermal coal.[122] Australia requested that the petition be dismissed on the ground that it cannot be held responsible for a global problem.[123] The UN Special Rapporteur on Human Rights and Environment, David Boyd together with his predecessor, John Knox, filed an amicus brief.[124]

The state party maintained that the communication is inadmissible because the alleged violations of treaties such as the Paris Agreement and the ICESCR are outside the scope of the present Covenant. Moreover, there is no basis for the authors' claim that climate treaties are relevant to the interpretation of the present Covenant because they have different aims and scopes and that "interpreting the Covenant through the Paris Agreement would be contrary to the fundamental principles of international law."[125] It further claimed that the authors have not substantiated their claim that they are victims under Article 1 of the Optional Protocol as there is no evidence that they face current or imminent threat of a violation of the rights they invoked. Neither have they shown any causal link between the alleged violations and state party's acts or omissions and are relying on a risk that has not yet materialized.[126] Moreover, none of the alleged failures to take mitigation measures fall within the scope of the Covenant and it is not possible under international human rights law to attribute climate change to the state party and legally, "it is not possible to trace causal links between the State party's contribution to climate change, its efforts to address climate change, and the alleged effects of climate change on the enjoyment of the authors' rights."[127]

Turning to the requirement to exhaust local remedies, the Committee noted the authors' statement that the highest court in Australia had ruled that "state organs do not owe a duty of care for failing to regulate environmental harm."[128] Because findings on admissibility whether covenant rights were breached cannot be disassociated from the merits of the case and in the absence of information from the state party whether effective domestic remedies existed at the relevant time, the Committee believed that article 5(2)(b) of the Covenant does not preclude it from examining the communication.

Examining the alleged violations, the Committee pointed out that Article 2 lays down a general obligation and does not give rise to a claim. With regard to the state party's claim that other international treaties are inadmissible because they lie outside the scope of the Covenant, the Committee observed that, while it is not competent to determine compliance with other treaties, to the extent that the authors are not seeking relief for their violations but rather refer to them in interpreting state party's obligations under the Covenant, the Committee considers that such interpretations relate to the merits of the authors' claim and therefore, Article 3 of the Optional Protocol is not an obstacle.[129]

With regard to mitigation measures, the Committee noted that the state party produces a large quantity of GHG emissions and also ranks high on world economic and social development indicators. In view of this, the Committee believed that the alleged actions or omissions fall under the state party's jurisdiction. With regard to adaptation measures, the Committee noted that the provisions invoked by the authors entailed positive obligations of state parties to ensure the protection of rights of individuals under their jurisdiction.

Regarding potential future damage and the notion of "victim," the Committee stated:

> individuals claiming to be a victim of a violation of a right protected under the Covenant must demonstrate either that a State party has, by act or omission, already impaired the exercise of their right or *that such impairment is imminent*...[130]

It noted that the authors presented information about the real predicaments they have actually experienced due to disruptive climate events and slow onset events which have compromised their ability to maintain their livelihoods, subsistence, and culture. The Committee considered that the risk of impairment of rights, based on adverse impacts experienced and ongoing, is more than a theoretical possibility and thus declared their claims admissible.

Regarding the merits, the Committee noted that Article 6 cannot be interpreted in a restrictive manner, and requires states to adopt positive measures to protect life; this obligation extends to reasonably foreseeable threats and life-threatening situations even if such threats do not result in the loss of life:

> Such threats may include adverse climate change impacts, and recalls that environmental degradation, climate change and unsustainable development constitute some of the most pressing and serious threats to the ability of present and future generations to enjoy the right to life.[131]

It observed that regional human rights tribunals have also established that environmental degradation can compromise the enjoyment of the right to life and that severe environmental degradation can affect an individual's well-being, thereby threatening the right to life. However, while authors have insecurity due to loss of predictability of weather patterns and culturally important food sources, they have not indicated that they face adverse impacts to their own health or are exposed to a situation of physical endangerment likely to threaten their life. Moreover, the authors' claims under Article 6 mainly relate to their culture which falls within Article 27 of the Covenant. Without robust national and international efforts, climate impact may expose individuals to a violation of Article 6 in the future. Relying on the *Teitiota case*, the Committee said: "...given that the risk of an entire country becoming submerged under water is such an extreme risk, the conditions of life in such a country may become incompatible with the right to life with dignity before the risk is realized."[132]

With regard to adaptation measures, the state party had not explained the delay in seawall construction, the reduction of marine resources used for food, and the loss of crops – elements that constitute the components of authors' private life, family and home. Thus, the Committee decided that by failing to implement adaptation measures to protect the authors' home, private life and family, the State party had violated article 17 of the Covenant. Based on the information provided, the Committee also decided that the state party violated article 27 of the Covenant. Having found violations of articles 17 and 27, the Committee did not deem it necessary to examine the claim under article 24(1). The Committee stated that pursuant to Article (2)(3)(a), the state party is under an obligation to provide an effective remedy:

> This requires it to make *full reparation* to individuals whose Covenant rights have been violated. Accordingly, the State party is obligated, inter alia, to provide adequate compensation, to the authors for the harm that they have suffered; engage in meaningful consultations with the authors' communities in order to conduct needs assessments; continue its implementation of measures necessary to secure the communities' continued safe existence on their respective islands; and monitor and review the effectiveness of the measures implemented and resolve any deficiencies as soon as practicable. *The State party is also under an obligation to take steps to prevent similar violations in the future.*[133]

This decision breaks new ground in international human rights law, not simply because it is the first decision to hold a state liable for the failure to take adequate mitigation and adaptation measures thereby violating the rights of its people. Other notable features include: referring to the Paris Agreement to interpret obligations under the Covenant; recognition that conditions on islands could become incompatible with a life of dignity long before the islands themselves become submerged; in the event of a violation, the state is under an obligation to make full reparation which includes adequate compensation; the need to engage in consultations with the injured communities to make needs assessments; continue implementation of measures to secure the safety of people; monitor and review the effectiveness of measures; and importantly, take steps to prevent similar violations in the future which means taking both mitigation and adaptation measures and fulfilling its commitments under the Paris Agreement.

5.5 Conclusion

This chapter discussed how the Human Rights Committee has addressed environmental degradation and pollution, climate change, and sustainable development within its mandate. One notable absence seems to be any references to SDGs. As discussed in other chapters, most treaty bodies, OHCHR and the UPR process all refer to SDGs. Their absence here seems quite conspicuous.

The chapter also discussed indigenous rights to the extent they implicated environmental degradation, especially their right to culture. Most of the communications involving indigenous rights arose under Article 27 of the ICCPR. As Mapping Report No 2 pointed out:

> Although on its face the protection of Article 27 is not limited to indigenous peoples, in this context the Committee has most often focused on environmental degradation and exploitation in the context of its impacts on them rather than other ethnic, religious or linguistic minorities.[134]

It is noteworthy that the Committee had referred to environmental degradation and pollution as far back as 1991, mostly in the context of the right to life. Compared to other treaty bodies, the number of instances where the Committee addressed environmental issues is fewer. However, because the Optional Protocol to the ICCPR has been in existence the longest, individual communications under the ICCPR outnumber those of other treaty bodies. The groundbreaking decision holding Australia liable for the damage caused to Torres Strait Islanders by climate change due to its acts or omissions and the duty to provide full reparation, will likely lead to similar claims in the future.[135] Likewise, the *Teitiota case* may have opened the door (slightly) to cases involving climate refugees. These two cases showed that the Committee will not shy away from deciding communications involving environmental (and climate) issues. However, the above discussion also revealed the jurisdictional hurdles potential victims must jump through especially with regard to global issues like climate change.

Notes

1 ICCPR (16 December 1966), art. 40, https://www.ohchr.org/sites/default/files/ccpr.pdf.
2 Ibid., art. 42.
3 As opposed to citizens. Some rights such the right to vote, are available only to citizens.
4 ICCPR, supra note 1, art. 2.
5 See Section 5.4 below.
6 First Optional Protocol to the ICCPR (16 December 1966), art. 1, https://www.ohchr.org/en/instruments-mechanisms/instruments/optional-protocol-international-covenant-civil-and-political.
7 Ibid., art. 2.
8 Ibid., art. 5.
9 *Mapping Human Rights Obligations Relating to the Enjoyment of a Safe, Clean, Healthy and Sustainable Environment: Individual Report on the International Covenant on Civil and Political Rights*, Report No. 2 (December 2013) ["Mapping Report No. 2"], https://www.ohchr.org/sites/default/files/Documents/Issues/Environment/Mappingreport/2.ICCPR-25-Feb.docx.
10 *General Comment No. 36*, CCPR/C/GC/36 (3 September 2019), available at: https://undocs.org/CCPR/C/GC/36.
11 GC No. 6 on the right to life adopted in 1982 and GC No. 14 on nuclear weapons and the right to life adopted in 1984. The latter provided that "designing, testing, manufacture, possession and deployment of nuclear weapons are among the greatest threats to the right to life which confront mankind today… The production, testing, possession, deployment and use of nuclear weapons should be prohibited and recognized as crimes against humanity."

12 GC No. 36, supra note 10, ¶ 21 (emphasis added). This has extended the rules of attributability to cover private entities where there is a reasonably foreseeable threat to life.
13 Ibid., ¶¶ 23–24.
14 Ibid., ¶ 26 (emphasis added).
15 See Teitiota case, discussed in Section 5.4. *Cf.* White House Report on the Impact of Climate Change on Migration (October 2021), available at: https://www.whitehouse.gov/wp-content/uploads/2021/10/Report-on-the-Impact-of-Climate-Change-on-Migration.pdf, that the US does not accept that the ICCPR includes obligations prohibiting refoulement, nor does it interpret article 6 as encompassing a positive duty to protect life in the face of all possible external threats. Moreover, it does not consider its international human rights obligations to require extending international protection to individuals fleeing the impacts of climate change.
16 GC No. 36, supra note 10, ¶ 62.
17 Ibid. (emphasis added).
18 The difference in nomenclature here and GC No. 14 which this GC replaced is quite striking. GC No. 14 clearly stated that production, testing, possession, deployment and use of nuclear weapons should be recognized as crimes against humanity whereas here the wording has been watered down.
19 GC No. 36, supra note 10, ¶ 66.
20 *General Comment adopted by the Human Rights Committee under Article 40, paragraph 4, of the International Covenant on Civil and Political Rights: Addendum*, CCPR/C/21/Rev. 1/Add.5 (26 April 1994), available at: https://undocs.org/CCPR/C/21/Rev.%201/Add.5.
21 Ibid., ¶ 3.2.
22 Ibid., ¶ 7 (footnotes omitted). Article 27 has been used extensively by indigenous (and other minority) groups to protect their rights, especially culture. See section 5.4 and Mapping Report No. 2, supra note 9.
23 Mapping Report No. 2, p. 15, supra note 9, quoting *Consideration of reports submitted by state parties under Article 40 of the Covenant: Colombia*, A/47/40(SUPP) (9 October 1992) ["Colombia report"], ¶¶ 378–379.
24 Ibid., p.16, citing Colombia report, supra note 23, ¶¶ 257–258.
25 *Concluding observations of the Human Rights Committee: Mexico*, CCPR/C/79/Add.109 (27 July 1999), ¶ 19.
26 *Concluding observations of the Human Rights Committee: Guyana*, CCPR/C/79/Add.121 (24 & 27 March 2000), ¶ 21.
27 *Concluding observations of the Human Rights Committee: Argentina*, CCPR/C/ARG/CO/4 (31 March 2010), ¶ 25.
28 *List of issues prepared in the absence of the initial report of Belize*, CCPR/C/BLZ/Q/1 (23 November 2012), ¶ 28.
29 *Concluding observations of the Human Rights Committee: Sweden* (24 April 2002) CCPR/CO/74/SWE ["Sweden report"], ¶ 15.
30 *Concluding observations on the third periodic report of Guinea*, CCPR/C/GIN/CO/3, (7 December 2018), ¶ 49.
31 Ibid., ¶ 50. It is noteworthy that the Committee points to the need to obtain free, prior and informed consent of local communities, not just indigenous communities with regard to natural resource management and projects that have a social and environmental impact whereas on the report of Dominica the Committee referred to the free, informed and prior consultation in relation to indigenous peoples.
32 *Concluding observations of the Human Rights Committee: Israel*, CCPR/C/ISR/CO/3 (3 September 2010), ¶ 18.
33 Ibid. The Committee relied on WHO standards here.
34 *Concluding observations of the Human Rights Committee: USA*, CCPR/C/USA/CO/3/Rev.1 (18 December 2006).

35 Adopted by the Commission on Human Rights in 1998, https://www.unhcr.org/en-us/protection/idps/43ce1cff2/guiding-principles-internal-displacement.html.
36 *Second periodic report of Austria*, CCPR/C/51/Add.2 (24–25 October 1991), ¶ 94.
37 Ibid., ¶ 95.
38 *Third periodic report of Poland*, CCPR/C/58/Add.10 and Add.13 (28–29 October 1991), ¶ 145.
39 *Second periodic report of Peru*, CCPR/C/51/Add.4 (31 March–2 April 1992), ¶ 321.
40 *Second periodic report of Belgium*, CCPR/C/57/Add.3 (7 April 1992), ¶ 408 (emphasis added). This is consistent with GC No. 6 which noted the positive aspects of the right to life and *E. H. P. v. Canada, Communication No. 67/1980*, CCPR/C/OP/1 at 20 (1984) ["Port Hope case"], http://hrlibrary.umn.edu/undocs/html/67–1980.htm, discussed in section 5.4.
41 *Third periodic report of the Ukrainian Soviet Socialist Republic*, CCPR/C/56/Add.B (5–6 November 1990), ¶ 199.
42 Ibid., ¶ 202. Panama was asked to describe its environmental policy under the provisions on right to life, *Second periodic report of Panama*, CCPR/C/42/Add.7 and 11 (4–5 April 1991), ¶ 431.
43 *Concluding observations in the absence of the initial report of Dominica*, CCPR/C/DMA/COAR/1 (24 April 2020), ¶ 24.
44 Ibid., ¶ 25.
45 *Concluding observations on the initial report of Cabo Verde*, CCPR/C/CPV/CO/1/Add.1 (3 December 2019), ¶ 17.
46 Ibid, ¶ 18. It is noteworthy that the Committee referred to the need to develop environmental standards, conduct EIAs, and adopt a precautionary approach to protect people including the most vulnerable.
47 Dominica observations, supra note 43, ¶ 47. Here the formulation adopted by the Committee is different from what is required under UNDRIP which requires the free, prior and informed consent (FPIC) of indigenous peoples in certain instances. *Cf.* Chapter 6 where the ESCR Committee has consistently referred to free, prior and informed consent of indigenous peoples and even local communities and Chapter 7 where the CEDAW Committee has applied this in relation to rural women.
48 Ibid., ¶ 48 (emphasis added). While the Committee refers to FPIC here, it calls upon the state *to consider* taking such measures.
49 *Concluding observations on the sixth periodic report of New Zealand*, CCPR/C/NZL/CO/6 (28 April 2016), ¶ 43.
50 Ibid., ¶¶ 45–46.
51 *Concluding observations of the Human Rights Committee: Russian Federation*, CCPR/C/RUS/CO/6 (24 November 2009).
52 Ibid.
53 *Concluding observations of the Human Rights Committee: Sweden*, CCPR/C/SWE/CO/6 (7 May 2009), ¶ 20.
54 Sweden report, supra note 29, ¶ 15.
55 Ibid.
56 *Third periodic report of Sweden*, CCPR/C/58/Add.7 (27–28 March 1991), ¶ 322.
57 USA report, supra note 34.
58 Ibid. See also the comments on the report of Ecuador where the Committee noted that following the largest uprising of indigenous peoples in Ecuador's history in 1990, a dialog was undertaken which resulted in far reaching reforms, including the donation of over 1 million hectares of land, *Third periodic report of Ecuador*, CCPR/C/58/Add.9 (6–7 November 1991), ¶ 222.
59 *Third periodic report of Colombia*, CCPR/C/64/Add.3 (2–3 April 1992), ¶ 382.
60 Ibid.
61 This section draws from author's book chapter (2019) "Environmental Rights and International Human Rights Covenants: What Standards Are Relevant?" in

Environmental Rights: Development of Standards, Turner, S., Shelton, D., Razzaque, J., McIntyre, O., & May, J., eds., Cambridge: Cambridge University Press, p. 17, doi: 10.1017/9781108612500.

62 Port Hope case, supra note 40.

63 Ibid.

64 Ibid.

65 Ibid.

66 *Communication No. 645/1995*, CCPR/C/57/D/645/1995 (30 July 1996).

67 GC No 6 elaborates on the right to life (replaced by GC No 36)

68 *CCPR General Comment No. 14: Article 6 (Right to Life) Nuclear Weapons and the Right to Life* (9 November 1984), https://tbinternet.ohchr.org/_layouts/15/treatybodyexternal/Download.aspx?symbolno=INT%2fCCPR%2fGEC%2f4723&Lang=en.

69 Ibid., ¶ 4.

70 Ibid., ¶ 6.

71 Communication No. 645/1995, supra note 66, ¶ 5.7.

72 Ibid., ¶ 5.6.

73 Ibid., ¶ 5.9.

74 *Susila Malani Dahanayake and 41 other Sri Lankan citizens v. Sri Lanka*, CCPR/C/87/D/1331/2004 (25 July 2006), https://undocs.org/CCPR/C/87/D/1331/2004, referred to in Mapping Report No. 2, supra note 9.

75 Ibid., ¶ 3.2.

76 Ibid., ¶ 4.7.

77 *E. P. et al. v. Colombia, Communication No. 318/1988*, U.N. Doc. CCPR/C/39/D/318/1988 (1990), http://hrlibrary.umn.edu/undocs/session39/318-1988.html.

78 Ibid., 8.

79 Communication No 167/1984 (26 March 1990), http://hrlibrary.umn.edu/undocs/session38/167–1984.html. U.N. Doc. Supp. No. 40, A/45/40, (1990) at 1.

80 Ibid., ¶ 11.2.

81 Ibid., ¶ 16.2.

82 Ibid., ¶ 33. In several other communications where the Committee did not find a violation of Article 27, it nonetheless acknowledged the link between the right to culture and environmental degradation. *Ilmari Länsman et al. v. Finland*, CCPR/C/52/D/511/1992 (26 October 1994), https://tbinternet.ohchr.org/_layouts/treatybodyexternal/Download.aspx?symbolno=CCPR%2FC%2F52%2FD%2F511%2F1992&Lang=en; *Jouni Länsman et al v. Finland*, CCPR/C/587/D/671/1995 (30 October 1996), http://hrlibrary.umn.edu/undocs/html/671–1995.html; and CCPR/C/83/D/1023/2001 (17 March 2005), http://hrlibrary.umn.edu/undocs/1023-2001.html, where the Committee decided that logging activities authorized by the state were not extensive enough to threaten the survival of reindeer husbandry, Mapping Report No. 2, supra note 9, ¶ 19. See also *Mahuika v. New Zealand*, CCPR/C/70/D/547/1993 (27 October 2000), http://hrlibrary.umn.edu/undocs/547–1993.html and *Ángela Poma Poma v. Peru*, CCPR/C/95/D/1457/2006 (27 March 2009) ["Poma Poma case"], https://undocs.org/CCPR/C/95/D/1457/2006.

83 *Communication No 1453/2006*, CCPR/C/88/D/1453/2006 (18 October 2006), https://undocs.org/CCPR/C/88/D/1453/2006.

84 ICCPR, supra note 1, art. 17 embodies the right to privacy, family, home and correspondence.

85 Communication No 1453/2006, supra note 82, ¶ 3.1.

86 Ibid., ¶ 3.2.

87 Port Hope case, supra note 40.

88 Communication No 1453/2006, supra note 82, ¶ 5.2.

89 Poma Poma case, supra note 82.

90 ICCPR, supra note 1, art. 1(2) embodies the right of all peoples to freely dispose of their natural wealth and resources and provides that a people cannot be deprived of its own means of subsistence.

91 Ibid., art. 2(3)(a) embodies the right to an effective remedy.
92 Ibid., art. 14 embodies the right to equality before courts and tribunals.
93 Ibid., art. 17 embodies the right to privacy, family, home or correspondence.
94 Poma Poma case, supra note 82, ¶ 2.2.
95 Ibid., ¶ 3.1.
96 *General Comment No. 23: Article 27 (Rights of Minorities)*, CCPR/C/21/Rev.1/Add.5 (8 April 1994), https://undocs.org/CCPR/C/21/Rev.1/Add.5.
97 Poma Poma case, supra note 82, ¶ 7.2.
98 Ibid., ¶ 7.6 (emphasis added).
99 Ibid., ¶ 7.7.
100 This section draws from Atapattu, S., (2022) "Human Rights and the Environment," in Goodhart, M., ed., *Human Rights in Theory and Practice* (4th edn.), Oxford: Oxford University Press, p. 297 ["Atapattu (2022)"]; and (2022) "Migrating with Dignity: Protecting the Rights of 'Climate Refugees' with the *non-refoulement* Principle," in Behrman, S., & Kent, A., eds., *Climate Refugees: Global, Local and Critical Approaches*, Cambridge: Cambridge University Press, p. 127.
101 The term "climate refugees" is not a legally accepted term. It simply describes those who are compelled to move due to adverse consequences of climate change. At the Paris Conference in 2015 (COP21) the parties requested the Executive Committee of the Warsaw International Mechanism on Loss and Damage to establish a taskforce on "displacement related to the adverse impacts of climate change," Decision 1/CP.21, "Adoption of the Paris Agreement," FCCC/CP/2015/10/Add.1 (2015), ¶ 49, https://unfccc.int/resource/docs/2015/cop21/eng/10a01.pdf#page=2.
102 See Behrman, S. & Kent, A. (2020) "Climate-Induced Migration: Will Tribunals Save the Day?" *Hong Kong Journal of Law and Public Affairs*, vol. 2, no. 2, doi: 10.2139/ssrn.3682504; and Buchanan, K. (July 2015) "New Zealand: 'Climate Refugee' Case Overview," Law Library of Congress, Global Legal Research Directorate, available at: https://www.loc.gov/item/2016295703.
103 Convention Relating to the Status of Refugees, Geneva (1951) and Protocol of 1967, https://www.unhcr.org/4d934f5f9.pdf.
104 The "disappearance" of these states raises several legal issues, such as those relating to statehood, nationality, and the right to self-determination of the people. See Gerrard, M. & Wannier, G., eds. (2013) *Threatened Island Nations: Legal Implications of Rising Seas and a Changing Climate,* New York: Cambridge University Press; Atapattu, S. (2016) *Human Rights Approaches to Climate Change: Challenges and Opportunities*, London: Routledge, Chapter 9.
105 *Ioane Teitiota v. New Zealand*, CCPR/C/127/D/2728/2016 (24 October 2019), ¶ 2.2, https://undocs.org/CCPR/C/127/D/2728/2016.
106 Ibid., ¶ 2.1.
107 Ibid., ¶ 8.5.
108 Ibid.
109 Ibid.
110 GC No. 31, *The Nature of the General Legal Obligation Imposed on States Parties to the Covenant*, CCPR/C/21/Rev.1/Add.13 (26 May 2004), https://undocs.org/CCPR/C/21/Rev.1/Add.13.
111 Ibid., ¶ 9.4.
112 Ibid. The exact wording was used in the Torres Strait Islanders case.
113 Ibid., ¶ 9.5 (footnotes omitted).
114 Ibid., ¶ 9.12.
115 Ibid., ¶ 9.11 (emphasis added). Again, exact wording was used in the Torres Strait Islanders case.
116 Atapattu (2022), supra note 100.
117 This is the first climate change legal action against Australia alleging a violation of human rights and the first legal action filed by the inhabitants of a low-lying island

against a national government for inaction on climate change. See Cullen, M. (27 June 2019) "Climate Change and Human Rights: The Torres Strait Islanders' Claim to the UN Human Rights Committee," *Groningen Journal of International Law blog*, available at: https://www.grojil.org/blog2/2019/06/27/climate-change-and-human-rights-the-torres-strait-islanders-claim-to-the-un-human-rights-committee; see also, Cullen, M. (September 2018) "Eaten by the sea": Human Rights Claims for the Impacts of Climate Change upon Remote Subnational Communities," *Journal of Human Rights and Environment*, vol. 9, no. 2, pp. 171–193, doi: 10.4337/jhre.2018.02.03.

118 *Views adopted by the Committee under article 5(4) of the Optional Protocol, concerning Communication No. 3624/2019*, CCPR/C/135/D/3624/2019 (22 September 2022), Advance unedited version ["Communication No. 3624/2019"] https://tbinternet.ohchr.org/Treaties/CCPR/Shared%20Documents/AUS/CCPR_C_135_D_3624_2019_34335_E.docx.

119 Article 24(1) deals with non-discrimination of children.

120 Ibid., ¶ 3.1.

121 Ibid.

122 Ibid.

123 See *Sacchi v. Argentina, Decision adopted by the Committee on the Rights of the Child under the Optional Protocol to the Convention on the Rights of the Child on a communications procedure in respect of Communication No. 104/2019*, CRC/C/88/D/104/2019 (8 October 2021), https://undocs.org/CRC/C/88/D/104/2019, discussed in Chapter 8, where the CRC Committee stressed that the global nature of climate change does not absolve states of their *individual responsibility* to address climate change (emphasis added).

124 Faa, M. (10 December 2020) "Torres Strait Islander complaint against climate change inaction wins backing of UN legal experts," *Pacific Beat*, https://www.abc.net.au/news/2020-12-11/torres-strait-islander-complaint-against-climate-change-inaction/12972926.

125 Communication No. 3624/2019, supra note 118, ¶ 4.1.

126 Ibid., ¶ 4.2.

127 Ibid., ¶ 4.3.

128 Ibid., ¶ 7.3.

129 First Optional Protocol, supra note 6, art. 3 refers to anonymous or frivolous claims or claims that are an abuse of the rights of submission.

130 Communication No. 3624/2019, supra note 118, ¶ 7.9 (emphasis added, footnotes omitted).

131 Ibid., ¶ 8.3. See identical language in the *Teitiota case.*

132 Ibid., ¶ 8.7. Identical language was used in the *Teitiota case.*

133 Ibid., ¶ 11 (emphasis added). Several Committee members added individual opinions. Member Duncan Laki Muhumuza while agreeing with the Committee decision, relying on the *Urgenda* decision, believed that there is a violation of Article 6 because the State party failed to prevent foreseeable loss of life from the impact of climate change.

134 Mapping Report No. 2, supra note 9, ¶ 24.

135 See Erin Daly, "The UNHRC's Torres Strait Islands Decision: A Major Advance and a Roadmap for the Future" (3 October 2022), https://gnhre.org/community/the-unhrcs-torres-strait-islands-decision-a-major-advance-and-a-roadmap-for-the-future.

References

Atapattu, S. (2016) *Human Rights Approaches to Climate Change: Challenges and Opportunities*, London: Routledge.

Atapattu, S. (2019) "Environmental Rights and International Human Rights Covenants: What Standards Are Relevant?" in *Environmental Rights: Development of Standards*,

Turner, S., Shelton, D., Razzaque, J., McIntyre, O., & May, J., eds., Cambridge: Cambridge University Press, doi: 10.1017/9781108612500.

Atapattu, S. (2022a) "Human Rights and the Environment," in Goodhart, M., ed., *Human Rights in Theory and Practice* (4th edn.), 297–319, Oxford: Oxford University Press.

Atapattu, S. (2022b) "Migrating with Dignity: Protecting the Rights of 'Climate Refugees' with the *non-refoulement* Principle," in *Climate Refugees: Global, Local and Critical Approaches*, Behrman, S., & Kent, A., eds., Cambridge: Cambridge University Press, p. 127, doi: 10.1017/9781108902991.007.

Behrman, S. & Kent, A. (2020) "Climate-Induced Migration: Will Tribunals Save the Day?" *Hong Kong Journal of Law and Public Affairs*, vol. 2, no. 2, doi: 10.2139/ssrn.3682504.

Bordes and Temeharo v. France (1996) *Communication No. 645/1995*, CCPR/C/57/D/645/1995, http://hrlibrary.umn.edu/undocs/html/DEC64557.htm.

Brun v. France (2006) *Communication No 1453/2006*, CCPR/C/88/D/1453/2006, https://undocs.org/CCPR/C/88/D/1453/2006.

Buchanan, K. (2015) *New Zealand: 'Climate Refugee' Case Overview*, Law Library of Congress, Global Legal Research Directorate, https://www.loc.gov/item/2016295703.

CCPR (1982) *General Comment No. 6*.

CCPR (1984) *CCPR General Comment No. 14*, https://tbinternet.ohchr.org/_layouts/15/treatybodyexternal/Download.aspx?symbolno=INT%2fCCPR%2fGEC%2f4723&Lang=en.

Communication No. 645/1995 (1996) CCPR/C/57/D/645/1995.

Concluding observations in the absence of the initial report of Dominica (2020) CCPR/C/DMA/COAR/1.

Concluding observations of the Human Rights Committee: Argentina (2010) CCPR/C/ARG/CO/4.

Concluding observations of the Human Rights Committee: Colombia (1992) A/47/40(SUPP).

Concluding observations of the Human Rights Committee: Guyana (2000) CCPR/C/79/Add.121.

Concluding observations of the Human Rights Committee: Israel (2010) CCPR/C/ISR/CO/3.

Concluding observations of the Human Rights Committee: Mexico (1999) CCPR/C/79/Add.109.

Concluding observations of the Human Rights Committee: Russian Federation (2009) CCPR/C/RUS/CO/6.

Concluding observations of the Human Rights Committee: Sweden (2002) CCPR/CO/74/SWE.

Concluding observations of the Human Rights Committee: Sweden (2009) CCPR/C/SWE/CO/6.

Concluding observations of the Human Rights Committee: USA (2006) CCPR/C/USA/CO/3/Rev.1.

Concluding observations on the initial report of Cabo Verde (2019) CCPR/C/CPV/CO/1/Add.1.

Concluding observations on the sixth periodic report of New Zealand (2016) CCPR/C/NZL/CO/6.

Concluding observations on the third periodic report of Guinea (2018) CCPR/C/GIN/CO/3.

Cullen, M. (2018) "Eaten by the sea": Human Rights Claims for the Impacts of Climate Change upon Remote Subnational Communities," *Journal of Human Rights and Environment*, vol. 9, no. 2, pp. 171–193, doi: 10.4337/jhre.2018.02.03.

Cullen, M. (2019) "Climate Change and Human Rights: The Torres Strait Islanders' Claim to the UN Human Rights Committee," *Groningen Journal of International Law Blog*, https://www.grojil.org/blog2/2019/06/27/climate-change-and-human-rights-the-torres-strait-islanders-claim-to-the-un-human-rights-committee.

Dahanayake et al. v. Sri Lanka (2006) *Communication No. 1331/2004*, CCPR/C/87/D/1331/2004, https://undocs.org/CCPR/C/87/D/1331/2004.

Daly, E. (2022) *The UNHRC's Torres Strait Islands Decision: A Major Advance and a Roadmap for the Future*, https://gnhre.org/community/the-unhrcs-torres-strait-islands-decision-a-major-advance-and-a-roadmap-for-the-future.

Decision 1/CP.21, Adoption of the Paris Agreement (2015) FCCC/CP/2015/10/Add.1, https://unfccc.int/resource/docs/2015/cop21/eng/10a01.pdf#page=2.

E. H. P. v. Canada (1984) *Communication No. 67/1980*, CCPR/C/OP/1, http://hrlibrary.umn.edu/undocs/html/67-1980.htm.

E. P. et al. v. Colombia (1990) *Communication No. 318/1988*, CCPR/C/39/D/318/1988, http://hrlibrary.umn.edu/undocs/session39/318-1988.html.

Faa, M. (2020) "Torres Strait Islander Complaint against Climate Change Inaction Wins Backing of UN Legal Experts," *Pacific Beat*, https://www.abc.net.au/news/2020-12-11/torres-strait-islander-complaint-against-climate-change-inaction/12972926.

First Optional Protocol to the ICCPR (1966), art. 1, https://www.ohchr.org/en/instruments-mechanisms/instruments/optional-protocol-international-covenant-civil-and-political.

General Comment No. 23 (1994) CCPR/C/21/Rev.1/Add.5.

General Comment No. 31 (2004) CCPR/C/21/Rev.1/Add.13.

General Comment No. 36 (2019) CCPR/C/GC/36.

Gerrard, M. & Wannier, G., eds. (2013) *Threatened Island Nations: Legal Implications of Rising Seas and a Changing Climate*, New York: Cambridge University Press.

Guiding Principles on Internal Displacement (1998), https://www.unhcr.org/en-us/protection/idps/43ce1cff2/guiding-principles-internal-displacement.html.

ICCPR (1966) https://www.ohchr.org/sites/default/files/ccpr.pdf.

Länsman et al. v. Finland (1994) *Communication No. 511/1992*, CCPR/C/52/D/511/1992, https://tbinternet.ohchr.org/_layouts/treatybodyexternal/Download.aspx?symbolno=CCPR%2FC%2F52%2FD%2F511%2F1992&Lang=en.

Länsman et al. v. Finland (1996) *Communication No. 671/1995*, CCPR/C/587/D/671/1995, http://hrlibrary.umn.edu/undocs/html/671–1995.html.

Länsman et al. v. Finland (2005) *Communication No. 1023/2001*, CCPR/C/83/D/1023/2001, http://hrlibrary.umn.edu/undocs/1023–2001.html.

List of issues prepared in the absence of the initial report of Belize (2012) CCPR/C/BLZ/Q/1.

Mahuika v. New Zealand (2000) *Communication No. 547/1993*, CCPR/C/70/D/547/1993, http://hrlibrary.umn.edu/undocs/547–1993.html.

Mapping Human Rights obligations relating to the enjoyment of a safe, clean, healthy and sustainable environment: Individual report on the International Covenant on Civil and Political Rights, Report No. 2, (2013) https://www.ohchr.org/sites/default/files/Documents/Issues/Environment/Mappingreport/2.ICCPR-25-Feb.docx.

Ominayak and the Lubicon Lake Band v. Canada (1990) *Communication No 167/1984, Supp. No. 40*, A/45/40, http://hrlibrary.umn.edu/undocs/session38/167–1984.html.

Poma Poma v. Peru (2009) *Communication No. 1457/2006*, CCPR/C/95/D/1457/2006, https://undocs.org/CCPR/C/95/D/1457/2006.

Sacchi v. Argentina (2021) *Communication No. 104/2019*, CRC/C/88/D/104/2019, https://undocs.org/CRC/C/88/D/104/2019.

Second periodic report of Austria (1991) CCPR/C/51/Add.2.

Second periodic report of Belgium (1992) CCPR/C/57/Add.3.

Second periodic report of Panama (1991) CCPR/C/42/Add.7 and 11.
Second periodic report of Peru (1992) CCPR/C/51/Add.4.
Third periodic report of Colombia (1992) CCPR/C/64/Add.3.
Third periodic report of Ecuador (1991) CCPR/C/58/Add.9.
Third periodic report of Poland (1991) CCPR/C/58/Add.10 and Add.13.
Third periodic report of Sweden (1991) CCPR/C/58/Add.7.
Third periodic report of the Ukrainian Soviet Socialist Republic (1990) CCPR/C/56/Add.B.
Teitiota v. New Zealand (2019) *Communication No. 2728/2016*, CCPR/C/127/D/2728/2016, https://undocs.org/CCPR/C/127/D/2728/2016.
Views adopted by the Committee under article 5(4) of the Optional Protocol, concerning Communication No. 3624/2019 (2022) CCPR/C/135/D/3624/2019, advance unedited version, https://tbinternet.ohchr.org/Treaties/CCPR/Shared%20Documents/AUS/CCPR_C_135_D_3624_2019_34335_E.docx.
White House (2021) *Report on the Impact of Climate Change on Migration*, https://www.whitehouse.gov/wp-content/uploads/2021/10/Report-on-the-Impact-of-Climate-Change-on-Migration.pdf.

6 Committee on Economic, Social and Cultural Rights

6.1 Introduction

The human rights treaty that comes closest to environmental issues is the International Covenant on Economic, Social, and Cultural Rights (ICESCR)[1] even though it does not refer to environmental pollution. Many of the rights recognized in the Covenant rely on a healthy and sustainable environment for their realization and cannot be enjoyed in a degraded environment. The rights protected under the ICESCR that are closely related to a healthy and sustainable environment include: the right to work (Article 7); social security (Article 9); adequate standard of living, including adequate food, clothing and housing and the right to be free from hunger (Article 11); right to physical and mental health (Article 12); and the right to take part in cultural life (Article 15).

Unlike civil and political rights, economic, social, and cultural rights (ESC rights) are to be realized progressively, subject to available resources.[2] This has led some states and scholars to argue that ESC rights are not justiciable, thereby creating a hierarchy of rights.[3] Many older constitutions embody only civil and political rights while newer constitutions embody ESC rights, including environmental rights[4] and even rights of nature.[5] Parties have undertaken to ensure the equal right of men and women to the enjoyment of all ESC rights embodied in the Covenant. Thus, principles of equality and non-discrimination underlie the entire Covenant.[6]

The Committee on Economic, Social and Cultural Rights (ESCR Committee) oversees the implementation of the ICESCR. It was established by a resolution of the UN Economic and Social Council (ECOSOC) in May 1985 which renamed the Working Group established by ECOSOC in 1978 as the "Committee on Economic, Social and Cultural Rights"[7] to carry out the monitoring functions assigned to ECOSOC under Part IV of the ICESCR. The Committee is a body of independent experts. However, unlike the UN Human Rights Committee whose mandate is included in the ICCPR itself, the ESCR Committee was established by an ECOSOC resolution.[8] Despite this distinction, its mandate, procedure, duties, and powers are very similar to those of the Human Rights Committee.

This chapter discusses how the ESCR Committee has addressed environmental issues, climate change, and sustainable development including the SDGs. It uses

DOI: 10.4324/9781003128847-9

the individual report on the ICESCR prepared for the UN Independent Expert on Human Rights and the Environment, John Knox, as part of his mapping exercise in 2013 as background for this chapter.[9]

6.2 General Comments

Several General Comments adopted by the ESCR Committee are relevant to the present discussion. GC No 3 elaborates on the nature of state parties' obligations.[10] It notes that Article 2 has a dynamic relationship with other provisions of the Covenant and describes the nature of the legal obligations undertaken by states which include both obligations of conduct and obligations of result.[11] While the differences between the nature of obligations in the two Covenants are often highlighted, the similarities are rarely pointed out. The ICESCR provides for the progressive realization of rights, but it imposes obligations which are of *immediate effect*. Of these, two are of particular importance: (a) the principle of non-discrimination; and (b) the obligation to take steps toward the full realization of rights:

> Thus while the full realization of the relevant rights may be achieved progressively, steps towards that goal must be taken within a reasonably short time after the Covenant's entry into force for the States concerned. Such steps should be deliberate, concrete and targeted as clearly as possible towards meeting the obligations recognized in the Covenant.[12]

States are required to take all appropriate means, including the adoption of legislative measures to implement the obligations under the treaty. Legislative measures are often indispensable for the fulfillment of obligations, especially those relating to non-discrimination but it is not the only measure. Rather, "all appropriate means" must be given its full and natural meaning. While states must decide for themselves which means are most appropriate under the circumstances, the ultimate determination whether all appropriate measures have been taken rests with the Committee. These measures may include judicial remedies, access to remedies and standing, as well as administrate, financial, educational, and social measures.[13]

The GC notes that the term "progressive realization" is used to describe the intent; it recognizes that full realization of ESC rights will not be achieved in a short period of time. In this sense, the obligation differs from that in Article 2 of the ICCPR which embodies an obligation to ensure their immediate realization. However, this should not be misinterpreted as depriving the obligation of all meaningful content:

> [t]he phrase must be read in the light of the overall objective, indeed the raison d'être, of the Covenant which is to establish clear obligations for States parties in respect of the full realization of the rights in question. It thus imposes an obligation to move as expeditiously and effectively as possible towards that

> goal. Moreover, any deliberately retrogressive measures in that regard would require the most careful consideration and would need to be fully justified by reference to the totality of the rights provided for in the Covenant and in the context of the full use of the maximum available resources.[14]

The Committee believes that *a minimum core obligation* to ensure the minimum essential levels of each right is incumbent upon every state party.[15] Similarly, any assessment as to whether a state has discharged its minimum core obligation must consider the resource constraints in that country. With regard to taking steps "to the maximum of its available resources," it must demonstrate that "every effort has been made to use all resources that are at its disposition in an effort to satisfy, *as a matter of priority*, those minimum obligations."[16]

Finally, the Committee noted that the phrase "maximum available resources" was intended to refer to both the resources existing within a state and those available from the international community through international cooperation and assistance. Relying on Articles 55 and 56 of the UN Charter, the General Comment emphasized that international cooperation for development and therefore for the realization of ESC rights, *is an obligation of all States*, particularly of those states which are in a position to assist others.[17]

Of the 25 General Comments so far adopted by the ESCR Committee, several refer to environmental pollution and/or the need for environmental protection. GC No 4 on the right to adequate housing notes that "adequacy" is determined in part by social, economic, cultural, climatic, ecological and other factors and identified certain aspects that must be taken into account: (a) legal security of tenure; (b) availability of services, materials, facilities and infrastructure; (c) affordability; (d) habitability; (e) accessibility; (f) location; and (g) cultural adequacy. The Committee elaborated on some of these aspects:

> An adequate house must contain certain facilities essential for health, security, comfort and nutrition. All beneficiaries of the right to adequate housing should have sustainable access to natural and common resources, safe drinking water, energy for cooking, heating and lighting, sanitation and washing facilities, means of food storage, refuse disposal, site drainage and emergency services.[18]

In addition, it encourages state parties to apply the Health Principles of Housing prepared by the WHO as housing is the environmental factor most associated with disease. With regard to accessibility, the GC refers to the need to provide full and sustainable access to adequate housing resources to disadvantaged groups who include, among others, victims of natural disasters and people living in disaster-prone areas.

GC No 12 on the Right to Food recognizes that adequate food is crucial for the enjoyment of all rights.[19] Regarding the normative content of Article 11, it elaborates that the right to adequate food should not be interpreted narrowly and while it is to be realized progressively, states have a core obligation to take necessary action to alleviate hunger even during natural and other disasters. Moreover, the concept of adequacy underlies a number of factors that must be taken into account when determining whether particular foods or diets that are accessible is the most appropriate and introduced the notion of sustainability:

> The notion of sustainability is intrinsically linked to the notion of adequate food or food security, implying food being accessible for both present and future generations. The precise meaning of "adequacy" is to a large extent determined by prevailing social, economic, cultural, climatic, ecological and other conditions, while "sustainability" incorporates the notion of long-term availability and accessibility.[20]

It called upon states to formulate a strategy to implement the right to food, address issues such as sustainable management and use of natural resources for food; and protect vulnerable populations even where states face severe resource constraints whether caused by economic adjustment, recession, climatic conditions or other factors.

GC No 14 elaborates on the right to health and recognizes that it is indispensable for the exercise of other human rights, including the rights to food, housing, work, education, dignity, and procedural rights.[21] In a direct reference to a healthy environment, the General Comment notes that:

> [t]he right to health embraces a wide range of socio-economic factors that promote conditions in which people can lead a healthy life, and extends to the underlying determinants of health, such as food and nutrition, housing, access to safe and potable water and adequate sanitation, safe and healthy working conditions, and *a healthy environment*.[22]

The Committee recognizes that the notion of health has undergone substantial change and more determinants of health are being taken into consideration such as resource distribution and gender differences. Thus, the underlying determinants of health include access to safe and potable water and adequate sanitation, an adequate supply of safe food, nutrition and housing, and healthy occupational and environmental conditions.[23] Moreover, the participation of the population in health-related decision-making at all levels is an important aspect.

Interpreting Article 12(2)(b), under the caption "The right to healthy natural and workplace environment," the Committee noted that the improvement of "all aspects of environmental and industrial hygiene" comprises, *inter alia*, preventive measures in respect of occupational accidents and diseases; ensuring an adequate supply of safe and potable water and basic sanitation; reducing the population's exposure to harmful substances such as radiation, chemicals, and other detrimental environmental conditions that impact upon human health.[24]

The General Comment further noted the importance of protecting vital medicinal plants, animals, and minerals necessary for the full enjoyment of health of indigenous peoples. Moreover, development-related activities that lead to the displacement of indigenous peoples from their traditional territories and environment, denying them their source of nutrition, and breaking the symbiotic relationship with their lands, have a detrimental impact on their health.

With regard to the specific legal obligations of states, GC No 12 notes that states "should refrain from unlawfully polluting air, water and soil, e.g. through industrial waste from State-owned facilities, from using or testing nuclear, biological or chemical weapons if such testing results in the release of substances harmful to human health…"[25] With regard to the violations of the obligation to protect, the Comment refers to the failure to enact laws to prevent the pollution of water, air, and soil by extractive and manufacturing industries.[26] It further recognized the need to give priority to vulnerable and marginalized groups when providing medical and financial aid, and distributing resources such as potable water, food, and medical supplies.

GC No 15 on the Right to Water[27] adopted in 2002 also refers to environmental conditions. This is the first time that water was recognized as a human right. It recognizes that water is a limited natural resource and a public good and that the human right to water is indispensable for leading a life of dignity. Similar to rights to health and food, it is a prerequisite for the realization of other human rights. Article 11 on an adequate standard of living provides the legal basis for the right to water which has been recognized in other instruments as well.

GC No 15 recognizes the importance of ensuring sustainable access to water resources for agriculture to realize the right to food, paying attention to disadvantaged and marginalized farmers, including women farmers. In addition, environmental hygiene, as an aspect of the right to health, requires taking steps to prevent unsafe and toxic water conditions. Thus, states must ensure that natural water resources are protected from contamination by harmful substances.[28] It provides that water should be treated as a social and cultural good, not primarily as an economic good. The realization of the right must be sustainable, "ensuring that the right can be realized for present and future generations."[29] The quantity of water available for each person should correspond to the WHO guidelines; with regard to quality, water must be safe, free from micro-organisms, chemical substances and radiological hazards that constitute a threat to health. Moreover, the right to water should be realized without discrimination with special attention being paid to marginalized groups, including women, children, rural people, nomadic communities, and indigenous peoples. States must ensure that access to traditional water sources in rural areas should be protected from unlawful encroachment and pollution.[30] Likewise, indigenous peoples' access to water resources on their ancestral lands must be protected from encroachment and unlawful pollution.[31]

Regarding the nature of states' obligations, the obligation to respect requires states to refrain from arbitrarily interfering with traditional arrangements for water allocation and unlawfully diminishing or polluting water, for example, by testing weapons.[32] Relying on principles of humanitarian law, the Committee noted that during armed conflicts, emergency situations, and natural disasters, state parties must protect objects indispensable for survival of the civilian population, including drinking water installations and protection of the natural environment against widespread, long-term, and severe damage.[33] With regard to the obligation to protect, states are required to prevent third parties from interfering in the enjoyment of the right which includes adopting necessary legislation and other measures to restrain third parties from "polluting and inequitably extracting from water resources, including natural sources, wells and other water distribution systems."[34] With regard to the obligation to fulfill, states are required to protect water sources and facilitate sustainable access to water, particularly in rural and deprived urban areas. Moreover, states must adopt comprehensive strategies and programs to ensure sufficient and safe water for present and future generations[35] which may include: (a) reducing depletion of water resources through unsustainable practices; (b) reducing contamination of watersheds and water-related ecosystems (c) monitoring water reserves; (d) ensuring that proposed developments do not interfere with access to water; and (e) assessing the impacts of actions "that may impinge upon water availability and natural-ecosystems watersheds, such as climate changes, desertification and increased soil salinity, deforestation and loss of biodiversity…."[36]

Another relevant General Comment is No 21 on the right to cultural life[37] which may be exercised individually or collectively. Culture encompasses, *inter alia*, ways of life, rites and ceremonies, natural and man-made environments.[38] There are several necessary conditions for the full realization of the right based on equality and non-discrimination: availability, accessibility, acceptability, adaptability and appropriateness. Availability includes the shared open spaces essential to cultural interaction, such as parks, and natural areas including seas, lakes, rivers, mountains, forests and nature reserves and their flora and fauna, which give nations their character and biodiversity.[39]

Similar to other rights, the right to culture is also closely linked to the enjoyment of other rights. Thus, states have a duty to implement their obligations under Art 15(1)(a) together with other obligations. While account must be taken of national and regional particularities and historical, cultural, and religious backgrounds, it is the duty of states to protect *all* human rights. Thus, "no one may invoke cultural diversity to infringe upon human rights guaranteed by international law, nor to limit their scope."[40] The Committee recognizes women, children, older persons, persons with disabilities, minorities, migrants, indigenous peoples, and people living in poverty as requiring special protection. The provisions on indigenous peoples are worth reproducing here which recognize the strong communal dimension of indigenous peoples' cultural life:[41]

> Indigenous peoples' cultural values and rights associated with their ancestral lands and their relationship with nature should be regarded with respect and protected, in order to prevent the degradation of their particular way of life, including their means of subsistence, the loss of their natural resources and, ultimately, their cultural identity. States parties must therefore take measures …. protect the rights of indigenous peoples to own … and use their communal lands, territories and resources, and, where they have been otherwise

inhabited or used without their free and informed consent, take steps to return these lands and territories.[42]

It further recognizes the importance of their traditional knowledge, access to genetic resources, seeds, medicines and knowledge of fauna and flora and the need to respect the principle of free, prior, and informed consent (FPIC) of indigenous peoples in all matters covered by their specific rights.[43]

The Committee notes that the States are obliged to protect indigenous peoples from illegal and unjust exploitation of their lands, territories and resources by state entities or private or transnational enterprises.[44] States must encourage the participation of minorities, indigenous peoples, and other communities in the design and implementation of laws and policies that affect them and obtain their free, informed, and prior consent when their cultural resources and cultural expression, are at risk.[45]

The ESCR Committee is currently working on two General Comments that are relevant to the present discussion: draft GC No 26 on land and ESC rights,[46] and GC on sustainable development and ESC rights.[47] The draft GC No 26 is currently open for comments. The draft notes that access to land is an important precondition for the realization of several rights in the Covenant including adequate food, water and housing, and health. Moreover, the sustainable use of natural resources depends on how people have access to land and how land use is governed.[48] With environmental degradation and climate change, the availability of land is diminishing, while the increasing populations require more food. Land tenure systems are becoming important in this context. However, competition for land is increasing in both rural and urban areas with competing uses, large-scale development projects, and urbanization. Other factors that compound the problem include unsustainable agronomic practices, unsustainable land management, climate change, growing demand for agricultural commodities for food, bioenergy, fiber and feedstock, and large-scale industrial agriculture. Moreover, land has cultural value. The GC refers to the obligations of states, especially non-discrimination and equality, participation, consultation and transparency, and obligations to respect, protect and fulfill including extraterritorial obligations to respect, protect, and fulfill.

The draft GC notes that the impact of climate change on access to land will be severe in many countries and sea level rise will impact housing, agriculture, and fisheries. Plus, extreme weather events such as droughts and floods will affect access to land. States should thus comply with their commitments under the Paris Agreement and design adaptation policies taking into consideration land use changes induced by climate change, and use maximum available resources to address the impacts of climate change, particularly on disadvantaged groups.[49] Main contributors to climate change should assist those countries that are most affected but are unable to cope with the impacts. States must also support mitigation and adaptation measures with environmental and social safeguards to ensure that human rights are not affected, provide access to information and facilitate consultations, and respect the free, prior, and informed consent of indigenous peoples.[50]

With regard to the second proposed General Comment,[51] the Committee notes that although sustainable development is not explicitly mentioned in the Covenant, its significance was recognized in the 1992 Rio Declaration on Environment and Development[52] and is at the heart of the 2030 Agenda and SDGs. The drafting group has identified ten key themes to be explored in the GC including: natural resource governance, equity in access and limitations in the use; environmental degradation and biodiversity loss; climate change; gender equality; indigenous peoples, peasants and others in rural areas; private actors; and the relationship between sustainable development and key concepts in the Covenant.[53] It has prepared an issue paper as a discussion guide and will continue regional consultations with academics, civil society, practitioners and experts.

6.3 Concluding Observations

By ratifying the ICESCR, state parties have agreed to submit reports on the measures they have adopted and the progress made in protecting the rights recognized in the Covenant as well as the difficulties in fulfilling the obligations and the factors that contribute to those difficulties. This section discusses the Concluding Observations of the ESCR Committee as they relate to climate change, sustainable development including SDGs and environmental pollution. In addition, reference will be made to indigenous rights and the free, prior, and informed consent principle, environmental impact assessments, disposal of natural resources and preservation of forests and related issues.

6.3.1 *Climate Change*

The majority of the Concluding Observations from around 1999 make references to climate change, at times in relation to indigenous peoples' rights[54] and the SDGs.[55] In a few instances, reference was made to climate change and extraterritorial obligations.[56] In this section, we will discuss some of the noteworthy statements that the Committee made in its concluding observations.

On the report of Norway, the Committee expressed concern about the licenses Norway had issued for the exploration and exploitation of petroleum and natural gas reserves in the Arctic Ocean and the Barents Sea and their impact on global warming.[57] It noted Norway's contribution to the Green Climate Fund and its efforts to meet its 2030 targets to reduce its GHG emissions. The Committee recommended that Norway:

> [i]ntensify its efforts to achieve its nationally determined contribution under the Paris Agreement to reduce emissions by at least 50 per cent and towards 55 per cent compared to 1990 levels by 2030, and to promote alternative and renewable energy sources. It also recommends that the State party reconsider its decision to increase oil and natural gas exploitation and take its human

> rights obligations as a primary consideration in its natural resource exploitation and export policies.[58]

The reference to the Paris Agreement, the recommendation to reduce GHG emissions by 50% compared to 1990 levels by 2030, promote alternative energy sources, and to reconsider the decision to increase oil and gas exploitation are significant especially given that they are made by a human rights body.

Similar sentiments were expressed in its Concluding Observations on the report of Switzerland.[59] The Committee expressed concern that Switzerland was not making an effort to meet the GHG reduction target for 2020 and that the emission reduction target of 50% by 2030 compared to 1990 is not compatible with the mitigation objectives set by the international community. It recommended that:

> the State party intensify its efforts to meet its greenhouse gas emission reduction targets for 2020 and raise the target for 2030 so that it is consistent with the commitment to limit temperature rise to 1.5°C. The Committee also recommends that the State party take the measures necessary to reduce public and private investment in the fossil fuel industry and to ensure that such investment is consistent with the need to reduce greenhouse gas emissions.[60]

The Committee discussed climate change and extraterritorial obligations on the report of Ecuador.[61] While taking note of the mitigation and adaptation measures included in Ecuador's first NDC, the Committee expressed concern that the increase in extractive activities "runs counter to the State party's commitments under the Paris Agreement"[62] and would have a negative impact on the enjoyment of ESC rights by people and future generations. It recommended that Ecuador reconsider the increase in oil development and large-scale mining *in light of its commitments under the Paris Agreement*.[63] This recommendation is significant because of the reference to the Paris Agreement. The Committee also encouraged Ecuador to promote alternative and renewable energy sources, reduce GHG emissions, and set national targets with time-bound benchmarks, drawing attention to its 2018 statement on climate change.[64]

Similarly, the Committee referred to climate change and extraterritorial obligations on the report of Argentina.[65] It expressed concern over the plans for large-scale exploitation of unconventional fossil fuels through hydraulic fracturing because this would consume a significant percentage of the entire global carbon budget for achieving the 1.5°C target in the Paris Agreement. It further expressed concern that the hydraulic fracturing plan would run counter to the commitments under the Paris Agreement and would have a negative impact on global warming as well as on the enjoyment of ESC rights by people and future generations. The Committee thus recommended that Argentina reconsider the large-scale exploitation of unconventional fossil fuels to ensure compliance with the obligations under the Covenant in light of the commitments under the Paris

Agreement. It encouraged Argentina to promote alternative and renewable energy sources, reduce GHG emissions and set national targets with time-bound benchmarks[66] and drew attention to its 2018 statement on climate change.

On the report of Mauritius, the Committee noted that natural disasters, in the face of climate change, have had a serious impact on the enjoyment of Covenant rights, especially for those affected by sea level rise in the southern part of the country.[67] It recognized Mauritius' negligible contribution to climate change and recommended: strengthening its response to disasters and disaster risk reduction methods; adopting a human rights-based approach and establishing monitoring mechanisms; allocating financial resources; strengthening the disaster preparedness of communities, especially those living in the south; and seeking international support to mobilize financial and technical support "which *it is entitled* in mitigating and responding to the effects of climate change."[68]

Several points are of note here: the Committee recognized the negligible contribution of Mauritius to climate change; the need to adopt a rights-based approach to disasters and risk reduction methods; that Mauritius (and by implication other developing countries whose contribution to climate change is negligible) is entitled to receive financial and technical support. It is, however, unfortunate that the Committee continues to refer to these severe weather events as "natural disasters."

On the report of Cabo Verde, the Committee expressed concern about the impact that natural disasters and cyclical droughts have on the most vulnerable populations and their Covenant rights, which could be exacerbated by climate change.[69] It recommended strengthening policies and strategies for mitigating the impact of disasters and reducing the vulnerability of communities and whose livelihood is dependent on climatic conditions; improving infrastructure and structural safety of houses; and updating contingency and relocation plans in consultation with the communities concerned.[70] While the reference again to "natural" disasters is unfortunate, the emphasis on community participation on relocation plans and the impact on vulnerable communities is important.

The Committee expressed concern that climate change would affect the enjoyment of ESC rights by people in the Russian Federation[71] and noted that climate change in the Arctic region will also affect other parts of the territory and impact people including indigenous peoples living in that region. It encouraged the state party to monitor and provide information on the impact of climate change on the enjoyment of ESC rights and recommended increasing efforts to reduce GHG emissions and set national targets with time-bound benchmarks.[72]

The Committee made extensive comments and recommendations relating to climate change on the report of Australia:

> The Committee is concerned about the continued increase of carbon dioxide emissions in the State party … despite the State party's commitments as a developed country under the United Nations Framework Convention on Climate Change and the Kyoto Protocol, as well as its nationally determined contribution under the Paris Agreement. The Committee is also concerned

> that environmental protection has decreased in recent years as shown by the repeal of the emissions trading scheme in 2013, and the State party's ongoing support to new coal mines and coal-fired power stations. The Committee is also concerned that climate change is disproportionately affecting the enjoyment of Covenant rights by indigenous peoples.[73]

The Committee recommended revising climate change and energy policies, taking immediate measures to reduce absolute emissions of GHGs, and pursue alternative and renewable energy production. It also encouraged the state party to review its support of coal mines and coal exports and recommended addressing the impact of climate change on indigenous peoples and engaging them in policy design and implementation.[74]

On the report of Canada, the Committee voiced its concern that climate change is affecting the enjoyment of ESC rights by indigenous peoples[75] and noted that environmental regulations have been weakened, notably by the enactment of the Budget Bill C38 (2012) and in the context of extractive industries. The Committee recommended: addressing the impact of climate change on indigenous peoples and engaging them in policy design and implementation; ensuring that before the use of non-conventional fossil energies, affected communities are consulted and impact assessments carried out; pursuing alternative and renewable energy production; strengthening legislation and regulations in accordance with international human rights obligations; and ensuring that environmental impact assessments are carried out in relation to extractive industries.[76]

On several occasions, the Committee commended member states for their positive action on climate change. It commended Cambodia for launching a project for carbon credits for community forestry under the Clean Development Mechanism and the REDD program of the UNFCCC.[77] Similarly, it commended Argentina for adopting a law to protect and safeguard glaciers and periglacial areas[78] while the Committee welcomed the adoption of legislation on climate protection by Ukraine and giving effect to the Kyoto Protocol and the UNFCCC.[79] The Committee welcomed the establishment of the Climate Change Trust and the Climate Change Resilience Fund, and the adoption of the Climate Change Strategy and Action Plan of 2009 in Bangladesh.[80]

The Committee expressed concern about the impact of climate change on the enjoyment of ESC rights by people in Bangladesh, *despite the negligible contribution of the State party*.[81] It recommended that the state party integrate human rights into plans on climate change and disaster risk reduction with the participation of the affected communities and civil society. Moreover, the state party should strengthen international cooperation in order to mobilize financial and technological support *to which it is entitled in mitigating and responding to* the effects of climate change.[82] On the report of Mongolia, the Committee acknowledged the needs of people exacerbated by the extreme climatic conditions[83] and noted the growing food insecurity among vulnerable groups such as children, older persons, female headed households, pensioners, and small herders compounded by drought and harsh winter, which killed more than 2 million heads of livestock.[84]

6.3.2 *Environmental Pollution, Environmental Degradation, and Environmental Impact of Mining Operations and Extractive Industries*

The Committee addressed environmental pollution in several of its Concluding Observations, with issues ranging from mining operations to the use of pesticides, and from access to safe drinking water[85] to air pollution. A large number of comments had been made in relation to the impact of mining, deforestation, and development activities on indigenous peoples and their lands. These are discussed later in the chapter.

On the report of Guinea,[86] while noting the revisions to the 2013 Mining Code, the Committee expressed concern about the negative effects of extractive activities on the environment and the health of local communities and about populations being displaced without adequate compensation by mining and hydroelectric companies.[87] The Committee recommended: ensuring compliance with the Mining Code and that communities share in the economic benefits of the mining industry; preparing independent studies on the effects of extractive and hydroelectric activities on ESC rights prior to undertaking investment projects; guaranteeing the quality of water sources, and holding those in mining activities accountable for polluting water sources; and where property has been expropriated, ensuring that landowners and those with usage rights are consulted and paid adequate compensation.

On the report of Senegal,[88] the Committee commended the inclusion of the rights to a healthy environment, natural resources and land in their Constitutional Act of 2016 but expressed concern about the lack of measures taken to address atmospheric pollution, particularly in Dakar, dumping of household waste, and the discharge of wastewater in urban and peri-urban areas, and the pollution from the use and disposal of plastic bags.[89] The Committee encouraged the state party to take immediate steps to address environmental hazards that affect the health of the population by regulating polluting vehicles; improving wastewater monitoring and management; and enforcing the law on plastic bags.[90]

The Committee voiced concern about reports that Israel had given licenses to Israeli and multinational companies for oil and gas extraction and renewable energy projects in occupied territory without consulting the affected communities while prohibiting Syrians and Palestinians from accessing and developing their natural resources.[91] The Committee recommended immediately ceasing the issue of licenses for the exploitation of natural resources in the occupied territories. In addition, the Committee expressed concern about the hazardous impact of aerial herbicide sprayed by companies hired by the Ministry of Defense adjacent to the fence between Israel and Gaza, especially the impact on crops and soil. It further expressed concern about the restrictions imposed on Palestinians to access their agricultural land, water sources, irrigation facilities, and marine resources, and the confiscation of fishing boats, depriving Palestinians of their means of subsistence.[92] The Committee recommended carrying out an assessment of the impact of herbicide spraying on the livelihoods, health, food security of Palestinians and the environment and, based on the precautionary principle, cease such spraying and ensure that farmers and fishermen have free access to their land, irrigation facilities and marine resources.[93]

On the report on Kazakhstan, the Committee recommended preparing human rights and environmental impact assessments before entering into investment and trade agreements or licensing investments.[94] The need to conduct social, environmental and human rights impact assessments prior to granting licenses for extractive activities and providing fair and adequate reparation to those negatively affected by extractive activities were among the recommendations made on the report of the Russian Federation.[95]

The Committee voiced concern at the excessive amount of fluoride and boron in the water supply system in Estonia and the high radon content in groundwater in northern Estonia reported to be closely associated with cancer which may put those cleaning the filers at risk.[96] The Committee, drawing attention to GC No 15, recommended: ensuring that the content of fluoride and boron remains below the limit in all water supply systems; providing those exposed to radon-contaminated water with alternative drinking water sources; revisiting legal regulations on radon in drinking water to ensure compliance with EU and WHO standards; and enforcing laws on water treatment and monitor compliance.[97]

The Committee expressed concern at the lack of instruments in Cameroon to ensure the protection of ESC rights when granting permits for projects for economic development or exploitation of natural resources.[98] It referred to reports on the negative impacts of these projects on the traditional lifestyles of population groups, including indigenous peoples, and on their access to land, food supply, and an adequate standard of living. The Committee recommended adopting guidelines for evaluating the impact of development projects on ESC rights and the environment, including those implemented by private actors, and especially those involving the territories and natural resources of indigenous peoples.[99]

On the report of Mali, the Committee expressed concern about the negative impacts of mining operations that cause irreversible damage to the environment and infringe the right to health and an adequate standard of living of affected communities.[100] It recommended developing guidelines for assessing the impact of mining projects on human rights and the environment; providing compensation to those affected by mining activities and ensuring they benefit from the revenue from such activities; and demanding mining companies to take steps to prevent water and air pollution and soil degradation resulting from their activities and to reclaim damaged areas.[101]

Similar concerns were voiced on the report of Argentina:

> The Committee is concerned about the negative impact that the use of specific methods of unconventional oil and gas exploitation, such as hydraulic fracturing, may have on the environment, water, health and the risk of earthquakes, and about the fact that the local impact of these forms of exploitation has not been adequately assessed and the local populations have not been duly consulted.[102]

The Committee recommended that the state party adopt a regulatory framework for hydraulic fracturing, including impact assessments, consultations with the communities with appropriate documentation of its effects on air and water pollution, radioactive emissions, occupational health and safety, effects on public health, noise and light pollution, potential to trigger seismic activity and threats to agriculture, soil quality, and climatic system.[103] Furthermore, the Committee expressed deep concern about the increased use of pesticides and herbicides that contain glyphosate, despite their serious adverse effects on human health and the environment and despite WHO's identification of glyphosate as a probable carcinogen. It recommended adopting the precautionary principle as part of its regulatory framework with regard to the use of pesticides and herbicides, in particular those that contain glyphosate to avoid negative health impacts and environmental degradation.[104]

The Committee addressed exploitation of natural resources on the report of Colombia and commended the State party for setting up consultations with communities that may be affected by projects that exploit natural resources, including agro-industrial projects.[105] However, despite the opposition of the communities, the projects go ahead and the Committee voiced concerned about the adverse effects of these activities on the environment, including deforestation which has a negative impact on the enjoyment of ESC rights.[106] It recommended that the state party give due weight to the public concerns; undertake social and environmental impact studies on activities that exploit natural resources; ensure that licensing agreements with private entities include measures to mitigate their impact on the enjoyment of ESC rights; provide adequate compensation to affected communities; and take appropriate measure to preserve forests.[107] Moreover, the Committee expressed concern over the disproportionate use of water for mining activities, and the pollution of rivers which has resulted in serious violations of rights to water, food, and health and caused damage to the environment.[108] It recommended, drawing attention to GC No 15, that the state party take measures to protect water resources and ensure the use of water by mining industry does not jeopardize access to safe drinking water, particularly for affected communities and that water is properly treated by the mining companies.[109]

Soil and groundwater pollution caused by landfill methods of waste disposal as well as extensive use of agricultural chemicals and persistent organic pollutants, including pesticides and insecticides were among the topics addressed by the Committee on the report of the Republic of Moldova.[110] It recommended that the state party use environmentally-friendly methods to manage, monitor, collect, and treat waste as an alternative to landfill; adopt appropriate waste-recycling programs; and promote sustainable consumption and production patterns. It urged the state party to regulate and monitor the use of harmful chemicals in agriculture.[111]

The Committee expressed concern about the high use of agrochemicals seriously affecting the health of farming communities on the report of Sri Lanka.[112] Noting that although glyphosate pesticide was officially banned in 2015, it is still used in plantations, the Committee recommended urgently addressing the very high use of agrochemicals, taking steps to enforce the ban, and extending the ban to cover all agrochemicals that adversely affect the health of the population and the environment.[113]

On the report of the Netherlands the Committee addressed activities leading to pollution both domestically and abroad.[114] While welcoming the adoption of the national action plan on business and human rights, the Committee expressed regret that the plan only contains guidelines for companies operating abroad and does not provide for formal monitoring mechanisms. It referred to the reports of damage to people's homes in Groningen due to gas extraction, the polluting oil refinery in Curacao, serious damage to the environment, and to the livelihoods of indigenous peoples by a company domiciled in the state party. The Committee recommended, referring to GC No 24, that the state party take measures to ensure the safety and mental health of the people residing in the area of gas extraction in Groningen, provide compensation to the victims, prevent future damage, expedite an overhaul of the oil refinery industry in Curacao to avert pollution, and remove the legal obstacles to holding companies accountable for violations of ESC rights resulting from their operations domestically and abroad.[115]

On the report of Burkina Faso, the Committee expressed alarm at the impact of businesses on the rights of rural communities, small farmers and agro-pastoralists that exploit the natural resources of their territories, especially the serious impact on rights to health, environment, education, housing and water, particularly in the Essakane community, where the mining company has not compensated the displaced populations.[116] The Committee recommended implementing the Environmental and Social Management Plan to ensure that no land is transferred to investors or used for economic development projects before conducting a full evaluation of the impact on the ESC rights of peoples and their free, prior, and informed consent obtained.[117] It further recommended that the Mining and Environment Code is enforced and that those who have their rights violated have access to remedies, as required under the Constitution.[118] The Committee also expressed concern at the lack of access to safe drinking water, particularly in rural areas, and the lack of a national strategy for the sanitation system, in particular with regard to the treatment and disposal of solid waste and wastewater. Drawing attention to GC No 15 and statement on the right to sanitation, the Committee recommended that safe and affordable drinking water and sanitation is provided for all, especially those in rural areas.[119]

The Committee expressed concern about continued land-grabbing in the Philippines and that the Comprehensive Agrarian Reform Program Extension with Reforms has been phased out, leaving many farmers landless while women farmers were among the minority who benefited.[120] The Committee recommended taking measures to stop land-grabbing, facilitating the distribution of land to landless farmers, and ensuring that women are not discriminated against in land distribution.[121] On the report of Kenya, the Committee voiced its concern at the lack of implementation of ESC rights of those internally displaced by disasters, development projects, and environment preservation projects.[122]

On the reports of Kyrgyzstan, the Committee was concerned about pollution and ground contamination[123] and the adverse effects on health and the environment caused by the high level of industrial pollution and by a wide range of sources and activities: ground contamination from uranium tailings, toxic waste dumps, and burial sites for pesticides that are persistent organic pollutants; import of obsolete, prohibited, or poor-quality pesticides; the location of a residential area near burial grounds for cattle with anthrax; and mining, including the operation of the world's last-known mercury mine in Khaidarkan. Concerned that people's awareness about health and environmental risks of radioactive and toxic substances was very low,[124] the Committee made extensive and detailed recommendations: relocating uranium tailings, pesticides that are persistent organic pollutants, and radioactive and hazardous waste; closing the mine in Khaidarkan and providing access to other economic activities in the region; ensuring that residential areas are not located near contaminated ground; reducing industrial pollution and enforcing prohibited pesticides; establishing normative frameworks on radiation and nuclear safety, pollution, and chemicals management, in line with international standards; and raising awareness on risks of radioactive and toxic substances and safety measures.[125]

On the report of Mongolia, the Committee discussed the issue of herders who lost their livestock and are engaged in artisanal and small-scale mining activities which are not regularized and have poor working conditions, involving the use of dangerous chemicals such as mercury.[126] The Committee recommended taking measures, including a legal framework, to protect their rights especially enforcing regulations on mercury-free processing plants and securing access to land and licenses for them. Concerned about the large amount of pesticides in vegetables, the Committee recommended taking measures to prevent and investigate cases of pesticides in food.[127] Regarding safe drinking water and sanitation, the Committee expressed concern about the disparity between rural and urban areas and about soil and water contamination caused by mining activities.[128] Referring to GC No 15 and the statement on sanitation, the Committee urged the state party to address this disparity and the impact of mining on water and soil.[129] Furthermore, the Committee was extremely concerned about the levels of air pollution leading to respiratory, heart, and lung diseases and recommended that the state party take measures to address air pollution and its health impacts and introduce new energy efficient heating systems.[130]

The Committee expressed concern at the shortcomings in the implementation of environmental regulations and industrial activity in Thailand.[131] Reminding the state party's obligations under Article 12(2)(b) of the Covenant, the Committee recommended that it regulate environmental protection and ensure enforcement of its environmental legislation, to prevent harmful health effects. On the report of Montenegro, the Committee expressed serious concern over Roma communities that live in deplorable conditions in environmentally polluted areas without access to electricity, running water, and sewage disposal.[132]

The Committee made extensive comments on water and sanitation, mining, and land-grabbing on the report of Indonesia.[133] It expressed concern that, in rural areas, about one quarter of the population does not have access to safe drinking water and sanitation despite the adoption of the National Policy for Community-Based Water Supply and Environmental Sanitation. Referring to GC No 15, the Committee called upon the state party to improve access to safe and clean drinking water and sanitation, particularly in rural areas. The Committee further expressed concern at human rights violations in mining and plantations sectors, including rights to livelihood, food, water, and labor and cultural rights and that the free, prior, and informed consent of affected communities is not always sought. However, even where consultations have taken place, their informed decisions have not been guaranteed.

The Committee noted that extractive projects have not brought about tangible benefits for local communities; and that there is inadequate monitoring for their human rights and environmental impact. Moreover, affected communities have not been afforded remedies, and together with human rights defenders, they have been subject to violence and persecution. The Committee called upon the state party to review legislation and practices in the mining and plantations sectors and guarantee legal assistance to communities during consultations on extractive projects with a view to ensuring their free, prior, and informed consent; ensure that license agreements are monitored for human rights and environmental impact during their implementation; guarantee legal assistance to communities complaining about human rights violations, and investigate allegations of breach of license agreements, and revoke licenses, as necessary; ensure that tangible benefits and their distribution are defined in license agreements; and engage in dialogue with human rights defenders, investigate allegations of reprisals, and bring perpetrators to justice.[134] Expressing concern at the large number of land disputes and cases of land-grabbing in the country, the Committee voiced its concern that the Presidential Regulation on Procurement of Land for Realizing Development for Public Interest has rendered individuals and communities vulnerable to land-grabbing. It urged the state party to adopt a land policy and establish an institution to oversee the settlement of land disputes; review laws and regulations which make communities vulnerable to land-grabbing; and get the involvement of the national human rights institutions and civil society.[135]

On the report of Uzbekistan, the Committee expressed concern that a large part of the population lacks access to sanitation and safe drinking water, especially regions affected by drought and the Aral Sea catastrophe. It noted that the pollution of water, an inadequate sewage system, and the contamination of soil due to the extensive use of chemicals in agriculture, have contributed to the increase in waterborne diseases and morbidity, and that airborne diseases are the second largest cause of mortality in the country.[136] The Committee urged the state party to invest more resources to improve safe water supply and sanitation, especially in rural areas, and to take measures to protect water, soil, and air from contamination and prevent and treat water and airborne diseases, and cooperate on water management with neighboring countries to find a viable solution to the Aral Sea catastrophe.[137]

The Committee expressed concern about the adverse environmental effects of industrial pollution and food contamination, and their impact on an adequate standard of living and health on the report of China.[138] It noted that, despite the measures to mitigate ecological degradation, environmental pollution and food contamination continue, their implementation and monitoring remain inadequate, and administrative authorities and private companies are not held accountable for violating environmental legislation. Urging the state party to address environmental threats that affect the health and the standard of living of the population, and the enjoyment of their ESC rights, the Committee recommended the enforcement of environmental regulations, imposing sanctions, and providing adequate compensation to those affected.[139]

On the report of Kuwait, the Committee noted the functions of the Directorate for the Environment, but regretted lack of information on how verbal complaints about pollution from residents near industrial sites are dealt with.[140] It recommended that the state party take measures, including engagement with residents and civil society organizations, to find solutions to exposure to air pollution and other environmental concerns. While the state party has taken steps to protect archaeological sites on Failaka Island, the Committee was concerned at the risk posed by development projects to these sites. It recommended implementing laws and regulations to protect historical sites, assessing the impact of development projects on their conservation, and facilitating access to cultural heritage by the general population.[141]

On the report of Iran, the Committee commented on the adverse environmental effects of the river diversion program, sugar-cane farming and industrial pollution in Khuzestan province and their negative impact on an adequate standard of living and health.[142] It recommended taking urgent steps to address the impact of river diversions and industrial activity in Khuzestan, including environmental pollution and water shortages, on agriculture and human health.[143]

The Committee voiced its concern about the environmental impacts of mining and agribusiness projects and their effects on water in rural areas on the report of Ecuador[144] and recommended that the state party establish measures to protect the environment and the right to water. It also expressed concern about corporate land purchases and their impact on landownership by campesinos, the living conditions in high-risk zones where basic services are not guaranteed, and about the effects of "land trafficking."[145] The Committee recommended developing land titling plans to safeguard campesinos' ownership of land and establishing mechanisms to prevent forced sales in rural areas. Moreover, when resettling families located in high-risk areas the state party should respect the right to due process by providing the families full information about the conditions of their resettlement.[146]

The Committee expressed concern about the construction and operation of the Gilgel Gibe III hydroelectric dam and its impact on the traditional practices and means of subsistence of indigenous peoples who rely on the Omo River, endangering local food security in Ethiopia.[147] It urged the state party to address the adverse social and environmental impact of the dam and to initiate, prior to the construction of hydroelectric projects, comprehensive impact assessments and consultations with affected communities, and providing opportunities to present views and influence decision-making.[148]

The Committee discussed the relationship between free trade/economic agreements and ESC rights on the report of the Dominican Republic.[149] It noted that information about the impact of the US-Dominican Republic Central America Free Trade Agreement (CAFTA) and the Economic Partnership agreements with the EU on ESC rights was not available and that *investment agreements are being approved before environmental and social impact assessments are carried out.*[150] It encouraged the state party to take into account its obligations under the Covenant when negotiating with investors and agreements on economic partnerships to ensure that ESC rights, particularly of disadvantaged and marginalized groups, are not undermined and that environmental and social impact assessments are carried out before investment agreements are approved.[151]

The impact of mining activities on the environment and ESC rights was the subject of Committee's comments on the report of the Democratic Republic of the Congo.[152] It expressed concern that, despite the adoption of a mining code (2002) and a mining plan (2004), and the review of all mining contracts, illegal exploitation and mismanagement of natural resources continue with the involvement of foreign companies. The Committee noted that in the resource-rich province of Katanga, mining activities continue to the detriment of the rights of people who remain extremely poor and are deprived of basic social services and infrastructure. It voiced concern about the lack of transparency surrounding the revision of mining contracts and the granting of new contracts, such as the exclusive concessions granted for uranium extraction.[153] The Committee urged the state party to take measures to ensure that its natural resources are not subjected to illegal exploitation and mismanagement; review the mining contracts in a transparent and participatory way; repeal contracts which are detrimental to the Congolese people; and ensure that future contracts are concluded in a transparent manner. The Committee encouraged the state party to implement the Extractive Industries Transparency Initiative (EITI); adopt measures to control the export of minerals; impose sanctions on those involved in the illicit trade in natural resources; allocate revenue from the mining sector for the development of the province of Katanga; and provide its inhabitants with basic social services and infrastructure to improve their living conditions.[154]

The Committee discussed deforestation in Cambodia and noted that according to the Mid-Term Review of the National Strategic Development Plan 2006–2010 a logging moratorium has been imposed and over 200,000 hectares of forest land have been reclaimed from land-grabbing and encroachment.[155] However, it expressed deep concern over the FAO global forest survey that estimated that the state party had lost 29% of its tropical forest cover over the last five years, with the continuing destruction of the Prey Long forest. The rapid increase in economic land concessions even within protected zones is a major factor in the degradation of natural resources, adversely affecting the ecology and biodiversity, resulting in the displacement of indigenous peoples from their lands without just compensation and resettlement, and the loss of livelihood for rural communities who depend on land and forest resources for their survival.[156] The Committee urged the state party to review its policy on converting protected zones into economic concessions, and conduct environmental and social impact assessments for all economic activities including mining and oil explorations, including consultations with relevant stakeholders and communities with due regard to their right to participate. Moreover, the granting of economic concessions must ensure sustainable development for all Cambodians who should share in the benefits of progress and the state party should provide, in its next periodic report, information on the progress made.[157]

On the report of Nigeria, the Committee expressed alarm at the devastation that oil exploration has caused to environment and the quality of life in those areas, including Ogoniland where oil has been extracted without due regard for the health and well-being of the people and their environment.[158]

The Committee discussed forced evictions associated with development projects such as Three Gorges Project in China, forced evictions and demolitions in connection with the 2008 Olympic Games[159] and of historic structures in Lhasa, Tibet. The Committee recommended undertaking consultations with affected residents before implementing development projects[160] and requested the state party to include, in its next periodic report, information on its environmental policies, in particular, on reducing atmospheric pollution, and to evaluate the environmental impact of large infrastructure development projects.[161]

On the report of Ecuador, the Committee expressed concern about the health and environmental impacts of natural resources extraction by companies at the expense of land and cultural rights of indigenous communities and the equilibrium of the ecosystem.[162] The Committee recommended implementing legislative and administrative measures to avoid violations of environmental laws and rights by transnational companies.[163] It urged the state party to implement laws on occupational health and safety of workers, and to adopt measures to protect workers from the use of toxic substances in the banana-growing and small mines sectors.[164]

The Committee expressed concern at the destruction of the environment by widespread pollution on the report of the Russian Federation[165] and the high level of contamination of both domestically produced and imported foodstuffs. For domestic production, the contamination and pollution appear to be caused by the improper use of pesticides and the improper disposal of heavy metals and oil spills. For imported food, the contamination appears to be caused by the illegal practices of some food importers and noted that the government must ensure that such food does not reach the market.[166] The Committee also expressed alarm at the extent of environmental problems and the industrial leakage of harmful waste products is so severe in some regions that they could be considered environmental disaster areas.[167] Moreover, lack of adequate funds to modernize the water delivery system has affected access to clean water. The Committee recommended taking action to protect indigenous peoples from exploitation by oil and gas companies and ensuring access to traditional and other sources of food. It pointed out that the food supply is linked to a seriously polluted environment and lack of investment in infrastructure to improve the water supply, and the state party should examine these linkages and take action to clean up the environment and prevent enterprises from engaging in further pollution, especially the food chain. The state party should improve the water supply system as a matter of priority and take action against enterprises which have imported contaminated food.[168]

On occasion, the Committee praised state parties for undertaking positive measures. It commended Tunisia's efforts on environmental protection, including the framework of the Ninth Economic and Social Development Plan (1997–2001) and the increased budget which will be used, *inter alia*, for the development of equipment to combat pollution, waste management, the use of waste water in agricultural production, and for combating desertification.[169] Similarly, the Committee noted that Egypt had undertaken new measures to introduce environment- and health-friendly fuel, beginning with the public transportation systems in major cities where pollution is a grave threat to life and health.[170] Likewise, the Committee noted the efforts of Mauritius to ensure better education, health, nutrition, and a cleaner environment.[171]

6.3.3 *Sustainable Development and Sustainable Development Goals*

The Committee referred to sustainable development and SDGs on several occasions. For example, on the report of Guinea, the Committee took note of the Accelerated Food and Nutrition Security and Sustainable Agriculture Programme 2016–2020 but noted that a large number of people are facing food insecurity. Moreover, access to safe drinking water and sanitation remains a problem, particularly in rural areas.[172] The Committee recommended, referring to the UN Declaration on the Rights of Peasants and Other People Working in Rural Areas,[173] to guarantee all peasants access to programs and agroecological solutions and build resilience of agriculture to environmental shocks, taking account of SDG target 1.5.[174] Referring to GC No 12 the Committee recommended ensuring access to safe drinking water and sanitation throughout the country.

The Committee referred to SDGs in the context of social security on the report of South Africa and noted that social protection floors is an essential element of the right to social security and SDGs.[175] On the issue of addressing poverty in Burkina Faso, the Committee recommended redoubling efforts to combat poverty, "by ensuring that the Strategy for Accelerated Growth and Sustainable Development is implemented from a rights-based perspective" and giving priority to the most disadvantaged and marginalized groups.[176] The Committee referred to sustainable development in the context of development assistance on the report of France.[177] Expressing concern that its official development assistance had fallen short of the target of 0.7% of gross national product (GNP), the Committee voiced concern that due diligence requirements such as those relating to sustainable development and to the mitigation of environmental and social risks, do not provide protection for the ESC rights. The Committee recommended developing methodological tools for analyzing the impact of operations funded by development agencies on the enjoyment of ESC rights.[178]

On the report of Mali, the Committee noted the adoption of the Framework for Economic Recovery and Sustainable Development,[179] the adoption of National Action Plan on Achieving the Sustainable Development Goals, and the National

Programme on Socioeconomic Development (2011–2030) in Turkmenistan,[180] and welcomed the success achieved in the promotion of sustainable *human* development in Tunisia[181] and stated with regard to Agenda 2030:

> The Committee recommends that the State party take fully into account its obligations under the Covenant and ensure the full enjoyment of the rights enshrined therein in the implementation of the 2030 Agenda for Sustainable Development at the national level, with international assistance and cooperation when needed. Achievement of the Sustainable Development Goals would be significantly facilitated by the State party establishing independent mechanisms to monitor progress and treating beneficiaries of public programmes as rights holders who can claim entitlements. Implementing the Goals on the basis of the principles of participation, accountability and non-discrimination would ensure that no one is left behind. In this regard, the Committee draws the attention of the State party to its statement on the pledge to leave no one behind (E/C.12/2019/1).[182]

It is noteworthy that this paragraph appears in 46 of the concluding observations adopted since 2015. A different formulation appears in the report on Finland:

> The Committee notes the State party's assessment regarding the progress in the implementation of the 2030 Agenda for Sustainable Development in areas relating to social sustainability, the economy and work. The Committee also renews its recommendation to take fully into account obligations under the Covenant and the objective of full realization of the rights enshrined therein in the implementation of the 2030 Agenda at the national level, including in the recovery from the COVID-19 pandemic. Moreover, the Committee recommends that the State party support the global commitment of the decade of action to achieve the Sustainable Development Goals.[183]

6.3.4 Indigenous Rights

The Committee discussed indigenous peoples and their rights in several concluding observations. On the report of Ecuador, the Committee voiced concern about the increase in mining concessions awarded in indigenous territories and relaxing the rules governing extractive activities in the buffer zone of the Yasuní National Park protected area, which is home to indigenous peoples, who live in voluntary isolation.[184] The Committee recommended: ensuring indigenous peoples' legal security to their traditional lands and natural resources and adequate consultation and obtaining their free, prior, and informed consent on the establishment of protected areas; and preventing hydrocarbon activities in the Yasuní National Park protected area and its buffer zone.[185] The Committee expressed further concern at the environmental impact of large-scale mining and other extractive activities and at the lack of information on measures to ensure the right to water, especially in the context of the spraying and extractive activities at the northern border. It recommended taking measures to protect communities affected by environmental degradation, such as rural, Afro-descendant and indigenous communities in Esmeraldas, and guaranteeing their enjoyment of the Covenant rights.[186]

The Committee expressed concern about the failure to respect the right of indigenous peoples and communities of African descent to prior consultation on decisions that affect them and recommended: updating regulations in consultation with indigenous peoples to ensure the right to be consulted and free, prior, and informed consent, in accordance with international human rights standards; guaranteeing the prior consultation of people of African descent; enforcing the judgments of the provincial courts; and establishing, in consultation with indigenous peoples, a mechanism to implement the recommendations of the Special Rapporteur on Indigenous Peoples.[187]

The Committee welcomed the adoption of the Traditional and Khoi-San Leadership Bill in South Africa which recognized the Khoi and the San as indigenous peoples.[188] However, it noted that indigenous peoples continue to be marginalized and discriminated against, affecting their ESC rights. The Committee recommended ensuring the enjoyment by indigenous peoples of ESC rights, their participation in the implementation of the above Bill and other matters concerning them, and ratifying ILO Convention No 169.[189]

The Committee expressed concern at the lack of compliance with the results of the land survey of indigenous communities, the demarcation of their territories and the prohibition of evictions from their lands on the report of Argentina.[190] No mechanisms have been established for the titling of land traditionally occupied by indigenous peoples and more than 120,000 hectares, many of which are in protected forests, were cleared in 2017 despite complaints from indigenous communities. The Committee recommended the implementation of relevant laws, completing the demarcation process and granting community land titles to indigenous communities; with the participation of indigenous peoples, undertake an assessment of the risk to ESC rights, particularly the right to health, before authorizing the exploitation of natural resources; providing the National Institute of Indigenous Affairs its own budget and human resources to complete the land survey; and expediting the demarcation of indigenous territories in the Great Reserve of El Impenetrable in Chaco Province.[191] The Committee expressed concern about the repeated violation of the right of indigenous peoples to be consulted and to free, prior, and informed consent, particularly in relation to extractive activities and lack of information about reparations when their rights have been infringed.[192] The Committee recommended:

> [t]hat the State party ensure that indigenous peoples are systematically consulted for the purpose of obtaining their free, prior and informed consent at the national and provincial levels before concessions are granted to State-owned companies or third parties for the economic exploitation of the lands and territories traditionally occupied by those peoples. In particular, the Committee recommends that, for the implementation of the right to be consulted and to free, prior and informed consent, the State party use

> the protocols drawn up and agreed upon with indigenous peoples, in order to ensure that factors specific to each people and each case are taken into account.[193]

On the report of New Zealand, the Committee expressed concern that the Treaty of Waitangi is neither legally enforceable nor referred to in the Constitution Act,[194] and the recommendations of the Waitangi Tribunal are frequently ignored by the Government.[195] Limited effort had been made to ensure the participation of Māori in decision-making concerning laws that impact their rights, including land and water rights. Moreover, the principle of free, prior, and informed consent is not systematically implemented, especially in relation to development and extractive activities carried out on Māori territories. The Committee's recommendations included: implementing, in partnership with Māori representative institutions, the recommendations of the Constitutional Advisory Panel and the proposals in the 2016 Matike Mai Aotearoa report; implementing the recommendations of the Waitangi Tribunal including its landmark report Ko Aotearoa Tēnei; developing a national strategy to bring legislation and public policy in line with UNDRIP; considering the Waitangi Tribunal recommendations on engaging with Māori, and adopting mechanisms to ensure their participation in decisions affecting their rights. Moreover, the state party should include provisions on transparency and public participation in its trade strategy, and ensure that its climate change policies are developed in partnership with Māori, through their effective participation in the Climate Commission; obtain the free, prior, and informed consent of indigenous peoples, especially with regard to extractive and development activities, and conduct social, environmental, and human rights impact assessments prior to granting licenses for extractive and development activities; and consider ratifying ILO Convention No 169.[196]

Noting that climate change is disproportionately affecting ESC rights of indigenous peoples in Australia, the Committee recommended addressing the impact of climate change on indigenous peoples and engaging with them in policy and program design and implementation.[197] The Committee voiced its concern that indigenous peoples lack constitutional recognition[198] and experience disadvantage across all socioeconomic indicators; and that consultations with them on programs and policies that affect them are inadequate and there is insufficient compliance with the free, prior, and informed consent principle in relation to the White Paper on the development of Northern Australia, with regard to the approval of extractive projects on indigenous lands. The Committee recommended constitutional recognition of indigenous peoples; implementing programs to respect, protect, and realize the rights of indigenous peoples, in consultation with indigenous representative bodies and civil society; incorporating the free, prior, and informed consent principle in the Native Title Act 1993 and other legislation; applying the principles in UNDRIP; and ratifying ILO Convention No 169.[199]

The Committee, while noting the efforts made by the Philippines to protect the rights of indigenous peoples, expressed concern at the conflicts between the protection of indigenous peoples' ancestral lands and the provisions of the 1995

Mining Act and the 1974 Forestry Reform Code, and the delay in adopting the National Land Use Bill.[200] It was also concerned about the failure to uphold the right to free, prior, and informed consent of indigenous peoples regarding their lands and territories and to implement the mandatory representation of indigenous peoples in local decision-making bodies, and the displacement of indigenous peoples due, *inter alia*, to extractive and logging operations. The recommendations included: implementing the 1997 Indigenous Peoples' Rights Act to ensure that, in accordance with UNDRIP, indigenous peoples' rights to their lands and resources are protected; their free, prior, and informed consent is obtained in respect of any legislation, policy or project affecting their lands and resources and before granting licenses to private companies; adopting the National Land Use Bill; ratifying ILO Convention No 169;[201] and ensuring that indigenous peoples are represented on local decision-making bodies, such as local mining boards and development units.[202]

The Committee expressed concern about Sami land rights on the report of Sweden[203] and noted that the Sami people encountered barriers to their rights including access to their ancestral lands and their traditional ways of living which was exacerbated by the increase in extractive and development projects on Sami lands. The Committee recommended ensuring that all Sami, including non-reindeer-herding Sami, enjoy equal access to water and land and adopting legislative changes to avoid distinctions among the Sami people; obtaining their free, prior, and informed consent on decisions that affect them and provide legal assistance; reviewing legislation especially the Minerals Act, the Minerals Strategy, and the Environmental Code that regulate activities including development projects and extractive industries; adopting the Nordic Sami Convention and domestic solutions; and ratifying ILO Convention No 169.[204]

While the Committee welcomed the drafting of legislation on free, prior and informed consultation in Honduras, it expressed concern that indigenous peoples were not properly involved in that process and their right of prior consultation with a view to obtaining their free, prior and informed consent was not respected when concessions are granted for the exploitation of natural resources or other development projects.[205] Noting that indigenous peoples' right to dispose of their lands and natural resources was not protected, the Committee recommended involving indigenous peoples in preparing the draft law; ensuring that this legislation complies with international standards, including ILO Convention No 169 and UNDRIP; consulting with indigenous peoples with a view to obtaining their free, prior, and informed consent in decisions that affect their ESC rights; and taking measures to safeguard their right to dispose of their lands and natural resources.[206] The Committee also expressed concern about natural resource exploitation projects that are causing irreparable damage to the environment and the health of people, in particular of indigenous and Afro-Honduran peoples and recommended adopting regulations to evaluate the social and environmental impact of natural resource exploitation projects, consulting affected communities and ensuring that they receive compensation for losses incurred and a share of the profits from these activities.[207]

The Committee reiterated its concern about Ecuador's failure to obtain free, prior, and informed consent of indigenous peoples for natural resource development projects that affect them and that the Executive Decree of August 2012 was issued without consulting them.[208] It urged the state party to engage in consultations regarding mining and hydrocarbon resource exploration and development and provide opportunities to implement cultural safeguards and compensatory remedies. It recommended suspending the implementation of Executive Decree, working with indigenous peoples to design the law, and holding prior consultations on the proposed legislation. The Committee urged the State party to comply with the Inter-American Court of Human Rights' decision in *Kichwa de Sarayaku v. Ecuador*[209] and expressed concern at the lack of mechanisms for consensus-building around development activities that could reconcile them with the world views of indigenous peoples and nationalities.[210]

On the report of Argentina, the Committee expressed concern about the negative consequences of lithium exploitation in Salinas Grandes on the environment, access to water, way of life, and subsistence of indigenous communities.[211] The Committee called upon the state party to ensure the protection of indigenous communities from mining exploration and exploitation projects. With regard to Salinas Grandes, the Committee urged the state party to comply with the decision of the Supreme Court and referred to its statement on the obligations of States parties regarding the corporate sector and ESC rights.[212] The Committee also voiced concern about the increased use of chemical pesticides and transgenic soya seeds in regions traditionally inhabited by indigenous communities and their impact on the food security of these communities. Noting that the scale of deforestation has forced indigenous peoples to leave their traditional territories and many activities are carried out without prior consultation with the affected groups, the Committee recommended the full implementation of laws on the protection of non-renewable resources with a view to combating deforestation.[213]

6.3.5 Human Rights Defenders

The Committee referred to human rights defenders on a few occasions. It voiced concern about the security of human rights defenders in Ecuador, especially those defending ESC rights, environmental rights, and the rights to territory and natural resources.[214] It recommended, referring to its 2017 statement on defenders,[215] that the state party adopt measures to protect defenders including indigenous peoples, people of African descent, and Montubio people and prevent violations of their rights, particularly in the context of extractive activities.[216]

The Committee referred to reports of human rights defenders being threatened and harassed especially those working to protect ESC rights and the environment in South Africa and recommended that the state party provide them a safe environment to protect ESC rights by ensuring that all cases of intimidation, harassment and violence against human rights defenders are investigated and the perpetrators brought to justice.[217] The Committee voiced concern about the criminal convictions of social and indigenous leaders who protested against bills on water management and development projects that would impact natural reserves such as that of Lake Kimsakocha. It recommended adopting safeguards, inter alia, in relation to the freedom of assembly and the right to participate in peaceful demonstrations.[218]

6.4 Resolutions, Statements, and Press Releases

In response to the IPCC's 1.5 Degree report[219] the Committee issued a statement welcoming the report and noting that it demonstrated that "climate change constitutes a massive threat to the enjoyment of economic, social and cultural rights."[220] It noted that while states have already made pledges, their human rights obligations should guide them in designing and implementing measures to address climate change. It further noted that climate change already affects, in particular, rights to health, food, water and sanitation, and will worsen in the future. It is anticipated that climate change will result in nearly 38,000 deaths per year as of 2030 and nearly 100,000 deaths per year as of 2050, with the largest impacts to be felt in Southeast Asia.[221]

The Committee issued a statement on the Pledge to "Leave no one Behind,"[222] a central principle that guides the 2030 Agenda and the 17 SDGs. Noting that 2030 Agenda is guided by the principles of the UN Charter and other human rights instruments, the Committee pointed out that the ICESCR is a fundamental pillar[223] which provides protection of ESC rights for all, especially disadvantaged and marginalized groups, and is based on the inherent dignity of the human person. Noting that systemic discrimination arises in a variety of contexts and due to a variety of factors and circumstances, the Committee recognized that affected groups include, among others, people living in poverty; *nations and communities vulnerable to climate change and environmental pollution and degradation*; indigenous peoples; women; refugees, internally displaced persons and migrants; persons with disabilities; the elderly; and children.[224] It stressed that by respecting, protecting, and fulfilling the rights in the Covenant, States will be able to fulfill their pledge to ensure that no one is left behind.[225]

The Committee held a Day of General Discussion on October 14, 2019 and prepared an Issues Paper on "state obligations under the International Covenant on Economic, Social and Cultural Rights and governance of land tenure."[226] Panel 6 was devoted to land under changing environmental conditions and climate change. Referring to the IPCC 1.5 Degree report, the Committee pointed out that how we use land has a decisive impact on climate mitigation and the Summary for Policymakers recognized the relationship between governance of land tenure and the ability for land to better fulfill its role in climate regulation.[227] The Committee also referred to the Intergovernmental Science-Policy Platform on Biodiversity and Ecosystem Services (IPBES) which referred to the importance of using customary practices and indigenous and local knowledge within local, tribal, or indigenous communities for sustainable land management.

6.5 Conclusion

There is a clear link between ESC rights and the quality of the environment. The enjoyment of many of the rights in the Covenant would be jeopardized in a degraded environment. Likewise, climate change and unsustainable development have a detrimental impact on the enjoyment of these rights. This chapter discussed how the ESCR Committee has addressed environmental pollution, climate change, sustainable development and SDGs, indigenous rights, and human rights defenders. As the discussion revealed, the Committee has made extensive comments and recommendations in relation to these issues and ESC rights. Of particular note are the references to the free, prior, and informed consent principle, and the need to hold consultations with a view to obtaining FPIC, not just in relation to indigenous peoples, but also in relation to other communities affected by decisions and activities especially by extractive industries carried out by private and multinational companies.

In relation to climate change, the Committee referred on several occasions to the commitments under the Paris Agreement and urged states to stop supporting and expanding fossil fuel exploitation as they run counter to the commitments under the Paris Agreement. It also addressed extraterritorial obligations of states and the need to control the activities of multinational companies abroad. While the Committee referred to national targets with "time bound benchmarks" on several reports, it did not elaborate on this.

The Committee stressed the need to adopt environmental principles, including the precautionary principle and environmental (and social and human rights) impact assessments on several occasions and admonished states for approving investment treaties before environmental and social impacts were carried out. It also referred to the need to comply with WHO standards, ILO standards, and international standards relating to indigenous peoples, especially those in UNDRIP. Significantly, the Committee stressed the need to provide adequate (and occasionally fair) compensation to individuals and communities whose rights were violated and to ensure that communities, especially indigenous communities, benefit from profits from activities on their lands.

Notes

1 ICESCR, adopted by UNGA resolution 2200A (XXI) on 16 December 1966, entered into force in 1976, https://www.ohchr.org/sites/default/files/cescr.pdf.
2 Ibid., art. 2(1).
3 See de Schutter, O. (2019) *International Human Rights Law: Cases, Materials, Commentary* (3rd edn.), Cambridge: Cambridge University Press, p. 297; and Tomuschat, C. (2014) *Human Rights: Between Idealism and Realism* (3rd edn.), Oxford: Oxford University Press, p. 139. While the Official UN position is that all rights are universal, interdependent, interrelated and indivisible (Vienna Declaration, 1993), literature often refers to "generations" of rights thereby implying a hierarchy.
4 Examples include Nepal, Switzerland, Portugal, Angola, and Bolivia. See Boyd, D. (2011) *The Environmental Rights Revolution: A Global Study of Constitutions, Human Rights, and the Environment*, Vancouver: UBC Press; Gellers, J. (2017) *The Global*

Emergence of Constitutional Environmental Rights, London: Routledge; May, J. & Daly, E. (2015) *Global Environmental Constitutionalism*, Cambridge: Cambridge University Press.

5 Examples include Ecuador and Bolivia.

6 ICESCR, supra note 1, art. 2.

7 Resolution 1985/17 (28 May 1985), https://documents-dds-ny.un.org/doc/RESOLUTION/GEN/NR0/663/73/IMG/NR066373.pdf?OpenElement; see https://www.ohchr.org/en/hrbodies/cescr/pages/cescrindex.aspx.

8 Under Part IV, state parties are required to submit their country reports to the UN Secretary General who then transmits copies to ECOSOC as well as to any specialized agencies. ECOSOC may transmit reports to the Commission on Human Rights (which was replaced by the HRC) for study and general recommendations on reports submitted by states.

9 *Mapping Human Rights Obligations Relating to the Enjoyment of a Safe, Clean, Healthy and Sustainable Environment, Individual Report on the International Covenant on Economic, Social and Cultural Rights, Report No 1* (December 2013) ["Mapping Report No 1"], https://www.ohchr.org/en/special-procedures/sr-environment/mapping-report.

10 *General Comment No 3*, E/1991/23 (14 December 1990), https://www.refworld.org/pdfid/4538838e10.pdf.

11 These were proposed by the International Law Commission in their seminal work on state responsibility. See ILC, *Responsibility of States for Internationally Wrongful Acts* (2001), https://legal.un.org/ilc/texts/instruments/english/draft_articles/9_6_2001.pdf.

12 GC No 3, supra note 10, ¶ 2.

13 Ibid., ¶¶ 6–7.

14 Ibid., ¶ 9.

15 Ibid., ¶ 10 (emphasis added).

16 Ibid. (emphasis added).

17 Ibid., ¶ 14.

18 *General Comment No 4* (1991), ¶ 8, https://tbinternet.ohchr.org/_layouts/15/treatybodyexternal/Download.aspx?symbolno=INT%2fCESCR%2fGEC%2f4759&Lang=en.

19 *General Comment No 12* (12 May 1999), https://undocs.org/E/C.12/1999/5.

20 Ibid., ¶ 7.

21 *General Comment No 14*, E/C.12/2000/4 (11 August 2000).

22 Ibid., ¶ 4 (emphasis added).

23 Ibid., ¶ 11.

24 Ibid., ¶ 15. The Committee cited in support Principle 1 of the Stockholm Declaration, UNGA resolution 45/94 on a healthy environment for the well-being of individuals, Principle 1 of the Rio Declaration and Article 10 of the San Salvador Protocol to the American Convention of Human Rights.

25 Ibid., ¶ 34.

26 Ibid., ¶ 51.

27 *General Comment No 15*, E/C.12/2002/11 (20 January 2003), https://undocs.org/E/C.12/2002/11.

28 Ibid., ¶ 8.

29 Ibid., ¶ 11. For the definition of sustainability, the General Comment relies on the Rio Declaration on Environment and Development (1992) and Agenda 21 adopted at the Rio Conference on Environment and Development.

30 Ibid., ¶ 16(c).

31 Ibid., ¶ 16(d).

32 Ibid., ¶ 21.

33 Ibid., ¶ 22. Here the Committee relied on Geneva Conventions.

34 Ibid., ¶ 23.

35 Ibid., ¶ 28. Reference is made to Agenda 21 and the World Summit on Sustainable Development and Plan of Implementation.

36 Ibid. Reference was made to the Convention on Biological Diversity, the Convention to Combat Desertification and the UNFCCC in this regard. Moreover, any person or groups who have been denied their right to water should have access to effective remedies at both national and international levels (¶ 55).
37 *General Comment No 21*, E/C.12/GC/21 (21 December 2009).
38 Ibid., ¶ 13.
39 Ibid., ¶ 16(a).
40 Ibid., ¶ 18.
41 Ibid., ¶ 36 (footnotes omitted), relying on UNDRIP.
42 Ibid., ¶ 36 (footnotes omitted).
43 Ibid., ¶ 37 (footnotes omitted), relying on UNDRIP and ILO Convention No 169.
44 Ibid., ¶ 50(c).
45 Ibid., ¶ 55(e).
46 See OHCHR website: https://www.ohchr.org/en/calls-for-input/2021/call-written-contributions-draft-general-comment-no-26-land-and-economic.
47 See OHCHR website: https://www.ohchr.org/en/treaty-bodies/cescr/general-comment-sustainable-development-and-international-covenant-economic-social-and-cultural.
48 *General Comment No 26* (2021) on land and economic, social and cultural rights, E/C.12/69/R.2, ¶ 1, https://www.ohchr.org/en/documents/general-comments-and-recommendations/ec1269r2-draft-general-comment-no-26–2021-land-and.
49 Ibid., ¶ 54.
50 Ibid., ¶ 55 (footnotes omitted).
51 *General Comment on Sustainable Development and the ICESCR*, available at: https://www.ohchr.org/en/treaty-bodies/cescr/general-comment-sustainable-development-and-international-covenant-economic-social-and-cultural.
52 Rio Declaration on Environment and Development (1992), *Report of the United Nations Conference on Environment and Development*, A/CONF.151/26 (vol 1), https://www.un.org/en/development/desa/population/migration/generalassembly/docs/globalcompact/A_CONF.151_26_Vol.I_Declaration.pdf
53 See supra note 51.
54 See Section 6.3.4
55 See Section 6.3.3
56 *Concluding observations on the fourth periodic report of Ecuador*, E/C.12/ECU/CO/4 (14 November 2019) ["Ecuador observations"], discussed below.
57 *Concluding observations on the sixth periodic report of Norway*, E/C.12/NOR/CO/6 (2 April 2020), ¶ 10.
58 Ibid., ¶ 11.
59 *Concluding observations on the fourth periodic report of Switzerland*, E/C.12/CHE/CO/4 (18 November 2019), ¶ 18 ["Switzerland observations"]. The Committee made similar remarks on the report of Germany and expressed regret that Germany was not on course to meet its GHG emission reduction targets for 2020, despite its domestic commitment to reducing emissions by 55% by 2030 compared to 1990 levels. It recommended that Germany intensify efforts to reach its targets for 2020 and comply with its obligations under the Paris Agreement by submitting its 2030 targets as its NDC, *Concluding observations on the sixth periodic report of Germany*, E/C.12/DEU/CO/6 (27 November 2018), ¶¶ 18–19 ["Germany observations"].
60 Switzerland observations, supra note 59, ¶19. The Committee also referred to its 2018 statement on climate change and the Covenant and the 2019 joint statement issued by the CEDAW Committee, ESCR Committee, the Committee on Migrant Workers, CRC Committee and the Committee on Persons with Disabilities.
61 Ecuador observations, supra note 56.
62 Ibid., ¶ 11.
63 Ibid., ¶12 (emphasis added).

64 *Climate Change and the International Covenant on Economic, Social and Cultural Rights* (8 October 2018), https://www.ohchr.org/en/NewsEvents/Pages/DisplayNews.aspx?NewsID=23691&LangID=E.
65 *Concluding observations on the fourth periodic report of Argentina*, E/C.12/ARG/CO/4 (1 November 2018), ¶ 13 ["Argentina observations"].
66 Ibid., ¶ 14. The wording is identical to the comments made in relation to Ecuador.
67 *Concluding observations on the fifth periodic report of Mauritius*, E/C.12/MUS/CO/5 (5 April 2019) ["Mauritius observations"], ¶ 9.
68 Ibid., ¶ 10.
69 *Concluding observations on the initial report of Cabo Verde*, E/C.12/CPV/CO/1 (27 November 2018) ["Cabo Verde observations"], ¶ 8.
70 Ibid., ¶ 9.
71 *Concluding observations on the sixth periodic report of the Russian Federation*, E/C.12/RUS/CO/6 (16 October 2017), ¶ 42 ["Russian Federation observations"].
72 Ibid., ¶ 43. While the Committee has made reference to "time-bound benchmarks" in several Concluding Observations, it is not clear what this phrase means. No other clarification is provided on this.
73 *Concluding observations on the fifth periodic report of Australia*, E/C.12/AUS/CO/5 (11 July 2017), ¶ 11 ["Australia observations"].
74 Ibid., ¶ 12.
75 *Concluding observations on the sixth periodic report of Canada*, E/C.12/CAN/CO/6 (23 March 2016), ¶ 53 ["Canada observations"].
76 Ibid., ¶ 5. On the report of Finland, the Committee expressed concern about the lack of adequate measures to address the adverse consequences of climate change on the Sami people. It recommended that Finland adopt measures to address the adverse impact of climate change on their land and resources. *Concluding observations on the sixth periodic report of Finland*, E/C.12/FIN/CO/6 (17 December 2014), ¶ 9(d) ["Finland observations"]. The Committee made a reference to climate change in the context of the right to food on the fifth periodic report of Sri Lanka, noting it as one of the factors contributing to high levels of malnutrition and wasting in the country. *Concluding observations on the fifth periodic report of Sri Lanka*, E/C.12/LKA/CO/5 (4 August 2017), ¶ 55, ["Sri Lanka observations"]. The Committee also referred to the need to prioritize investments in climate resilient agriculture and to address the situation of victims of the severe drought and heavy flooding that took place in 2016 and 2017.
77 *Concluding observations on the report of Cambodia*, E/C.12/KHM/CO/1 (12 June 2009).
78 *Concluding observations on the report of Argentina*, E/C.12/ARG/CO/3 (14 December 2011) ["Third Argentina observations"].
79 *Concluding observations on the report of Ukraine*, E/C.12/UKR/CO/5 (4 January 2008).
80 *Concluding observations on the initial report of Bangladesh*, E/C.12/BGD/CO/1 (18 April 2018) ["Bangladesh observations"].
81 Ibid., ¶ 13 (emphasis added).
82 Ibid., ¶ 14 (emphasis added).
83 *Concluding observations on the report of Mongolia*, E/C.12/1/Add.47 (1 September 2000), ¶ 10.
84 Ibid., ¶ 14.
85 *Concluding observations on third periodic report of Slovakia*, E/C.12/SVK/CO/3 (14 November 2019), ¶¶ 33–34, where the Committee expressed concern that a large number of Roma people lacked access to clean water. Drawing attention to GC No 15, the Committee recommended that the state party take measures to provide its population, especially those in rural areas with adequate and safe water and sanitation services; *Concluding observations on second periodic report of Turkmenistan*, E/C.12/TKM/CO/2 (31 October 2018), ¶ 31 ["Turkmenistan observations"], where the Committee expressed concern about the great disparity between rural and urban

areas with regard to access to safe drinking water and sanitation; *Concluding observations on fourth periodic report of New Zealand*, E/C.12/NZL/CO/4 (1 May 2018), ¶ 42 ["New Zealand observations"], where the Committee expressed concern at persistent challenges in access to safe drinking water leading to outbreaks of disease and recommended that the state party take immediate steps to address the obstacles to access to safe drinking water.

86 *Concluding observations on the initial report of Guinea*, E/C.12/GIN/CO/1 (30 March 2020) ["Guinea report"].

87 Ibid., ¶ 16.

88 *Concluding observations on third periodic report of Senegal*, E/C.12/SEN/CO/3, (13 November 2019) ["Senegal observations"].

89 Ibid., ¶ 34.

90 Ibid., ¶ 35.

91 *Concluding observations on the fourth periodic report of Israel*, E/C.12/ISR/CO/4, (12 November 2019) ["Israel observations"], ¶ 14.

92 Ibid., ¶ 45.

93 Ibid.

94 *Concluding observations on second periodic report of Kazakhstan*, E/C.12/KAZ/CO/2 (29 March 2019) ["Kazakhstan observations"], ¶ 52.

95 Russian Federation observations, supra note 71, ¶ 15(d).

96 *Concluding observations on third periodic report of Estonia*, E/C.12/EST/CO/3 (27 March 2019) ["Estonia observations"], ¶ 40.

97 Ibid., ¶ 41.

98 *Concluding observations on the fourth periodic report of Cameroon*, E/C.12/CMR/CO/4 (25 March 2019) ["Cameroon observations"], ¶ 16.

99 Ibid., ¶ 17.

100 *Concluding observations on the initial report of Mali*, E/C.12/MLI/CO/1 (6 November 2018), ¶ 43 ["Mali observations"].

101 Ibid., ¶ 44.

102 Argentina observations, supra note 65, ¶ 57.

103 Ibid., ¶ 58.

104 Ibid., ¶¶ 59–60.

105 *Concluding observations on the sixth periodic report of Colombia*, E/C.12/COL/CO/6 (19 October 2017), ¶ 15 ["Colombia observations"].

106 The Committee thus drew a clear link between environmental degradation and the enjoyment of rights. Likewise, it expressed deep concern about the continued deforestation in Brazil which impacts the enjoyment of ESC rights. It recommended taking measures to combat deforestation to ensure the enjoyment of ESC rights, especially by indigenous and vulnerable groups. See *Concluding observations on the report of Brazil*, E/C.12/BRA/CO/2 (12 June 2009), where the Committee noted the ratification of ILO 169 and the WHO Framework Convention on Tobacco Control, and *Concluding observations on the report of Honduras*, E/C.12/1/Add.57 (21 May 2001), where the Committee referred to the need to address excessive deforestation which is affecting the habitat of indigenous populations.

107 ¶ 16.

108 ¶ 59.

109 ¶ 60. Similar concerns were expressed on the report of the Republic of Korea regarding reports of contamination of rivers and the negative impact on the availability of safe drinking water. The Committee recommended, drawing attention to GC No 15, that the state party ensure the quality of water sources and the availability of safe potable drinking water for all. See *Concluding observations on the fourth periodic report of the Republic of Korea*, E/C.12/KOR/CO/4 (19 October 2017), ¶¶ 50–51.

110 *Concluding observations on the third periodic report of the Republic of Moldova*, E/C.12/MDA/CO/3 (19 October 2017) ["Moldova observations"].

111 ¶¶ 68–69.
112 Sri Lanka observations, supra note 76, ¶¶ 61–62.
113 Ibid. Similar concerns were expressed about the harmful effects of the insufficiently controlled use of agrochemicals on water quality and water resources on its comments on the report of Uruguay. The Committee recommended taking steps to control the use of agrochemicals, monitoring their public health effects and implementing the National Drinking Water and Sanitation Plan to protect the water resources and water quality. See *Concluding observations on the fifth periodic report of Uruguay*, E/C.12/URY/CO/5 (20 July 2017), ¶¶ 44–45 ["Uruguay observations"].
114 *Concluding observations on the sixth periodic report of the Netherlands*, E/C.12/NLD/CO/6 (6 July 2017).
115 Ibid., ¶¶ 11–13.
116 *Concluding observations on the initial report of Burkina Faso*, E/C.12/BFA/CO/1 (12 July 2016), ¶ 13 ["Burkina Faso observations"]. Similar concerns were expressed on the report of Niger where the Committee referred to the negative impact of extractive activities on the environment and on the health and rights of workers, rural communities, smallholder farmers and agro-pastoralists living where the mining of uranium, gold and oil takes place. It expressed alarm at: (a) the levels of uranium contamination; (b) the heavy water consumption associated with extractive activities and the contamination of water sources; and (c) land expropriation in the public interest in mining areas. The Committee recommended, relying on GC Nos 24 and 15, the strict application of legal provisions relating to the exploitation of natural resources, especially those on environmental protection, workers' health and rights of communities in mining areas; investigate the impact of extractive activities on the enjoyment of ESC rights; conduct inspections of mining sites; guarantee the quality of water sources, establish the liability of companies involved in mining activities that pollute the water sources; and pay adequate compensation where property has been expropriated. *Concluding observations on the initial report of the Niger*, E/C.12/NER/CO/1 (4 June 2018), ¶¶17–18 ["Niger observations"].
117 The free, prior and informed consent is usually required in relation to indigenous peoples. The Committee seems to have extended this requirement to *all communities* whose rights have been affected by development projects which is a significant development.
118 Burkina Faso observations, supra note 116, ¶ 14.
119 Ibid., ¶¶ 40–41. Similar concerns were expressed on the report of Togo where the Committee referred to inadequate monitoring of water quality and the pollution in urban areas caused by deficiencies in waste and sewage treatment and disposal. It also expressed concern about the harmful environmental and social impact of natural resource extraction activities. The Committee recommended strengthening the legal framework on environmental protection and social rights in relation to mineral exploitation and ensuring that mining companies establish provisions that allow tangible benefits for the realization of ESC rights of the communities. The Committee urged the state party to develop public sanitation, waste and sewage treatment services and safe drinking water distribution systems, particularly in rural areas, and to clean up sites that have become polluted by wastewater. It invited the state party to provide information in its next periodic report on the impact of the national action plan for water management and sanitation, *Concluding observations on the initial report of Togo*, E/C.12/TGO/CO/1 (3 June 2013), ¶¶ 27, 33. See also *Concluding observations on the initial report of Mauritania*, E/C.12/MRT/CO/1 (10 December 2012), ¶ 29, where similar concerns and recommendations were made. On the report of Ethiopia, the Committee noted that a large number of households in rural areas and those living in refugee camps do not have access to safe drinking water and sanitation, and over half of the households have to make a long journey to fetch drinking water. *Concluding observations on the report of Ethiopia*, E/C.12/ETH/CO/1–3 (31 May 2012), ¶ 23.

120 *Concluding observations on the combined fifth and sixth periodic reports of the Philippines*, E/C.12/PHL/CO/5–6 (26 October 2016), ¶ 45, ["Philippines observations"].

121 Ibid., ¶ 46.

122 *Concluding observations on the combined second to fifth periodic reports of Kenya*, E/C.12/KEN/CO/2–5 (6 April 2016), ¶ 11.

123 *Concluding observations on the combined second and third periodic reports of Kyrgyzstan*, E/C.12/KGZ/CO/2–3 (7 July 2015).

124 Ibid., ¶ 25.

125 Ibid.

126 *Concluding observations on the fourth periodic report of Mongolia*, E/C.12/MNG/CO/4 (7 July 2015), ¶ 16.

127 Ibid., ¶ 24.

128 Similar concerns were expressed on the report of Chile where the Committee referred to the disproportionate and unsustainable use of water by mining companies. It recommended ensuring access to drinking water and sanitation services for the most disadvantaged and marginalized groups, particularly in rural areas; ensuring moderate use of water by the mining industry; and adopting standards for the processing of wastewater in mining - *Concluding observations on the fourth periodic report of Chile*, E/C.12/CHL/CO/4 (7 July 2015), ¶ 27 ["Chile observations"]. See also *Concluding observations on the combined third to fifth periodic reports of Romania*, E/C.12/ROU/CO/3–5 (9 December 2014) ["Romania observations"], ¶ 20, where the Committee referred to poor access to sewage and sanitation systems in rural areas and informal settlements, and deficiencies in sewage treatments causing water pollution. The Committee called upon the state party to ensure universal access to safe drinking water and sanitation facilities particularly for the most disadvantaged groups and those in rural areas. On the report of Poland, the Committee requested the state party to provide in its next periodic report information on: the results of measures taken to protect the environment and improve industrial hygiene; and the right to water and sanitation, including the results of regulatory measures on water and sewage, *Concluding observations on the report of Poland*, E/C.12/POL/CO/5 (2 December 2009), ¶ 34.

129 Ibid., ¶ 25. Similar concerns were expressed on the report of the Philippines regarding the adverse effects of economic activities connected with the exploitation of natural resources, especially mining operations carried out in indigenous territories. The Committee noted the legal framework and institutional mechanisms established for the protection of the environment and the improvement of environmental and industrial hygiene, *Concluding observations on the report of Philippines*, E/C.12/PHL/CO/4 (1 December 2008), ["Second-Fourth Philippines observations"].

130 Ibid., ¶ 26.

131 *Concluding observations on the combined initial and second periodic reports of Thailand*, E/C.12/THA/CO/1–2 (19 June 2015), ¶ 31 ["Thailand observations"].

132 *Concluding observations on the initial report of Montenegro*, E/C.12/MNE/CO/1 (15 December 2014), ¶ 22. Similar concerns were expressed on the report of Romania where the majority of Roma continue live in substandard housing conditions, without safe drinking water or sanitation facilities, electricity, heating, sewage, and waste disposal, Romania observations, supra note 128, ¶ 18; see also the comments on the report of Ukraine where similar concerns were expressed regarding Roma people living in informal settlements and camps that lack basic infrastructure and services. The Committee referred to inadequate and heavily polluted living condition in the Dominican Republic, *Concluding observations on the report of the Dominican Republic*, E/C.12/1994/15 (19 December 1994), ¶ 14. The Committee requested information on the measures taken by Syria on environmental protection and how development was reconciled with environmental protection. The representative of the state party

stated that the industrial sector was required to conduct environmental impact studies and to take measures to address pollution, *Concluding observations on the report of the Syrian Arab Republic*, E/1992/23 E/C.12/1991/A (1992) ¶ 181.

133 *Concluding observations on the initial report of Indonesia*, E/C.12/IDN/CO/1 (19 June 2014).

134 Ibid., ¶ 28.

135 Ibid., ¶ 29.

136 *Concluding observations on the second periodic report of Uzbekistan*, E/C.12/UZB/CO/2 (13 June 2014).

137 Ibid., ¶ 25. The Committee referred to GC No 15 and the statement on sanitation. On the report of Kazakhstan the Committee referred to the pollution of the Aral Sea and the environmental pollution of the former nuclear test site of Semipalatinsk and voiced concern about air pollution and accumulation of waste and contamination of soil and water by industrial waste, agricultural pollutants and chemicals. The Committee urged taking immediate steps, including through regional cooperation, to address environmental hazards that affect the health of the population, and allocate more resources to enforce its environmental legislation, provide information on remedies available to those who have contracted illnesses due to environmental pollution, *Concluding observations on the report of Kazakhstan*, E/C.12/KAZ/CO/1 (7 June 2010), ¶ 35. On a previous report of Uzbekistan the Committee recognized that the effects of the Aral Sea ecological catastrophe affected the enjoyment of ESC rights by the population. It urged the State party to find a regional solution to this catastrophe, including through international technical cooperation, and to take measures to ensure the enjoyment of ESC rights, especially, the right to health. It expressed concern about the degree of environmental degradation in the country and its health impact especially on women and children. *Concluding observations on the report of Uzbekistan*, E/C.12/UZB/CO/12 (4 January 2006), ¶¶ 9, 47.

138 *Concluding observations on the second periodic report of China, including Hong Kong, China, and Macao, China*, E/C.12/CHN/CO/2 (13 June 2014).

139 Ibid., ¶ 32.

140 *Concluding observations on the second periodic report of Kuwait*, E/C.12/KWT/CO/2 (19 December 2013).

141 Ibid., ¶ 31.

142 *Concluding observations on the second periodic report of the Islamic Republic of Iran*, E/C.12/IRN/CO/2 (10 June 2013), ¶ 26.

143 Ibid.

144 *Concluding observations on the third periodic report of Ecuador*, E/C.12/ECU/CO/3 (13 December 2012), ¶ 25 ["Third Ecuador observations"]. The negative impact of extractive and mining activities on the environment and on the right to health, such as the serious public health problems in mining towns was discussed on the report of Mauritania. The Committee expressed concern about the insufficient regulatory measures and their weak enforcement and that mining activities have not generated much employment for local communities. The Committee recommended implementing EITI; ensuring that sanctions are applied for breaching environmental clauses in extractive and mining contracts; taking corrective measures to address environmental and health hazards caused by these activities; ensuring that the free, prior and informed consent of the population is obtained in decision-making; and ensuring that these activities bring tangible benefits to the enjoyment of ESC rights by the population. See *Concluding observations on the initial report of Mauritania*, E/C.12/MRT/CO/1 (10 December 2012), ¶ 8.

145 This is the first time that the Committee referred to "land trafficking" as opposed to "land grabbing," apparently referring to a term used by the state party in its report.

146 Third Ecuador observations, supra note 144, ¶ 26.

147 *Concluding observations on the combined initial, second and third periodic reports of Ethiopia*, E/C.12/ETH/CO/1–3 (31 May 2012), ¶ 24 ["Ethiopia observations"]. Similar concerns were expressed at the potential impact of the Ilisu Dam under construction and other dams on the enjoyment of ESC rights in Turkey, especially with regard to forced evictions, resettlements, displacement, and compensation of people affected, as well as their environmental and cultural impact. The Committee urged adopting a rights-based approach to infrastructure development projects, and to undertaking a review of its legislation and regulations on evictions, resettlement and compensation, especially the Ilisu dam, in line with GC No 7. See *Concluding observations on the initial report of Turkey*, E/C.12/TUR/CO/1 (12 July 2011), ¶ 26.
148 Ethiopia observations, supra note 147, ¶ 24.
149 *Concluding observations on the third periodic report of the Dominican Republic*, E/C.12/DOM/CO/3 (26 November 2010), ¶ 8.
150 Ibid. (emphasis added).
151 Ibid.
152 *Concluding observations on the combined second to fourth periodic reports of the Democratic Republic of the Congo*, E/C.12/COD/CO/4 (16 December 2009).
153 Ibid., ¶ 13.
154 Ibid.
155 *Concluding observations on the combined initial and second to fourth periodic reports of Cambodia*, E/C.12/KHM/CO/1 (12 June 2009), ¶ 6.
156 Ibid., ¶ 15.
157 Ibid.
158 *Concluding observations on the initial report of Nigeria*, E/C.12/1/Add.23 (16 June 1998), ¶ 29. Similar concerns were expressed on the report of Azerbaijan, where the Committee called upon the state party to regulate the oil industry, especially regarding its environmental impact, *Concluding observations on the initial report of Azerbaijan*, E/C.12/1/Add.20 (22 December 1997), ¶¶ 14, 28.
159 *Concluding observations on the initial report of the People's Republic of China* (including Hong Kong and Macao), E/C.12/1/Add.107 (13 May 2005), ¶ 31.
160 Ibid., ¶ 61.
161 Ibid., ¶ 63.
162 *Concluding observations on the second periodic report of Ecuador*, E/C.12/1/Add.100, 7 June 2004, ¶ 12 ["Second Ecuador observations"].
163 Ibid., ¶ 35.
164 Ibid., ¶ 41. Similar concerns were expressed on the report of Honduras about the negative effects of the use of pollutants and toxic substances in agricultural and industrial sectors, such as banana growing and gold-mining, and on the environment, affecting the health of workers and those living in the vicinity. The Committee noted that environmental impact studies conducted are not reviewed by independent bodies and recommended: (a) implementing existing legislative and administrative measures to avoid violations of environmental and labor laws by transnational companies; (b) adopting legislation to protect workers from health hazards from the use of toxic substances – such as pesticides and cyanide – in the banana-growing and gold-mining industries; (c) publicizing applications for mining concessions where the mining will take place, and opposition be allowed within three months; and (d) adopting measures to address environmental and health impacts of pollutants and toxic substances in agricultural and industrial sectors, such as banana growing and gold mining and establishing a mechanism to review the environmental impact studies conducted by these sectors, *Concluding observations on the initial report of Honduras*, E/C.12/1/Add.57 (21 May 2001), ¶¶ 44–46. Similarly, on the report of Estonia, the Committee recommended enforcing domestic and international standards for environmental protection to prevent harmful health effects, *Concluding observations on the initial report of Estonia*, E/C.12/1/Add.85 (19 December 2002), ¶ 54.

165 ECOSOC, *Concluding observations on the third report of the Russian Federation*, E/C.12/1/Add.13 (20 May 1997), ¶ 14 ["Third Russian Federation observations"].
166 Ibid.
167 Ibid., ¶ 25.
168 Ibid., ¶ 38.
169 *Concluding observations on the second periodic report of Tunisia*, E/C.12/1/Add.36 (14 May 1999) ["Tunisia observations"], ¶ 9.
170 *Concluding observations on the initial report of Egypt*, E/C.12/1/Add.44 (23 May 2000), ¶ 7.
171 *Concluding observations on the initial report of Mauritius*, E/C.12/1995/14 (28 December 1995), ¶ 8.
172 Guinea report, supra note 86, ¶ 39.
173 *Resolution 39/12, United Nations Declaration on the Rights of Peasants and Other People Working in Rural Areas*, A/HRC/RES/39/12 (8 October 2018), https://undocs.org/A/HRC/RES/39/12.
174 Guinea report, supra note 86, ¶ 40. Target 1.5 refers to building the resilience of the poor and vulnerable to climate-related extreme events and other economic, social and environmental shocks and disasters.
175 *Concluding observations on the initial report of South Africa*, E/C.12/ZAF/CO/1 (29 November 2018), ¶ 49 ["South Africa observations"]. The Committee referred to GC No 19 and its statement on social protection floors adopted in 2015. Similar recommendations were made on the Niger observations, supra note 116, ¶ 43. The Committee drew attention to GC No 19, its statement on social protection floors and SDGs on several reports: Bangladesh observations, supra note 80, ¶ 46; Colombia observations, supra note 105, ¶ 42; Uruguay observations, supra note 113, ¶ 31.
176 Burkina Faso observations, supra note 116, ¶ 33.
177 *Concluding observations on the fourth periodic report of France*, E/C.12/FRA/CO/4 (13 July 2016), ¶ 7.
178 Ibid.
179 Mali observations, supra note 100, ¶ 3.
180 Turkmenistan observations, supra note 85, ¶ 3. The Committee noted the adoption of a policy framework for sustainable development of indigenous peoples in the North Siberia and the far east of the Russian Federation but expressed concern about the lack of outcomes of the new policy, action plan and target program. *Concluding observations on the fifth periodic report of the Russian Federation*, E/C.12/RUS/CO/5 (1 June 2011), ¶ 7 ["Fifth Russian Federation observations"].
181 Tunisia observations, supra note 169, ¶ 5.
182 Ibid., ¶ 50. Identical language appears in several reports: Switzerland observations, supra note 59, ¶ 60, Ecuador observations, supra note 56, ¶ 65; Senegal observations, supra note 88, ¶ 46; *Concluding observations on sixth periodic report of Denmark*, E/C.12/DNK/CO/6 (12 November 2019), ¶ 75; Israel observations, supra note 91, ¶ 72; Mauritius observations, supra note 67, ¶ 65; Kazakhstan observations, supra note 94, ¶ 53; *Concluding observations on the sixth periodic report of Bulgaria*, E/C.12/BGR/CO/6 (29 March 2019), ¶ 53; Estonia observations, supra note 96, ¶ 56; Cameroon observations, supra note 98, ¶ 68; South Africa observations, supra note 175, ¶ 80; Cabo Verde observations, supra note 69, ¶ 70; Germany observations, supra note 59, ¶ 63; Mali observations, supra note 100, ¶ 56; Argentina observations, supra note 65, ¶ 65; Turkmenistan observations, supra note 85, ¶ 48; Niger observations, supra note 116, ¶ 63; *Concluding observations on the initial report of Central African Republic*, E/C.12/CAF/CO/1 (4 May 2018), ¶ 44; New Zealand observations, supra note 85, ¶ 53; *Concluding observations on the sixth periodic report of* Spain, E/C.12/ESP/CO/6 (25 April 2018), ¶ 52; Bangladesh observations, supra note 80, ¶ 74; Colombia observations, supra note 105, ¶ 71; Moldova observations, supra note 110, ¶ 78; Russian Federation observations, supra note 71, ¶ 62; Sri Lanka observations, supra note 76,

¶ 74; Uruguay observations, supra note 113, ¶ 61; *Concluding observations on the initial report of Pakistan*, E/C.12/PAK/CO/1 (20 July 2017), ¶ 91; *Concluding observations on the report of* Australia, ¶ 61; *Concluding observations on the combined second and third periodic reports of Liechtenstein*, E/C.12/LIE/CO/2–3 (3 July 2017), ¶ 35; *Concluding observations on the sixth periodic report of Cyprus*, E/C.12/CYP/CO/6 (28 October 2016), ¶ 47; Philippines observations, supra note 120, ¶ 61; *Concluding observations on the sixth periodic report of Poland*, E/C.12/POL/CO/6 (26 October 2016), ¶ 62; *Concluding observations on the fifth periodic report of Costa Rica*, E/C.12/CRI/CO/5 (21 October 2016), ¶ 67; *Concluding observations on the fourth periodic report of the Dominican Republic*, E/C.12/DOM/CO/4 (21 October 2016), ¶ 70 (slightly different wording).

183 *Concluding observations on the seventh periodic report of Finland*, E/C.12/FIN/CO/7 (30 March 2021), ¶ 52. The Committee referred to MDGs in its recommendations on the report of Ecuador and the need to pay attention to the disparities in urban and rural areas when implementing anti-poverty programs. It recommended that the state party strive to achieve the MDGs taking advantage of the technical advisory services that the UNDP provides, Third Ecuador observations, supra note 144, ¶ 23.

184 Ecuador observations, supra note 56, ¶ 15.

185 Ibid., ¶ 16.

186 Ibid., ¶ 18.

187 Ibid. In addition, the Committee addressed the environmental impact of mining and extractive activities and recommended taking measures in favor of the communities affected by environmental degradation and guarantee their enjoyment of ESC rights.

188 South Africa observations, supra note 175, ¶¶ 14–15.

189 Ibid., ¶ 15. ILO, Indigenous and Tribal Peoples Convention 1989 (No 169) ["ILO Convention 169"], https://www.ilo.org/dyn/normlex/en/f?p=NORMLEXPUB:55:0::NO::P55_TYPE,P55_LANG,P55_DOCUMENT,P55_NODE:REV,en,C169,/Document.

190 Argentina observations, supra note 65, ¶ 18.

191 Ibid., ¶ 19.

192 Ibid., ¶ 20.

193 Ibid., ¶ 21. The Committee also referred to the negative impact of unconventional oil and gas exploitation such as hydraulic fracking.

194 On the report of Bangladesh the Committee noted that while the Constitution recognizes tribes, "minor" races and ethnic sects and communities, there is absence of legal recognition of those who self-identify as indigenous peoples; the expropriation of ancestral land of indigenous peoples where free, prior and informed consent was not obtained; and at the lack of mechanisms for affected indigenous communities to take part in decision-making processes, Bangladesh observations, supra note 80, ¶¶ 15–16. The recommendations were similar to those of New Zealand including ratifying ILO Convention No 169. See also, Colombia observations, supra note 105, ¶¶ 17–18; Russian Federation observations, supra note 71, ¶¶ 14–15, where the Committee recommended providing groups negatively affected by extractive activities, including the Shor people, with fair and adequate remedies and reparation, referring to the recommendations made by the Committee on the Elimination of Racial Discrimination in 2017.

195 New Zealand observations, supra note 85, ¶ 8.

196 Ibid., ¶ 9.

197 Australia observations, supra note 73, ¶¶ 11–12.

198 The Committee referred to the lack of recognition of indigenous peoples in Angola and reports of discrimination in access to food, water, health and education and that development activities are impeding indigenous peoples from accessing their lands; and lack of legal framework for consultations prior to embarking on projects. It recommended adopting legislation to guarantee the rights of indigenous peoples to their

land and to freely dispose of their natural wealth and resources; seeking the free, prior and informed consent of indigenous peoples before granting licenses for economic activities on indigenous territories; and ensuring that licensing agreements with businesses provide for adequate compensation for the affected communities, *Concluding observations on the fourth and fifth periodic report of* Angola, E/C.12/AGO/CO/4–5 (15 July 2016), ¶¶ 19–20 ["Angola observations"].

199 Ibid., ¶ 16.

200 Philippines observations, supra note 120, ¶ 13.

201 Similar recommendation was made on the report of Angola, Angola observations, supra note 198, ¶ 20.

202 Ibid., ¶ 14.

203 *Concluding observations on the sixth periodic report of Sweden*, E/C.12/SWE/CO/6 (14 July 2016), ¶ 13 ["Sweden observations"]. The rights of Sami people were discussed on the report of Finland where the Committee recommended obtaining their free, prior and informed consent before granting licenses to private companies; ensuring that licensing agreements provide for adequate compensation to affected communities and taking measures to address adverse impact of climate change on the Sami people and speeding up ratification of ILO Convention No 169, Finland observations, supra note 76, ¶ 9. The Committee called upon the Russian Federation to ensure that licensing agreements with private entities provide for adequate compensation for the affected communities and that free, informed consent of indigenous communities be obtained prior to granting licenses to private companies for economic activities on indigenous territories, and urged ratifying ILO Convention No 169. Fifth Russian Federation observations, supra note 180, ¶ 7. The Committee used strong language on the 1997 report of the Russian Federation: "It is alarmed at reports that the economic rights of indigenous peoples are exploited with impunity by oil and gas companies which sign agreements under circumstances which are clearly illegal, and that the State party has not taken adequate steps to protect the indigenous peoples from such exploitation." Third Russian Federation observations, supra note 165, ¶ 14.

204 Sweden observations, supra note 203, ¶ 14. Similar concerns were expressed on the report of Canada where the Committee noted that the right to right to free, prior and informed consent of indigenous peoples to changes to their territories is not incorporated in legislation and that formal mechanisms to consult with them particularly in relation to extractive industries are lacking. The Committee recommended recognizing the right to free, prior and informed consent of indigenous peoples in its laws; establishing mechanisms to ensure their participation in decisions relating to development projects near their territories; engaging them in drafting legislation that affects them; applying principles in UNDRIP; and considering ratifying ILO No 169, Canada observations, supra note 75, ¶¶ 13–14, 19–20. See also *Concluding observations on the initial report of Namibia*, E/C.12/NAM/CO/1 (23 March 2016), ¶¶ 15–16, where reference was made to respecting FPIC, expediting the ratification of ILO No 169 and implementing the recommendations of the Special Rapporteur on Indigenous Peoples; *Concluding observations on the combined second to fourth periodic reports of Guyana*, E/C.12/GUY/CO/2–4 (28 October 2015), ¶¶ 14–17; Chile observations, supra note 128, ¶¶ 8–9 where the Committee urged Chile to comply with its commitment to guarantee recognition of indigenous peoples under the new Constitution, ensure FPIC and adopt regulations to assess the social and environmental impact of projects exploiting natural resources, especially within indigenous territories and ensuring that licensing agreements provide for adequate compensation for affected communities, especially indigenous peoples; Thailand observations, supra note 131, where extensive comments were made with regard to indigenous peoples, their territories and natural resources, adverse impacts of activities involving exploitation of natural resources, including large-scale projects such as the Map Ta Phut Industrial

estate and lack of consultations with affected communities. The Committee recommended adopting a rights-based approach to development projects and establishing participatory mechanisms to seek FPIC of affected communities.

205 *Concluding observations on the second periodic report of Honduras*, E/C.12/HND/CO/2 (11 July 2016), ¶ 11.

206 Ibid., ¶ 12.

207 Ibid., ¶¶ 45–46.

208 Third Ecuador observations, supra note 144, ¶ 9.

209 Ibid.

210 Ibid., ¶ 10. Lack of consultations and granting concessions for economic exploitation of indigenous lands and territories without obtaining the free, prior informed consent were expressed on the report of Argentina. The Committee recommended providing just and fair compensation to indigenous communities and ensuring their protection during mining projects. It requested information in the next periodic report about measures to protect traditional knowledge and cultural heritage of indigenous peoples including their ancestral land, Third Argentina observations, supra note 78, ¶ 25. On the report of Chad, the Committee expressed concern about the adverse effects of mining operations and oil exploration in indigenous territories, carried out in violation of indigenous peoples' right to their ancestral lands and natural resources. The Committee urged the State party to carry out environmental and social impact assessments of economic activities, particularly mining and oil exploration, to consult with the communities concerned, and encouraged ratifying ILO No 169, *Concluding observations on the report of Chad*, E/C.12/TCD/CO/3 (16 December 2009), ¶ 13. Similar concerns were expressed on the report of the Philippines with regard to the impact of mining operations on indigenous land and urged implementing the 1997 Indigenous Peoples Rights Act, Second-Fourth Philippines observations, supra note 129. On Ecuador's report, supra note 162, ¶ 12, the Committee noted that despite Constitutional guarantees of the right of indigenous people to own property communally, and to be consulted before natural resources are exploited in community property, these rights are not implemented. It urged the state party to ensure the participation of indigenous peoples on decisions that affect them and seek their consent prior to implementing natural resource-extracting projects in accordance with ILO Convention No 169. The Committee voiced concern about the impact of economic activities such as mining in the Imataca Forest Reserve and coal-mining, on the health, living environment and way of life of the indigenous populations in Venezuela and requested the state party to provide information in its next report on the measures taken to improve their situation, *Concluding observations on the report of Venezuela, E/C.12/1/Add.56* (21 May 2001), ¶¶ 12, 22.

211 Third Argentina observations, supra note 78.

212 Ibid., ¶ 9.

213 Ibid., ¶ 10.

214 Ecuador observations, supra note 56, ¶ 13.

215 *Statement on human rights defenders and economic, social and cultural rights*, E/C.12/2016/2 (29 March 2017).

216 Ibid., ¶ 14.

217 South Africa observations, supra note 175, ¶¶ 12–13.

218 Ibid., ¶ 10.

219 IPCC Special Report, *Global Warming of 1.5^{0}C (2018), Summary for Policymakers*, https://www.ipcc.ch/site/assets/uploads/sites/2/2022/06/SPM_version_report_LR.pdf.

220 *Climate change and the International Covenant on Economic, Social and Cultural Rights* (8 October 2018), https://www.ohchr.org/en/NewsEvents/Pages/DisplayNews.aspx?NewsID=23691&LangID=E.

221 Ibid.

222 *The pledge to leave no one behind: the International Covenant on Economic, Social and Cultural Rights and the 2030 Agenda for Sustainable Development*, E/C.12/2019/1 (5 April 2019).
223 Ibid., ¶ 4.
224 Ibid., ¶ 8 (emphasis added).
225 Ibid., ¶ 20.
226 *State Obligations under the International Covenant on Economic, Social and Cultural Rights and governance of land tenure* (4 October 2019), https://www.ohchr.org/sites/default/files/Documents/HRBodies/CESCR/Discussions/2019/Issues_paper.pdf. See also the proposed GC on land and ESC rights, discussed in Section 6.2.
227 Ibid., ¶ 28.

References

Boyd, D. (2011) *The Environmental Rights Revolution: A Global Study of Constitutions, Human Rights, and the Environment*, Vancouver: UBC Press.

Call for written contributions to the draft general comment (No. 26) on Land and Economic, Social and Cultural Rights (2021), https://www.ohchr.org/en/calls-for-input/2021/call-written-contributions-draft-general-comment-no-26-land-and-economic.

Climate change and the International Covenant on Economic, Social and Cultural Rights (2018), https://www.ohchr.org/en/NewsEvents/Pages/DisplayNews.aspx?NewsID=23691&LangID=E.

Concluding observations on the combined fifth and sixth periodic reports of the Philippines (2016), E/C.12/PHL/CO/5–6.

Concluding observations on the combined initial and second periodic reports of Thailand (2015), E/C.12/THA/CO/1–2.

Concluding observations on the combined initial and second to fourth periodic reports of Cambodia (2009), E/C.12/KHM/CO/1.

Concluding observations on the combined initial, second and third periodic reports of Ethiopia (2012), E/C.12/ETH/CO/1–3.

Concluding observations on the combined second and third periodic reports of Kyrgyzstan (2015), E/C.12/KGZ/CO/2–3.

Concluding observations on the combined second and third periodic reports of Liechtenstein, (2017), E/C.12/LIE/CO/2–3.

Concluding observations on the combined second to fifth periodic reports of Kenya (2016), E/C.12/KEN/CO/2–5.

Concluding observations on the combined second to fourth periodic reports of the Democratic Republic of the Congo (2009), E/C.12/COD/CO/4.

Concluding observations on the combined second to fourth periodic reports of Guyana, (2015), E/C.12/GUY/CO/2–4.

Concluding observations on the combined third to fifth periodic reports of Romania, (2014), E/C.12/ROU/CO/3–5.

Concluding observations on the fifth periodic report of Australia (2017), E/C.12/AUS/CO/5.

Concluding observations on the fifth periodic report of Costa Rica (2016), E/C.12/CRI/CO/5.

Concluding observations on the fifth periodic report of Mauritius (2019), E/C.12/MUS/CO/5.

Concluding observations on the fifth periodic report of the Russian Federation (2011), E/C.12/RUS/CO/5.

Concluding observations on the fifth periodic report of Sri Lanka (2017), E/C.12/LKA/CO/5.

Concluding observations on the fifth periodic report of Uruguay (2017), E/C.12/URY/CO/5.

Concluding observations on the fourth and fifth periodic report of Angola (2016), E/C.12/AGO/CO/4–5.
Concluding observations on the fourth periodic report of Argentina (2018), E/C.12/ARG/CO/4.
Concluding observations on the fourth periodic report of Cameroon (2019), E/C.12/CMR/CO/4.
Concluding observations on the fourth periodic report of Chile (2015), E/C.12/CHL/CO/4.
Concluding observations on the fourth periodic report of the Dominican Republic (2016), E/C.12/DOM/CO/4.
Concluding observations on the fourth periodic report of Ecuador (2019), E/C.12/ECU/CO/4.
Concluding observations on the fourth periodic report of France (2016), E/C.12/FRA/CO/4.
Concluding observations on the fourth periodic report of Israel (2019), E/C.12/ISR/CO/4.
Concluding observations on the fourth periodic report of Mongolia (2015), E/C.12/MNG/CO/4.
Concluding observations on the fourth periodic report of New Zealand (2018), E/C.12/NZL/CO/4.
Concluding observations on the fourth periodic report of the Republic of Korea, (2017), E/C.12/KOR/CO/4.
Concluding observations on the fourth periodic report of Switzerland (2019), E/C.12/CHE/CO/4.
Concluding observations on the initial report of Azerbaijan (1997), E/C.12/1/Add.20.
Concluding observations on the initial report of Bangladesh (2018), E/C.12/BGD/CO/1.
Concluding observations on the initial report of Burkina Faso (2016), E/C.12/BFA/CO/1.
Concluding observations on the initial report of Cabo Verde (2018), E/C.12/CPV/CO/1.
Concluding observations on the initial report of Central African Republic (2018), E/C.12/CAF/CO/1.
Concluding observations on the initial report of Egypt (2000), E/C.12/1/Add.44.
Concluding observations on the initial report of Estonia (2002). E/C.12/1/Add.85.
Concluding observations on the initial report of Guinea (2020). E/C.12/GIN/CO/1.
Concluding observations on the initial report of Indonesia (2014), E/C.12/IDN/CO/1.
Concluding observations on the initial report of Mali (2018), E/C.12/MLI/CO/1.
Concluding observations on the initial report of Mauritania (2012), E/C.12/MRT/CO/1.
Concluding observations on the initial report of Mauritius (1995), E/C.12/1995/14.
Concluding observations on the initial report of Montenegro (2014) E/C.12/MNE/CO/1.
Concluding observations on the initial report of Namibia (2016), E/C.12/NAM/CO/1.
Concluding observations on the initial report of the Niger (2018), E/C.12/NER/CO/1.
Concluding observations on the initial report of Nigeria (1998), E/C.12/1/Add.23.
Concluding observations on the initial report of Pakistan (2017), E/C.12/PAK/CO/1.
Concluding observations on the initial report of the People's Republic of China (including Hong Kong and Macao) (2005), E/C.12/1/Add.107.
Concluding observations on the initial report of South Africa (2018), E/C.12/ZAF/CO/1.
Concluding observations on the initial report of Togo (2013), E/C.12/TGO/CO/1.
Concluding observations on the initial report of Turkey (2011), E/C.12/TUR/CO/1.
Concluding observations on the report of Argentina (2011), E/C.12/ARG/CO/3.
Concluding observations on the report of Brazil (2009), E/C.12/BRA/CO/2.
Concluding observations on the report of Chad (2009), E/C.12/TCD/CO/3.
Concluding observations on the report of the Dominican Republic (1994), E/C.12/1994/15.
Concluding observations on the report of Honduras (2001), E/C.12/1/Add.57.
Concluding observations on the report of Kazakhstan (2010), E/C.12/KAZ/CO/1.
Concluding observations on the report of Mongolia (2000), E/C.12/1/Add.47.
Concluding observations on the report of Philippines (2008), E/C.12/PHL/CO/4.

Concluding observations on the report of Poland (2009), E/C.12/POL/CO/5.
Concluding observations on the report of the Syrian Arab Republic (1992), E/1992/23 E/C.12/1991/A.
Concluding observations on the report of Ukraine (2008), E/C.12/UKR/CO/5.
Concluding observations on the report of Uzbekistan (2006), E/C.12/UZB/CO/12.
Concluding observations on the report of Venezuela (2001), E/C.12/1/Add.56.
Concluding observations on the second periodic report of China, including Hong Kong, China, and Macao, China (2014) E/C.12/CHN/CO/2.
Concluding observations on the second periodic report of Ecuador (2004), E/C.12/1/Add.100.
Concluding observations on the second periodic report of Honduras (2016), E/C.12/HND/CO/2.
Concluding observations on the second periodic report of the Islamic Republic of Iran (2013), E/C.12/IRN/CO/2.
Concluding observations on the second periodic report of Kazakhstan (2019), E/C.12/KAZ/CO/2.
Concluding observations on the second periodic report of Kuwait (2013), E/C.12/KWT/CO/2.
Concluding observations on the second periodic report of Tunisia (1999), E/C.12/1/Add.36.
Concluding observations on the second periodic report of Turkmenistan (2018), E/C.12/TKM/CO/2.
Concluding observations on the second periodic report of Uzbekistan (2014), E/C.12/UZB/CO/2.
Concluding observations on the seventh periodic report of Finland (2021), E/C.12/FIN/CO/7.
Concluding observations on the sixth periodic report of Bulgaria (2019), E/C.12/BGR/CO/6.
Concluding observations on the sixth periodic report of Canada (2016), E/C.12/CAN/CO/6.
Concluding observations on the sixth periodic report of Colombia (2017), E/C.12/COL/CO/6.
Concluding observations on the sixth periodic report of Cyprus (2016), E/C.12/CYP/CO/6.
Concluding observations on the sixth periodic report of Denmark (2019), E/C.12/DNK/CO/6.
Concluding observations on the sixth periodic report of Finland (2014) E/C.12/FIN/CO/6.
Concluding observations on the sixth periodic report of Germany (2018), E/C.12/DEU/CO/6.
Concluding observations on the sixth periodic report of the Netherlands (2017), E/C.12/NLD/CO/6.
Concluding observations on the sixth periodic report of Norway (2020), E/C.12/NOR/CO/6.
Concluding observations on the sixth periodic report of Poland (2016), E/C.12/POL/CO/6.
Concluding observations on the sixth periodic report of the Russian Federation (2017), E/C.12/RUS/CO/6.
Concluding observations on the sixth periodic report of Spain (2018), E/C.12/ESP/CO/6.
Concluding observations on the sixth periodic report of Sweden (2016), E/C.12/SWE/CO/6.
Concluding observations on the third periodic report of the Dominican Republic (2010), E/C.12/DOM/CO/3.
Concluding observations on the third periodic report of Ecuador (2012), E/C.12/ECU/CO/3.
Concluding observations on the third periodic report of Estonia (2019), E/C.12/EST/CO/3.
Concluding observations on the third periodic report of the Republic of Moldova (2017), E/C.12/MDA/CO/3.
Concluding observations on the third report of the Russian Federation (1997), E/C.12/1/Add.13.
Concluding observations on the third periodic report of Senegal (2019), E/C.12/SEN/CO/3.
Concluding observations on the third periodic report of Slovakia (2019), E/C.12/SVK/CO/3.
de Schutter, O. (2019) *International Human Rights Law: Cases, Materials, Commentary* (3rd edn.), Cambridge: Cambridge University Press.
Gellers, J. (2017) *The Global Emergence of Constitutional Environmental Rights*, London: Routledge.

General Comment No. 3 (1990), E/1991/23.

General Comment No. 4 (1991), https://tbinternet.ohchr.org/_layouts/15/treatybody external/Download.aspx?symbolno=INT%2fCESCR%2fGEC%2f4759&Lang=en.

General Comment No. 12 (1999), E/C.12/1999/5.

General Comment No. 14 (2000), E/C.12/2000/4.

General Comment No. 15 (2003), E/C.12/2002/11.

General Comment No. 21 (2009), E/C.12/GC/21.

General Comment No. 26 (2021), E/C.12/69/R.2.

General Comment on Sustainable Development and the International Covenant on Economic, Social and Cultural Rights (2022), https://www.ohchr.org/en/treaty-bodies/cescr/general-comment-sustainable-development-and-international-covenant-economic-social-and-cultural.

ILC (2001) *Responsibility of States for Internationally Wrongful Acts*, https://legal.un.org/ilc/texts/instruments/english/draft_articles/9_6_2001.pdf.

ILO (1989) *Indigenous and Tribal Peoples Convention (No. 169)*, https://www.ilo.org/dyn/normlex/en/f?p=NORMLEXPUB:55:0::NO::P55_TYPE,P55_LANG,P55_DOCUMENT,P55_NODE:REV,en,C169,/Document.

IPCC (2018) Special Report, *Global Warming of 1.5⁰C, Summary for Policymakers*, https://www.ipcc.ch/site/assets/uploads/sites/2/2022/06/SPM_version_report_LR.pdf.

Mapping Human Rights Obligations Relating to the Enjoyment of a Safe, Clean, Healthy and Sustainable Environment, Individual Report on the International Covenant on Economic, Social and Cultural Rights, Report No. 1 (2013), https://www.ohchr.org/en/special-procedures/sr-environment/mapping-report.

May, J. & Daly, E. (2015) *Global Environmental Constitutionalism*, Cambridge: Cambridge University Press.

Resolution 39/12 (2018), A/HRC/RES/39/12.

Resolution 1985/17 (1985), https://documents-dds-ny.un.org/doc/RESOLUTION/GEN/NR0/663/73/IMG/NR066373.pdf?OpenElement.

Resolution 2200A (XXI), ICESCR (1966), https://www.ohchr.org/sites/default/files/cescr.pdf.

State Obligations under the International Covenant on Economic, Social and Cultural Rights and governance of land tenure, (2019), https://www.ohchr.org/sites/default/files/Documents/HRBodies/CESCR/Discussions/2019/Issues_paper.pdf.

Statement on human rights defenders and economic, social and cultural rights (2017), E/C.12/2016/2.

The pledge to leave no one behind: the ICESCR and the 2030 Agenda for Sustainable Development (2019), E/C.12/2019/1.

Tomuschat, C. (2014) *Human Rights: Between Idealism and Realism* (3rd edn.), Oxford: Oxford University Press.

Treaty Bodies: Committee on Economic, Social and Cultural Rights (2022), https://www.ohchr.org/en/hrbodies/cescr/pages/cescrindex.aspx.

World Conference on Human Rights (1993) Vienna Declaration and Programme of Action, https://www.ohchr.org/en/instruments-mechanisms/instruments/vienna-declaration-and-programme-action.

7 Committee on the Elimination of Discrimination against Women

7.1 Introduction and Mandate

The Convention on the Elimination of All Forms of Discrimination Against Women (CEDAW) was adopted in December 1979[1] and entered into force on September 3, 1981.[2] This is the main instrument governing discrimination against women and is a specialized treaty that is applicable in tandem with the International Bill of Rights.[3] Like many other human rights treaties of that era, it contains no provisions on environmental degradation and/or pollution, sustainable development, or climate change. The only provision that comes even close is Article 14 that addresses rights of rural women which requires states parties to ensure, *inter alia*, that rural women "enjoy adequate living conditions, particularly in relation to housing, sanitation, electricity and water supply, transport and communications."[4]

A Committee on the Elimination of Discrimination against Women ("CEDAW Committee") was established under Article 17 of the Convention. It consists of 23 experts on issues covered by the Convention who serve in their personal capacity.[5] Under Article 18, states have undertaken to submit to the UN Secretary General, for consideration by the Committee, a report on the measures they have adopted to give effect to the provisions of the Convention within one year after the treaty has entered into force for that state and thereafter at least every four years and whenever the Committee so requests. The Committee, through ECOSOC, reports annually to the UNGA on its activities and makes suggestions and general recommendations based on country reports.

An Optional Protocol to CEDAW was adopted in 1999 recognizing the competence of the Committee to receive and consider communications submitted in accordance with Article 2 of the Protocol:

> Communications may be submitted by or on behalf of individuals or groups of individuals, under the jurisdiction of a State Party, claiming to be victims of a violation of any of the rights set forth in the Convention by that State Party. Where a communication is submitted on behalf of individuals or groups of individuals, this shall be with their consent unless the author can justify acting on their behalf without such consent.[6]

DOI: 10.4324/9781003128847-10

Similar to other individual petitions, victims must exhaust all available domestic remedies, unless such remedies are unreasonably prolonged or unlikely to bring effective relief.[7]

Against this backdrop, this chapter discusses how the CEDAW Committee has addressed the issues covered in this book – environmental degradation/pollution, climate change, and sustainable development, including the SDGs. This chapter will use the individual report on CEDAW prepared for the Independent Expert on Human Rights and Environment[8] as the foundation which covers environmental issues until December 2013.

7.2 General Recommendations

While the Convention itself is silent on environmental protection and climate change, the Committee has adopted several recommendations that refer to these issues. General Recommendation (GR) No 21 on equality in marriage and family life[9] provides that "the voluntary regulation of population growth helps preserve the environment and achieve sustainable economic and social development."[10] GR No 23 on political and public life stresses the need for inclusion of women's participation in international bodies, and the need to integrate a gender perspective into their agenda.[11] It provides that many crucial decisions on global issues including nuclear disarmament, *development, and the environment* are taken with limited participation of women, in stark contrast to their participation at the non-governmental level.[12]

GR No 27 on the protection of human rights of older women[13] refers to climate change and natural disasters under areas of concern:

> Climate change impacts differently on women, especially older women who, due to their physiological differences, physical ability, age and gender, as well as social norms and roles and an inequitable distribution of aid and resources relating to social hierarchies, are particularly disadvantaged in the face of natural disasters. Their limited access to resources and decision-making processes increases their vulnerability to climate change.[14]

The Committee recommended that the state parties ensure that climate change and disaster risk reduction measures are gender-responsive and sensitive to the needs of older women and facilitate their participation in decision-making with regard to mitigation and adaptation.[15] While this recognition and the need to facilitate their participation in decision-making is significant, the Committee's reference to *natural disasters* is concerning as many of these severe weather events are triggered by climate change.

Similarly, GR No 34 on the rights of rural women[16] makes several references to sustainable development, SDGs, environment, and climate change. Recognizing the vital contributions of rural women, the recommendation refers to the SDGs, many of which address the situation of rural women and the need for specific attention to them and provides an important opportunity to advance process and outcome indicators.[17] The recommendation notes that discrimination against rural women cannot be understood without taking into account the macroeconomic roots of gender inequality. States fail to acknowledge the role of rural women in unpaid work, their contribution to GDP and, therefore, sustainable development. Moreover, environmental issues, including climate change and natural disasters, unsustainable use of natural resources, and poor waste management practices have detrimental impacts on the well-being of rural women.[18] The Committee thus recommended:

> States parties should address specific threats posed to rural women by climate change, natural disasters, land and soil degradation, water pollution, droughts, floods, desertification, pesticides and agrochemicals, extractive industries, monocultures, biopiracy and the loss of biodiversity, in particular agro-biodiversity. They should alleviate and mitigate those threats and ensure that rural women enjoy *a safe, clean and healthy environment*. They should effectively address the impact of such risks on rural women in the planning and implementation of all policies concerning the environment, climate change, disaster risk reduction, preparedness and management and ensure the full participation of rural women in designing, planning and implementing such policies.[19]

7.2.1 Sustainable Development and SDGs

Recognizing the contribution of rural women to the economy especially through the non-monetized sectors, GR No 34 notes that sustainable development must uphold their rights[20] and they must be regarded as drivers of sustainable development especially in relation to agriculture and rural development.[21] Toward this end, state parties should ensure that legal and policy frameworks relating to agriculture, water, forestry, livestock, fisheries, and aquaculture are gender-responsive and have adequate budgets. State parties should ensure that rural women are visible as stakeholders, decision makers, and beneficiaries, in line with various guidelines, GR No 23, and SDGs.[22] Moreover, states should establish gender units with senior-level staff in ministries relevant to rural development, provide adequate budgets, institutional procedures, accountability frameworks, and effective coordinating mechanisms. The recommendation calls upon states to strengthen rural economies, create local employment opportunities and livelihoods for rural women in the context of sustainable development.[23]

GR No 34 calls upon states to ensure rural women's right to employment by:

> Protecting the occupational health and safety of rural women by taking legislative and other measures to protect them against exposure to harmful chemicals. They should receive information about the health and environmental effects of the use of and exposure to chemicals, in particular hazardous chemicals, pesticides and other products used in agriculture and in extractive and other industries. States parties should develop and implement

public awareness programmes on those effects and on alternatives and ensure that no use, storage or disposal of hazardous materials or substances takes place without the explicit consent of rural women and their communities.[24]

7.2.2 Environmental Impact Assessments and Free, Prior, and Informed Consent

GR No 34, recognizing that rural women have a right to participate in decision-making at all levels but are inadequately represented in rural extension and water, forestry, and fishery services, calls upon states to ensure their active and informed participation in the development and implementation of all agricultural and rural development strategies, and in decision-making in relation to rural infrastructure and services, including water, sanitation, transportation, and energy. States should ensure that rural development projects are implemented only after gender and environmental impact assessments have been conducted with full participation of rural women, and after obtaining their free, prior, and informed consent. In addition, measures should be taken to mitigate adverse environmental and gender impacts.[25]

This provision is significant in several respects. *First*, it requires the preparation of *gender and* environmental impact assessments for rural development projects. *Second*, such assessments must be conducted with the full participation of rural women. *Third*, such development projects require the free, prior, and informed consent (FPIC) of rural women. This is significant as FPIC has hitherto been applied only in relation to indigenous peoples.[26] *Fourth*, the results of participatory assessments form the basis for any decisions regarding the implementation of such projects. *Finally*, effective measures should be taken to mitigate adverse environmental and gender impacts. While these provisions are laudable and break new ground, a few questions arise: it is not clear what the definition of "rural women" is, and is unlikely to be a homogenous group. Another unclear phrase is the reference to "participatory assessments."

The GR recognizes that rural women often have only limited rights over land and natural resources but their rights to land, natural resources including water, seeds, forests, and fisheries are fundamental rights. State parties should ensure through legislation that these rights are guaranteed on an equal basis with men. With regard to indigenous women in rural areas, states should ensure their equal access to ownership of and control over land, water, forests, fisheries, aquaculture, and other resources they have traditionally used. Additionally, states must enhance rural women's role in fisheries and aquaculture, their knowledge of sustainable use of fishery resources and promote their access to forests and sustainable forest resources, including safe access to fuelwood and non-wood forest resources.[27]

The Recommendation points out that industrial agriculture has been detrimental to rural women farmers and its consequences include soil degradation and erosion, water depletion, and use of cash crops instead of local food crops, use of genetically modified organisms and patenting of genetically altered crops whereas rural women are generally engaged in organic and sustainable farming practices.[28] GR No 34 also recognizes the circumstances that have led to the forced eviction of rural women from their lands such as global food, energy, financial, and environmental crises, leading to the sale or leasing of state-owned land to investors. This, in turn, has led to increased poverty and diminished access to land and natural resources, such as water, fuelwood, and medicinal plants. In addition, displacement affects rural women in multiple ways, including gender-based violence.[29]

Other relevant provisions include: implementing agricultural policies that support rural women farmers; promoting organic farming, protecting rural women from harmful pesticides and fertilizers; ensuring access to agricultural resources, including high-quality seeds, tools, knowledge and information, as well as equipment and resources for organic farming. In addition, state parties should: protect rural women's traditional and eco-friendly agricultural knowledge, and their right to preserve, use, and exchange traditional and native seeds; protect and conserve native and endemic plant species and varieties that are a source of food and medicine, and prevent patenting by national and transnational companies; prohibit contractual requirements on the mandatory purchase of seeds producing plants whose seeds are sterile (called "terminator seeds"), which prevent rural women from saving fertile seeds; ensure that land acquisitions do not violate their rights or result in forced eviction, and protect rural women from the acquisition of land by private companies, development projects, extractive industries, and megaprojects; and obtain their free and informed consent before approving any acquisitions or projects affecting rural lands and resources, including those relating to the lease and sale of land, land expropriation, and resettlement. Moreover, land acquisitions should be in line with international standards, and rural women should be adequately compensated.[30]

The Recommendation recognizes the need to ensure the realization of right to food within a framework of food sovereignty and that rural women and girls are among the most affected by water scarcity, aggravated by unequal access to natural resources and lack of infrastructure and services. It recognizes that they frequently walk long distances to fetch water, often spending up to four to five hours per day collecting water, carrying heavy containers, suffering illness from consuming unsafe water, and sometimes facing sexual violence and attacks.[31]

This GR is remarkable for its coverage of a wide range of issues applicable to rural women. Particularly noteworthy is the recognition of their right to a healthy environment, the rights to land, natural resources including water, seeds and forests, and fisheries as fundamental rights; the need to obtain their FPIC in relation to rural development projects and their right to participate in decision-making in relation to design, formulation, and implementation of projects.

7.2.3 *Climate Change*

Adopted in 2018 GR No 37 recognizes that climate change is exacerbating disasters globally by increasing both frequency and severity of climate hazards, increasing the vulnerability of communities.[32] Relying on the IPCC reports, the recommendation notes that a large proportion of extreme weather events is a result of climate change, and the human rights consequences of such disasters are affecting those populations that contributed least to the problem, disproportionately making them most vulnerable to the impacts. These groups include those living in poverty, young people, and future generations[33] and climate equity requires that these vulnerable groups including women and girls and countries are prioritized.

Women and girls are affected differently than men by climate change and disasters and experience greater risks, burdens, and impacts. In many contexts, gender inequalities impact the decision-making power of women and girls as well as access to resources such as food, water, land, and housing. Therefore, women and girls are more likely to be exposed to disaster-induced risks and are less able to adapt to changes in climatic conditions. Moreover,

> Although climate change mitigation and adaptation programmes may provide new employment and livelihood opportunities in sectors such as agricultural production, sustainable urban development and clean energy, failure to address the structural barriers faced by women in gaining access to their rights will increase gender-based inequalities and intersecting forms of discrimination.[34]

Moreover, mortality and morbidity levels are higher among women and girls during disasters who face an increased risk of gender-based violence during and after disasters. However, it is a mistake to categorize them as passive victims in need of protection as a negative gender stereotype that fails to recognize their contributions in the areas of disaster risk reduction, and climate mitigation and adaptation strategies:

> Well-designed disaster risk reduction and climate change initiatives that provide for the full and effective participation of women can advance substantive gender equality and the empowerment of women, while ensuring that sustainable development, disaster risk reduction and climate change objectives are achieved. It should be underlined that gender equality is a precondition for the realization of the Sustainable Development Goals.[35]

The GR provides guidance to states on the implementation of their obligations under CEDAW in relation to disasters and climate change and states that they must address in their reports the rights that may be affected by climate change and disasters, including extreme weather events such as floods and hurricanes, as well as slow-onset events such as melting of glaciers, droughts, and sea level rise.[36]

Stressing that it does not exhaustively cover the gender dimensions of climate mitigation and adaptation measures, the GR points out that it does not differentiate between disasters relating to climate change and other disasters. While a large proportion of disasters may be attributed to human-induced climate change, the recommendation is also applicable to other hazards and risks. It adopts a wide definition of "disasters" that include all hazards:

> [s]mall-scale and large-scale, frequent and infrequent, sudden- and slow-onset, caused by natural or human-made hazards, and related environmental, technological and biological hazards and risks, mentioned in the Sendai Framework for Disaster Risk Reduction 2015–2030, as well as any other chemical, nuclear and biological hazards and risks. Such hazards and risks include the testing and use of all types of weapons by State and non-State actors.[37]

Stating that human rights mechanisms have recognized the link between mitigation and adaptation in order to reduce increased disaster risk, the Committee points to the need to limit fossil fuel use and GHG emissions and harmful environmental impacts of extractive industries such as mining and fracking: "Any mitigation or adaptation measures should be designed and implemented in accordance with the human rights principles of substantive equality and non-discrimination, participation and empowerment, accountability and access to justice, transparency and the rule of law."[38]

The GR embodies international treaties and frameworks applicable to climate change, disaster prevention, and gender equality: CEDAW, Rio Declaration on Environment and Development (1992), the "Future We Want" outcome document (2012), UNFCCC (1992), and the Sendai Framework (2015) which emphasize, *inter alia*, gender equality, empowering women, participation of women in designing, implementing gender-sensitive disaster risk reduction policies, plans and programs, and building their capacity to secure alternative livelihoods in post-disaster situations.

Referring to the Lima Work Program on Gender adopted within the UNFCCC which established a plan for promoting gender balance and achieving gender-responsive climate policies, the Committee noted that the Conference of Parties agreed to advance the full and equal participation of women, promote gender-responsive climate policy, and mainstream a gender perspective into all elements of climate action.[39] It also referred to the Paris Agreement which embodies gender equality and the empowerment of women as principles and the need for adaptation plans to be, *inter-alia*, gender-responsive. Reference was also made to SDGs 3–6, 10, 11 and 13.[40]

The Committee referred to three cross-cutting principles in the Convention which are crucial and fundamental to drafting legislation, polices, and plans of action relating to disaster risk reduction and climate change: *equality and non-discrimination* with priority accorded to the most marginalized groups of women and girls including indigenous and refugee women; *participation and empowerment* ensuring that women have opportunities to participate in every stage of policy development, implementation and monitoring; and *accountability and access to justice* which require the provision of information and mechanisms to provide women and girls with remedies.[41]

Elaborating on these principles, the Committee pointed out that state parties have obligations under Article 2 of the Convention[42] to take measures to guarantee equality between women and men, including adopting participatory and gender-responsive policies and programs relating to disaster risk reduction and climate change. It recognized that intersecting forms of discrimination[43] may limit the access of particular groups of women to information, political power, and resources that would help mitigate the adverse impacts of disasters and climate change.[44]

In this regard, GRs Nos 32 (2014), No 33 (2015), 34 (2016), 35 (2017) and No 36 (2017) are also relevant. The Committee noted that GR No 37 does not contain an exhaustive list of every group whose rights should be integrated into laws, policies and programs on disaster risk reduction and climate change.

To ensure substantive equality between women and men in the context of disaster risk reduction and climate change, states parties must take specific steps to: eliminate all forms of discrimination against women in legislation, policies and programs, giving priority to addressing discrimination in relation to access to land and natural resources, and barriers to exercising their full legal capacity; and create effective mechanisms to guarantee that rights of women and girls are a primary consideration in devising measures at every level.

To ensure that women and girls are provided with equal opportunities to participate and engage in decision-making the Committee recommends that state parties adopt policies to achieve their equal participation in decision-making and development planning relating to disaster risk reduction and climate change; ensure equal representation of women in forums and mechanisms at all levels so that they can influence the development of policies, legislation and plans. States should extend these measures to include indigenous and other marginalized groups; and strengthen national institutions on gender-related issues and provide them with adequate resources and skills to prevent and respond to disasters and mitigate adverse effects of climate change.[45] Finally, women should be accorded equality before the law as women often face barriers to access justice and may encounter significant difficulties in claiming compensation and other forms of reparation and to adapt to climate change.[46]

The Recommendation recognized that gender-related dimensions of disaster risk reduction and the impacts of climate change are not well understood, and a lack of data disaggregated by sex, age, and ethnicity impede the development of appropriate strategies for disaster risk reduction and climate change. Recognizing that efforts have been made only recently to coordinate policies on gender equality, disaster risk reduction, climate change, and sustainable development, the Committee stated that:

> Programmes of action, budgets and strategies should be coordinated across sectors, including trade, development, energy, environment, water, climate science, agriculture, education, health and planning, and at levels of government, including local and subnational, national, regional and international,

> in order to ensure an effective and human rights-based approach to disaster risk reduction and climate change mitigation and adaptation.[47]

We now turn to a discussion of the relevant provisions in concluding observations issued by the CEDAW Committee.

7.3 Concluding Observations

As noted above, CEDAW does not include any provisions on environmental protection, pollution, climate change or sustainable development. However, many of its Concluding Observations refer to these issues, especially in the context of rural women and disaster risk reduction which is not surprising given the adoption of General Recommendations on these two topics. A perusal of the Concluding Observations from 2017 to 2020 revealed that SDGs received the highest number of mentions followed by climate change, especially in the context of disaster risk reduction. Other areas mentioned include access to water, renewable energy, indigenous women, rural women, mining, hazardous conditions of work, exploitation of natural resources, and the FPIC principle.

7.3.1 *Climate Change*

The Committee referred to climate change in many of its Concluding Observations, even as far back as 2007. Frequent references to climate change and disaster risk reduction, preparedness and management can be seen in more recent reports. On a few occasions the Committee even referred to climate displacement and statelessness.

On the reports of Kiribati,[48] the Committee welcomed the adoption of the joint implementation plan for climate change and disaster risk management, and the participation of women in the National Expert Group, in consultations on the development of plans and frameworks.[49] The Committee, however, expressed concern regarding: the limited participation of women in implementation of disaster risk management programs; the impact of seawater flooding of agricultural land and the pollution of wells on women's access to food, water, firewood and medicinal plants; and limited participation of women in migration policies. The Committee recommended ensuring women's participation in the implementation of climate change and disaster risk management initiatives; addressing the impact of climate change on women's access to resources and livelihoods; and reviewing the "migration with dignity" policy to ensure women's participation in employment opportunities abroad and respect women's agency and their mobility choices.[50]

On the report of Zimbabwe[51] the Committee noted that while the national climate policy adopted in 2016 included a gender component, events such as Cyclone Idai in March 2019 and floods disproportionately affected women and girls, which increased their risk of gender-based violence and food insecurity. It noted the absence of information on whether a gender perspective has been incorporated in the 2014 national climate change response strategy, and whether participation of women was sought during its development and implementation. Referring to GR No 37, the Committee recommended addressing the impact of Cyclone Idai on women and girls, and ensuring that a gender perspective is integrated into policies and programs on disaster risk reduction and climate change.[52]

On the report of Cambodia[53] the Committee expressed concern that women, especially rural women, are excluded from participation in the formulation and implementation of policies and action plans relating to disaster risk reduction, even though they are disproportionately affected as they are more likely to depend on agriculture than men.[54] It called upon the state party to ensure their effective participation, as agents of change and to include a gender perspective in these policies and plans taking into account their needs, especially of rural women.[55]

Disaster risk reduction and climate change was also mentioned on the report of Seychelles.[56] The Committee referred to the adoption of the Disaster Risk Management Act in 2014 and the establishment of the National Vulnerability Assessment Committee, to address the impact of climate change and disasters on women's rights, but expressed concern about the lack of a gender perspective in policies and programs on climate change.[57] Referring to GR No 37, the Committee recommended that the state party integrate a gender perspective into national policies and programs on climate change, with the participation of women.[58]

The Committee discussed the vulnerability of small island developing countries to climate change on the report of Cabo Verde.[59] It noted the efforts by the state party to invest in renewable energy and to adopt sex-disaggregated data; to implement gender mainstreaming strategies in the environment sector; and to adopt a national strategy for disaster risk reduction. It, however, expressed concern about the lack of participation of women in the development and implementation of these initiatives. In addition to referring to the need to ensure women's involvement in the development of legislation, national policies and programs on climate change, disaster response and disaster risk reduction, the Committee recommended taking measures to address hunger and food security of rural women due to impacts of climate change.[60]

The Committee noted the disproportionate impact of climate change on rural women and girls due to pre-existing inequalities on the report of Côte d'Ivoire.[61] Referring to its GR No 37, the Committee called on the state party to ensure the equal participation of rural women and girls in policymaking processes on disaster mitigation and climate change, and build on traditional, indigenous and local knowledge systems.[62] On the report of Macedonia,[63] the Committee referred to the impacts of climate change-related migration which affect agricultural production and cause internal migration of rural women. Referring to GR No 34, the Committee recommended adopting measures to ensure that rural women are represented in decision-making processes in the agriculture sector, including policies concerning disaster risk reduction, post-disaster management, and climate change.[64]

On the report of New Zealand, the Committee noted the steps taken to improve the situation of rural women and the development of its national climate adaptation plans[65] but noted that climate change has a disproportionate impact on women and there is a lack of information on the participation of women in decision-making in relation to policies and programs concerning rural development.[66] It recommended taking their greater vulnerability to natural disasters and climate change into account in adaptation processes and including traditionally male-dominated fields such as meteorology, disaster risk reduction, and climate change in girls' education.[67]

The Committee noted the adoption of gender equality as a cross-cutting principle in the Green State Development Strategy: Vision 2040 in Guyana but expressed concern about continuing to extract oil and gas there.[68] It pointed out that GHG emissions from these extractive industries could undermine the state party's obligations toward women because environmental degradation and disasters have a disproportionate impact on women, especially those in poverty.[69] Relying on GR No 37, the Committee recommended reviewing its climate change and energy policies, especially the extraction of oil and gas; and developing a disaster risk reduction strategy taking into account the negative impacts of climate change on gender equality and especially on those living in areas below sea level.[70] The Committee also expressed concern about rural and Amerindian women and girls who are disproportionately affected by climate change, oil and gas production, water pollution, and deforestation caused by mining activities.[71]

Gender and climate change was again addressed by the Committee on the reports of Mozambique.[72] It expressed concern about the disproportionate impact of Cyclone Idai and floods on women and girls because of their increased exposure to gender-based violence and food insecurity. Referring to GR No 37, the Committee recommended: ensuring that a gender perspective is integrated into the development and implementation of policies and programs on disaster risk reduction and climate change; assessing the impact of Cyclones Idai and Kenneth on women and girls; adopting a new action plan on gender, environment, and climate change; and ensure that women are included in the planning and implementation of policies on disaster management and disaster risk reduction.[73] The Committee further recommended incorporating a gender perspective in the 2010 environment and climate change action plan and strategy, and to ensure the participation of women in its development and implementation.[74]

On the report of Qatar,[75] the Committee welcomed the establishment of a national committee on climate change and clean development to give effect to the recommendations made under the UNFCCC and the Kyoto Protocol. It also welcomed the participation in the Global Dryland Alliance to address food insecurity and the negative impacts of climate change. It, however, expressed concern that the carbon emissions per capita in the state party are among the highest in the world, which has a significant negative impact on the rights of women and girls.[76] Referring to GR No 37, the Committee recommended increasing efforts to reduce carbon emissions and provide, in its next report, information on the participation of women in the development and implementation of initiatives on climate change, and the measures taken to incorporate a gender perspective into climate adaptation and mitigation efforts.[77]

The Committee made detailed recommendations with regard to the impact of climate change and natural disasters on women and in particular the impact of Hurricane Irma on the reports of Antigua and Barbuda.[78] It noted that in the aftermath of Hurricane Irma, the entire population had to be evacuated and it continues to pose challenges to the education, health, and livelihood of the affected women and girls. The Committee commended the efforts to support the affected women and girls but expressed concern that efforts to redistribute land in Barbuda may deprive women of communal land ownership.[79] The Committee recommended: addressing the impact of hurricanes on the environment and the education, health, and livelihood of the affected women and girls, and promoting their economic empowerment; developing a strategy to ensure their participation in recovery activities, in particular redistribution of land; allocating resources to the fund established to support those affected and making it accessible to affected women, in particular to disadvantaged groups, such as older women, women with disabilities, and women in the agricultural and fisheries sectors; and seeking assistance from the international community for post-disaster recovery.[80] The Committee welcomed the development of a gender-responsive disaster risk reduction plan and an early warning system and commended the efforts of the National Office of Disaster Services and the Department of Environment to engage women in planning and budgeting on disaster risk reduction, mitigation, and resilience. However, it expressed concern about the lack of information on the number of women in decision-making processes, given the vulnerability of the state party to the impacts of climate change and natural disasters, which disproportionately affect women.[81] Relying on GR No 37, the Committee recommended: ensuring that the rights of women and girls are a primary consideration in measures relating to climate change and disaster response especially, the needs of disadvantaged women; including a gender perspective on policies and programs on disaster response and disaster risk reduction; and ensuring the participation of women in policies and plans on disaster risk reduction, post-disaster management, and climate change and that women are represented in decision-making processes at all levels.[82]

The Committee discussed the impact of climate change and natural disasters on women on the report of the Bahamas. It commended that women occupy senior positions at the National Emergency Management Agency and that a gender assessment was conducted following Hurricane Matthew, [83] but noted the vulnerability of the state party to environmental and climate change and natural disasters which affect women disproportionately. The Committee believed that more could be done to include the special needs of women, in particular vulnerable groups, in disaster risk reduction preparedness, response, and recovery[84] Referring to GR No 37, the Committee recommended ensuring that women are represented in decision-making with regard to policies on disaster risk reduction, post-disaster management, and climate change, and using the results of gender assessments done after the hurricanes in policies and programs taking into account the special needs of women, including vulnerable groups.[85]

On the report of Lao People's Democratic Republic, the Committee expressed concern about the insufficient participation of women in developing policies and strategies on disaster risk reduction and climate change, given its vulnerability to flooding and drought.[86] Referring to GR No 37, the Committee recommended ensuring the participation of women in preparing and implementing national policies and programs on climate change, disaster response, and reduction, including a gender perspective, and increasing the number of women in decision-making bodies, including the National Committee for Natural Disaster Prevention and Control.[87]

The Committee referred to the adaptation policy framework to address climate change and disasters and incorporating a gender perspective on the report of Mauritius.[88] However, it expressed concern about the lack of information on the involvement of women in drafting legislation and policies and on the impact of programs aimed at engaging women as agents of change. Referring to GR No 37, the Committee recommended ensuring that women are involved in the development of legislation, policies, and programs on climate change, disaster response, and disaster risk reduction.[89] On the report of Cyprus, the Committee noted the risk assessment project on climate change but stated it required additional information on whether the state party plans to adopt a climate action plan and how it plans to ensure participation of women in its development and apply a gender perspective in relation to adaptation and mitigation measures.[90] The Committee commended Chile for the initiatives to address climate change through its national strategic plan for disaster risk management, and the inclusion of gender-focused training on disaster management.[91] It recommended expanding its strategic plan to areas that are most vulnerable to climate change, and ensure the participation of women in decision-making.[92]

The Committee made extensive comments on the gender dimension of climate change and disaster risk reduction on the report of Fiji.[93] Despite its leading role in climate negotiations as the presidency of the UNFCCC Conference of Parties, at the national level women are largely excluded from drafting and implementing policies and plans on climate change and disaster risk reduction, even though they are disproportionately affected and are the best agent for change.[94] It expressed concern about the risk that women face during and after disasters and the failure of relief measures to take their needs into account, and inadequate safeguards to protect them from sexual violence when displaced. The Committee expressed concern about the GHG emissions of fossil fuel companies operating in the country and recommended that: women participate in drafting plans and strategies for disaster preparedness and provide women-only shelters; set up public funds to support families in the wake of disasters and to provide basic necessities, including water, sanitation, food, and medication; ensure that private sector actors remedy the impact of their operations on women; and strengthen gender analysis and mainstream their concerns by getting their participation on adaptation and mitigation measures in their communities.[95]

Extensive comments were also made on the report of Marshall Islands.[96] The Committee noted the state party's efforts to address climate change and disasters, including the establishment of a green climate fund. It, however, expressed concern at the disproportionate impact of climate change, droughts, rising sea levels, and other weather-related disasters on women and girls which was compounded by the nuclear testing program of the United States, and caused severe environmental damage, including in the Bikini Atoll, a World Heritage Site.[97] It noted that no information had been provided on the participation of women in preparing the national policies and programs on climate change, disaster response, and disaster risk reduction, and whether a gender component was integrated into them. Referring to GR No 37, the Committee recommended seeking international assistance including climate finance from other countries, especially the United States "whose extraterritorial nuclear testing activities have exacerbated the adverse effects of climate change and natural disasters in the State party."[98] Moreover, the state party should ensure the participation of women in the preparation and implementation of national policies and programs on climate change,[99] disaster response, and disaster risk reduction and include a gender perspective in them. Women should have equal access to the green climate fund and training opportunities on climate mitigation and adaptation, including promoting the use of traditional knowledge in strategies for coping with natural disasters and climate change.[100]

On the report of the Republic of Korea, the Committee expressed concern that state party's energy policies on fossil fuel and coal-fired power plants, which result in GHG and other emissions, negatively affect women, especially pregnant women, and recommended reviewing energy and climate policies to ensure that they do not have an adverse impact on women and girls.[101]

On the report of Suriname, the Committee noted that, while the state party had adopted a climate action plan, no information was available on how women participated in the development of the plan and how a gender perspective was incorporated into the identification of mitigation and adaptation measures.[102] The Committee referred to a lack of gender perspective in the climate risk reduction strategies adopted by Suriname and that the second national communication to the UNFCCC submitted in 2016 did not take into account the vulnerability of women to natural disasters and the ability as agents of change. Referring to GR No 37, the Committee recommended that the state party ensure the participation of women, especially rural women, Maroon women, and indigenous women in the formulation of policies and actions plans on disaster risk reduction and climate change. It also recommended that such policies and plans and the third national communication include a gender perspective taking into account the specific needs of women.[103]

While commending Norway for its measures to address climate change, the Committee expressed concern about expanding extraction of oil and gas in the Arctic and that GHG emissions:

> [u]ndermine its obligations to ensure the substantive equality of women with men, as climate change disproportionately affects women, especially in situations of poverty, since they are more reliant on natural resources for their livelihoods than men and have lesser capacity to deal with natural hazards.[104]

The Committee recommended reviewing its climate change and energy policies, specifically its policy on the extraction of oil and gas, taking into account the disproportionate impact of climate change on women.[105]

Thus, while the Committee has referred to climate change and disasters, the disproportionate impact on women, and the need to include a gender perspective on national policies and action plans with their participation on several reports,[106] its comments on Tuvalu's report is particularly noteworthy as they deal with displacement in the context of climate change and possible statelessness.[107] It is significant that the Committee made comments as far back as 2009 and expressed concern that the state party did not have disaster management and mitigation plans in place to address potential internal and/or international displacement. The Committee recommended developing disaster management and mitigation plans in response to potential displacement and/or statelessness arising from environmental and climate change with women's participation. The Committee further encouraged the state party to seek assistance from the UN High Commissioner for Refugees and recommended integrating a gender perspective in sustainable development policies and plans.[108]

Noting the vulnerability of the state party to serious environmental and climate change, the Committee expressed concern that emigration of Tuvaluan citizens to neighboring countries has considerably increased and further internal or international displacement would give rise to statelessness. It noted that the impact of climate change, sea levels rise, and other climate-related disasters affect rural women disproportionately, as they rely on natural resources for their daily survival. The Committee was concerned about the lack of information about "those affected, including women, have been informed of or given opportunities to participate in decision-making processes to mitigate the impacts of climate change and in policymaking relating to them."[109] The Committee recommended developing:

> [d]isaster management and mitigation plans in response to potential displacement and/or statelessness arising from environmental and climate change and ensure that women, including those living on the outer islands, are included and may actively participate in planning and decision-making processes concerning their adoption.[110]

It recommended that a gender perspective be integrated into sustainable development policies and disaster risk reduction, post-disaster management, and climate change policies.

The Committee commended Bhutan for its leadership on climate change, its classification as a carbon-neutral country, and for the successful implementation of its NAPA to reduce risks and vulnerabilities from glacial lake outburst floods.[111] It, however, expressed concern at the lack of information on the participation of women in the development of policies and strategies on disaster risk reduction and climate adaptation. The Committee recommended that the state party provide information in the next report on the participation of women in the development and implementation of policies and strategies on disaster risk reduction and climate change adaptation and base them on a comprehensive gender analysis taking into account the needs of disadvantaged women, such as rural women.[112]

7.3.2 *Sustainable Development Goals*

SDGs featured on almost all the Concluding Observations issued by the Committee since SDGs were adopted[113] using identical language. The following paragraph appears in all Concluding Observations surveyed:

> The Committee welcomes the international support for the Sustainable Development Goals and calls for the realization of de jure (legal) and de facto (substantive) gender equality, in accordance with the provisions of the Convention, throughout the process of implementing the 2030 Agenda for Sustainable Development. The Committee recalls the importance of Goal 5 and of the mainstreaming of the principles of equality and non-discrimination throughout all 17 Goals. It urges the State party to recognize women as the driving force of the sustainable development of the State party and to adopt relevant policies and strategies to that effect.[114]

In addition to this common paragraph, some Concluding Observations contained further provisions that are relevant to the present discussion. Thus, on the report of Kiribati, the Committee referred to GR No 28 (2010) and SDG target 5.1, to end all forms of discrimination against women and girls everywhere. The Committee recommended adopting a definition of discrimination against women and girls in line with Article 1 of CEDAW, encompassing direct and indirect discrimination in the public and private spheres, recognizing intersecting forms of discrimination, and ensuring that discrimination based on gender is prohibited.[115] On the report of Iraq, the Committee referred to SDG target 5.A and recommended that the state party ensure coherence between the national development plan (2018–2022) and the SDGs.[116]

In its Concluding Observations on Andorra,[117] the Committee referred to the need for data collection and analysis and regretted the limited availability of statistical data disaggregated by sex, age, ethnicity, migration status, and disability that would enable the state party to determine the magnitude and nature of discrimination against women and girls which is necessary to evaluate progress achieved regarding the gender-related targets of the SDGs.[118] The Committee recommended enhancing the collection and analysis of such data and using measurable indicators to assess the progress made.[119]

In its earlier reports, the Committee simply referred to the need to achieve substantive gender equality in accordance with the provisions of CEDAW when implementing the 2030 Agenda.[120] Prior to its adoption, the Committee referred to the MDGs, noting that the full implementation of CEDAW is necessary to achieve MDGs and called for the integration of a gender perspective in all efforts toward the achievement of MDGs. The Committee called on state parties to include such information in their next periodic report.[121] Interestingly, in the reports issued in 2015, the Committee referred to the MDGs and the need to integrate those efforts into the post-2015 development framework.[122]

On the report of Norway,[123] in addition to calling for the realization of gender equality when implementing the 2030 Agenda,[124] the Committee, referring to SDG 5, commended the state party's effort to implement the SDGs, establishing an online knowledge platform, and implementing sustainable development policies, including measures to address climate change.[125]

7.3.3 *Environmental Pollution*

In addition to climate change, the Committee addressed environmental pollution in state parties in the context of gender equality. Thus, on the report of the Democratic Republic of the Congo,[126] the Committee referred to the environmental and sanitary impacts of industrial exploitation of copper and cobalt from the Frontier Mine, resulting in the drying up of the river, and affecting access to safe water and education. Its recommendations included: protecting women and girls working in the mining sector from exploitative and hazardous conditions of work; ensuring that they are represented in decision-making processes relating to mining activities; addressing environmental and health damage resulting from industrial exploitation of the Frontier Mine; and ensuring that they receive adequate compensation for the damage.[127]

On the report of Guyana,[128] the Committee went as far as recommending that the state party amend its environmental law to include a gender assessment in all environmental impact assessments and establish a mechanism to monitor their implementation. The state party should further guarantee rural women's contribution to national development through its Green State Development Strategy: Vision 2040. The Committee also stressed the need to ensure that rural women give their FPIC before any development, agro-industrial or extractive projects affecting their traditional lands and resources are initiated and that they can take advantage of benefit-sharing agreements in accordance with the ILO Convention No 169.[129] These recommendations are noteworthy for their reference to the FPIC principle in the context of rural women, which is typically confined to indigenous peoples.[130] Moreover, the recommendations refer to benefit-sharing agreements which are mostly seen under the Convention on Biological Diversity.[131]

The Committee endorsed the FPIC principle on the report of Angola:[132]

> [e]nsure the prior, free and informed consent of all women living in areas affected before the approval of any acquisition of land or the implementation of resource exploitation projects on rural land, including projects relating to the lease and sale of land, land expropriation and resettlement, and guarantee that they are adequately compensated.[133]

The Committee referred to the disproportionate impact on rural women of fracking and the exposure to hazardous and toxic chemicals and environmental pollution and climate change in its comments on the report of the United Kingdom.[134] Referring to GR No 34, the Committee recommended adopting a comprehensive ban on fracking.[135]

On the report of Marshall Islands,[136] the Committee pointed out that the nuclear testing program conducted by the US from 1946 to 1958 continues to pose severe environmental and health problems in the state party. Although the state party had approached the US through the Pacific Islands Forum and the Secretary General to get the recommendations made by the Special Rapporteur on Toxics[137] implemented, the Committee expressed concern that there is a lack of information on a comprehensive strategy to engage with the US to address the:

> [e]nvironmental damage and intergenerational health impacts, in particular on Marshallese women and girls, who disproportionately suffer from thyroid and other cancers as well as other reproductive health problems that are a cause of the large number of stillbirths and congenital birth defects in the State party.[138]

The Committee urged the state party to seek technical and financial assistance from the international community, the United Nations system and from the United States to: develop a strategy to implement the recommendations made by the Special Rapporteur on Toxics, to address the effects of nuclear weapon testing on the environment, health, and livelihood of the Marshallese people, in particular women and girls; and replenish the nuclear trust fund to provide compensation to the affected population.[139]

The Committee commended the Republic of Korea for its environmental policies and for aiming to reduce fine dust by 30% by 2022[140] and Suriname for its efforts to address the consequences of mining activities, including pollution from mercury.[141] However, the Committee expressed concern about the negative impact of activities of foreign-owned gold mining, petroleum extraction, and agribusiness companies and companies registered in Suriname and carrying out business abroad, especially because the existing legal framework does not hold foreign-owned companies accountable for violations of women's rights, particularly their land rights and right to health and for violations of environmental and labor standards.[142] The Committee recommended strengthening its legislation governing the conduct of companies registered in the state party, including the Mining Decree, to establish minimum standards for environmental protection, water quality, labor and health, with a view to protecting the rights of rural women, Maroon women, and indigenous women; and establishing a mechanism to regulate the activities of foreign-owned private mining companies and monitor their compliance with legislation and the human rights of women.[143]

The Committee also referred to the extreme poverty among rural women in Suriname and limited access to basic services including clean water, sanitation, and energy and expressed concern about the violation of land rights of indigenous and tribal women, despite three judgments of the Inter-American Court of Human Rights.[144] It expressed regret that rural women are excluded from decision-making processes, especially with regard to the use of land. It recommended implementing measures to ensure their equal participation including in relation to mining and logging concessions and rural and national development plans in line with GR No 34.[145]

The harmful impact of the use of pesticides, fertilizers, and agrochemicals on women's health, the need to promptly investigate complaints about their harmful use, and the need to guarantee rural and indigenous women's access to safe water and sanitation were expressed by the Committee on the report of Guatemala.[146] Similarly, on the report of Nauru, the Committee noted the environmental challenges, especially local food production, which affect the health and well-being of women and girls in Nauru. The Committee expressed concern that the negative consequences of phosphate mining were unaddressed and recommended taking measures to address the health concerns of women resulting from phosphate mining.[147]

The Committee expressed concern at the lack of mechanisms to assess the impact of environmental policies on women on the report of Federated States of Micronesia[148] and recommended establishing mechanisms to assess the impact of environmental policies on women. On the report of Mongolia, the Committee expressed concern on the negative environmental consequences of industrial operations, especially the mining sector which disproportionately affect rural women[149] and recommended that the state party:

> [i]nclude a gender perspective in national policies and action plans relating to climate change, disaster response and risk reduction, as well as to the negative environmental and socioeconomic consequences of industrial operations, principally those of the mining sector, focusing on women not only as victims but also as active participants in the formulation and implementation of such policies.[150]

The Committee noted the devastating impact of the drought and its impact on the harvest and income of rural women on the report of Haiti[151] and regretted that although women play an important role in food production, they have not been consulted or included in the development of agrarian policies. It recommended that rural women are represented in decision-making processes in the agriculture sector, enabling them to better address the impact of natural disasters and climate change.[152] On the report of the Solomon Islands, the Committee expressed concern over the adverse impact of large-scale development projects and logging and mining activities on women's control over land, and their limited share in the profits and income generated. It recommended establishing a legal framework to ensure that large-scale development projects do not undermine rural women's rights to land ownership and agricultural activities.[153] On the report of Iraq, the Committee referred to the increase in cancer cases, especially breast cancer among young women due to the deterioration of environmental conditions there. Noting that research indicated that the presence of toxic levels of mercury has led to an increase in birth defects and miscarriages, the Committee recommended adopting measures to address the worsening environmental conditions in order to reduce the incidence of birth defects in children and cancers and miscarriages in women.[154]

On the report of Cambodia, the Committee expressed concern that women are subjected to displacement due to large-scale land concessions and urban development and relocation sites frequently lack basic infrastructure and access to essential services, such as water, sanitation, and healthcare which affect women disproportionately.[155] The Committee urged the state party to ensure that land acquisitions for economic and other concessions follow due process and that adequate compensation is provided after a consultative process.[156] The Committee referred to the need to improve women's and girls' access to safe drinking water and adequate sanitation on the report of Djibouti and to encourage agro-pastoralist communities to settle near these water sources to increase food security and access to water for personal consumption and irrigation.[157]

On several occasions, the Committee referred to state parties' vulnerability to environmental threats. For example, the Committee took note of Tuvalu's vulnerability to environmental threats, including coastal erosion and rising sea levels as a result of climate change, and national disasters pose restrictions on movement and communication.[158] The Committee expressed concern about the impact of environmental degradation on rural women in Mozambique and recommended that such impact be evaluated and incorporated into solutions and policies to address the situation.[159] Likewise, it expressed concern about "the degree of environmental degradation in the country and its extremely negative impact on the health of the whole population, in particular women and children"[160] in Kazakhstan and urged the government to implement a sound environmental policy to protect the health of women and children. As far back as 1995, the Committee questioned Mauritius on how environmental degradation affected rural women which was especially visible in Rodrigues, mainly through soil erosion and deforestation. The Committee noted that the state party had initiated measures to sustain the environment involving terracing of land, reforestation, sensitization, waste management, and others.[161] The Committee requested more information from Poland on the activities of women in relation to environmental protection.[162]

7.3.4 *Renewable Energy*

On a few occasions, the Committee referred to the need to promote renewable energy. Thus, on the report of Cabo Verde,[163] the Committee noted the limited inclusion of rural women in the design and implementation of measures to promote renewable energies and sustainable use of the ocean. It recommended that rural women have access to electricity generated by renewable energy and to involve them in the design and implementation of measures to develop the sustainable use of the ocean ("the blue economy"). On the report of Nigeria, the

Committee noted the limited efforts of the state party to explore investment and employment opportunities for women in renewable energy while combating adverse effects of climate change.[164] The Committee recommended that the state party explore investment and employment opportunities for women in renewable energy in its efforts to implement SDGs 5, 7 and 13.[165]

7.3.5 *Women Human Rights Defenders*

On the report of Guatemala, the Committee referred to the need to protect women human rights defenders including indigenous women defending land rights and environmental resources from gender-based violence.[166] Similar observations were made in relation to Cambodia where female human rights defenders who advocate for women's land rights are subjected to intimidation and harassment by law enforcement personnel.[167] The Committee recommended that the state party investigate and prosecute cases of intimidation.

The Committee expressed serious concern about women human rights defenders in Thailand, especially those advocating for land rights, protection of the environment, and the rights of indigenous women and rural women[168] and recommended adopting effective measures to protect them and to enable them to freely undertake their work without violence or intimidation, and improving, in consultation with women human rights defenders, the Witness Protection Office within the Ministry of Justice.[169]

7.4 Conclusion

This chapter discussed how the CEDAW Committee has addressed environmental degradation, climate change and sustainable development, including SDGs within its mandate as these issues pertain to the enjoyment of rights by women and girls, including rural women and indigenous women. Several principles and themes can be identified from this discussion. First, one of the most important principles is the need to ensure women's involvement and participation in designing, developing, and implementing policies, programs, and strategies that affect them whether it is in relation to land use, climate change, renewable energy, mining projects, or other infrastructure projects. Second, the Committee has extended the FPIC principle to women, especially rural women, a principle typically applied in relation to indigenous people.[170] Third, the Committee advanced the notion of women benefiting from projects on lands that have been expropriated from them. Fourth, where their rights have been infringed, the Committee advanced the principle of adequate reparations and compensation.[171] Finally, the Committee has referred to the need to collect gender segregated data to ascertain the extent of gender discrimination and to inform policies to address it.

Similarly, GR No 34 on rural women is remarkable for the range of issues covered and the principles it embodies. First, it calls upon states to recognize and address the specific threats posed to rural women by environmental pollution, climate change, disasters,[172] agrochemicals, and loss of biodiversity. Second, it recognizes rural women's right to a healthy environment. Third, it calls upon states to address these threats in planning and implementing policies concerning the environment, climate change, and disaster risk reduction. Fourth, it calls upon states to ensure the full participation of rural women in designing, planning, and implementing such policies. Finally, it provides that states should ensure the protection and security of rural women and girls during all phases of disasters. Significantly, on several reports the Committee referred to the intersecting and multiple forms of discrimination facing women affected by conflict and natural disasters, as well as the intersecting challenges faced by rural indigenous women and the need to involve women, not as victims but as agents of change.

Notes

1 By UNGA resolution 34/180 (18 December 1979), available at: https://undocs.org/A/RES/34/180.
2 UN Convention on the Elimination of All Forms of Discrimination Against Women (CEDAW), https://www.ohchr.org/en/instruments-mechanisms/instruments/convention-elimination-all-forms-discrimination-against-women.
3 See Chapter 2.
4 CEDAW, supra note 2, art. 14(h).
5 Ibid., art. 17.
6 Optional Protocol to the Convention on the Elimination of All Forms of Discrimination against Women, (entered into force in December 2000), art. 2, https://www.ohchr.org/EN/ProfessionalInterest/Pages/OPCEDAW.aspx.
7 Ibid., art. 4.
8 *Mapping Human Rights Obligations Relating to the Enjoyment of a Safe, Clean, Healthy and Sustainable Environment, Individual report on the UN Convention on the Elimination of All Forms of Discrimination against Women, Report No. 4* (December 2013) ["Mapping Report No. 4"], https://www.ohchr.org/en/special-procedures/sr-environment/mapping-report.
9 Adopted by the Committee on the Elimination of Discrimination against Women (1994), *General Recommendation No. 21: Equality in Marriage and Family Relations*, https://tbinternet.ohchr.org/_layouts/15/treatybodyexternal/Download.aspx?symbolno=INT%2fCEDAW%2fGEC%2f4733&Lang=en. The CEDAW Committee refers to its "general comments" as "general recommendations." Legally, there is no distinction.
10 Ibid., ¶ 23.
11 Adopted at the Sixteenth Session (1997), *General Recommendation No. 21*, available at: https://tbinternet.ohchr.org/Treaties/CEDAW/Shared%20Documents/1_Global/INT_CEDAW_GEC_4736_E.pdf.
12 Ibid., ¶ 39 (emphasis added).
13 *General Recommendation No 27 on older women and protection of their human rights*, CEDAW/C/GC/27 (16 December 2010).
14 Ibid., ¶ 25.
15 Ibid., ¶ 35.
16 *General Recommendation No 34 (2016) on the rights of rural women*, CEDAW/C/GC/34, (7 March 2016).
17 Ibid., ¶ 2.
18 Ibid., ¶ 10.
19 Ibid., ¶ 12.
20 Ibid., ¶ 16. See *GR No 35 on gender-based violence against women, updating GR No 19*, CEDAW/C/GC/35, (14 July 2018), which refers to SDGs, particularly Goal 5 on

gender equality, and Goal 16 on promoting peaceful and inclusive societies for sustainable development, ¶ 35(b).

21 GR No 34, supra note 16, ¶ 35.

22 Ibid., ¶ 36. These are: Voluntary Guidelines on the Responsible Governance of Tenure of Land, Fisheries and Forests in the Context of National Food Security, and the Voluntary Guidelines for Securing Sustainable Small-Scale Fisheries in the Context of Food Security and Poverty Eradication.

23 Ibid., ¶ 51.

24 Ibid., ¶ 52(e).

25 Ibid., ¶ 54(e).

26 See Chapter 6 in this volume where the ESCR Committee also extended the FPIC principle to non-indigenous communities.

27 GR No 34, supra note 16, ¶ 59.

28 Ibid., ¶ 60.

29 See GR No 35, supra note 20, which recognized that gender-based violence is exacerbated by, *inter alia*, economic, cultural, social and environmental factors, as is seen in relation to displacement, migration and increased globalization of economic activities, including global supply chains, extractive and offshoring industry, armed conflict and terrorism.

30 GR No 34, supra note 16, ¶ 62.

31 Ibid., ¶ 82.

32 *General Recommendation No 37 on the gender-related dimensions of disaster risk reduction in the context of climate change*, CEDAW/C/GC/37 (13 March 2018).

33 Ibid., ¶ 1.

34 Ibid., ¶ 3.

35 Ibid., ¶ 7 (footnotes omitted).

36 Ibid., ¶ 11.

37 Ibid., ¶ 13.

38 Ibid., ¶ 14.

39 *Decision 18/CP.20* (2014), https://unfccc.int/resource/docs/2014/cop20/eng/10a03.pdf#page=35%22. See also *Decision 3/CP.25* (2019), https://unfccc.int/sites/default/files/resource/cp2019_13a01E.pdf.

40 SDGs are available at: https://sdgs.un.org.

41 GR No 37, supra note 32, ¶ 26.

42 Article 2 of CEDAW addresses discrimination against women and states parties have undertaken to, *inter alia*, adopt legislative and other measures prohibiting all forms of discrimination against women.

43 Intersectionality is a good lens to study the multiple and intersecting forms of discrimination as well as vulnerabilities. See Chapter 4, note 94.

44 GR No 37, supra note 32, ¶ 28.

45 Ibid., ¶¶ 32–36.

46 Ibid., ¶¶ 37–38.

47 Ibid., ¶ 41.

48 *Concluding observations on the combined initial, second and third periodic reports of Kiribati*, CEDAW/C/KIR/CO/1–3, (11 March 2020).

49 Ibid., ¶ 45.

50 Ibid., ¶ 46.

51 *Concluding observations on sixth periodic report of Zimbabwe*, CEDAW/C/ZWE/CO/6 (13 February 2020).

52 Ibid., ¶ 48.

53 *Concluding observations on sixth periodic report of Cambodia*, CEDAW/C/KHM/CO/6 (9 October 2019).

54 Ibid., ¶ 42.

55 Ibid., ¶ 43.

56 *Concluding observations on sixth periodic report of Seychelles*, CEDAW/C/SYC/CO/6 (25 October 2019).
57 Ibid., ¶ 45.
58 Ibid., ¶ 46. Similar comments were made on the report of Guyana. Referring to GR No 37, the Committee recommended developing a disaster risk reduction strategy that considers the negative effects on gender equality and the lives of women and their families, especially those living in areas below sea level. *Concluding observations on ninth periodic report of Guyana*, CEDAW/C/GUY/CO/9 (12 July 2019) ["Guyana observations"]. See also *Concluding observations on the eighth periodic report of the United Kingdom of Great Britain and Northern Ireland*, CEDAW/C/GBR/CO/8 (26 February 2019) ["United Kingdom observations"].
59 *Concluding observations on ninth periodic report of Cabo Verde*, CEDAW/C/CPV/CO/9 (11 July 2019) ["Cabo Verde observations"].
60 Ibid., ¶ 37.
61 *Concluding observations on fourth periodic report of Côte d'Ivoire*, CEDAW/C/CIV/CO/4 (5 July 2019).
62 Ibid., ¶ 48(c).
63 *Concluding Observations on the sixth periodic report of the former Yugoslav Republic of Macedonia*, CEDAW/C/MKD/CO/6 (1 November 2018). On the report of Thailand, the Committee noted that women, especially rural women, are excluded from participation in the formulation and implementation of policies and plans on climate change and disaster risk reduction. It recommended ensuring their effective participation not only as those disproportionately affected but also as agents of change and that these policies and plans include a gender perspective taking into account the needs of women, especially rural women, *Concluding observations on the combined sixth and seventh periodic reports of Thailand*, CEDAW/C/THA/CO/6–7 (5 July 2017), ¶¶ 46–47 ["Thailand observations"].
64 Macedonia observations, supra note 63, ¶ 42(d). See also the *Concluding observations on the fifth periodic report of Fiji*, CEDAW/C/FJI/CC/5 (23 February 2018), ¶ 49 ["Fiji observations"] where the Committee referred to the worsening situation of rural women due to the intensification of natural disasters and climate change affecting the agriculture sector and damaging crops. It recommended considering the greater vulnerability of rural women when drafting policies on disaster risk reduction and climate change.
65 *Concluding observations on the eighth periodic report of New Zealand*, CEDAW/C/NZL/CO/8 (12 July 2018).
66 Ibid., ¶ 41(d).
67 Ibid., ¶ 32(e).
68 Guyana observations, supra note 58.
69 Ibid., ¶ 41.
70 Ibid., ¶ 42.
71 Ibid., ¶ 43(d).
72 *Concluding observations on third to fifth periodic reports of Mozambique*, CEDAW/C/MOZ/CO/3–5 (3 July 2019).
73 Ibid., ¶ 44.
74 Ibid., ¶ 43.
75 *Concluding observations on second periodic report of Qatar*, CEDAW/C/QAT/CO/2 (2 July 2019).
76 Ibid., ¶ 43.
77 Ibid., ¶ 44.
78 *Concluding observations on combined fourth to seventh periodic reports of Antigua and Barbuda*, CEDAW/C/ATG/CO/4–7 (20 February 2019).
79 Ibid., ¶¶ 9–10.

80 Ibid., ¶ 10.

81 Ibid., ¶ 50.

82 Ibid., ¶ 51. Similar comments were made on the fourth periodic report of Botswana, CEDAW/C/BWA/CO/4 (1 March 2019), ¶ 46. The Committee noted that the Ministry of Environment, Wildlife and Tourism had adopted a gender mainstreaming approach in 2014 but there was lack of information on whether a gender perspective is reflected in policies and programs on disaster risk reduction and climate change and whether women have opportunities to participate in policymaking and decision-making. Referring to GR No 37, the Committee recommended ensuring that a gender perspective is integrated into national policies and programs that reflect the needs and concerns of women and that they are included in planning and decision-making. See also United Kingdom observations, supra note 58, ¶ 54, where the Committee made similar comments regarding the participation of rural women and girls.

83 *Concluding Observations on the sixth periodic report of the Bahamas*, CEDAW/C/BHS/CO/6 (25 October 2018), ¶ 47 ["Bahamas observations"]. Similar comments were made regarding women in leadership positions on the report of Barbados, but the Committee noted that the state party did not incorporate a gender perspective into its disaster risk reduction strategies. It recommended including women as active participants in the formulation and implementation of such policies, *Concluding observations on the combined fifth to eighth periodic reports of Barbados*, CEDAW/C/BRB/CO/5–8 (12 July 2017), ¶¶ 47–48 ["Barbados observations"].

84 Bahamas observations, supra note 83, ¶ 48. On the report of the Federated States of Micronesia, the Committee noted that the state party was the first Pacific Island state to adopt a climate change act, but referred to the disproportionate impact of climate change, rising sea levels and weather-related disasters on women, lack of information on women's participation in decision-making and the lack of a gender component in policies and programs. It recommended ensuring women's participation in policies and programs on climate change, disaster response and risk reduction, *Concluding observations on the combined initial to third periodic reports of the Federated States of Micronesia*, CEDAW/C/FSM/CO/1–3 (24 February 2017), ¶¶ 42–43 ["Micronesia observations"].

85 Bahamas observations, supra note 83, ¶ 48.

86 *Concluding observations on the combined eighth and ninth periodic reports of the Lao People's Democratic Republic*, CEDAW/C/LAO/CO/8–9 (2 November 2018).

87 Ibid., ¶ 52. See also *Concluding observations on the sixth periodic report of Samoa*, CEDAW/C/WSM/CO/6 (26 October 2018), ¶¶ 41–42, where the Committee made similar comments noting the vulnerability of the state party to climate change and the disproportionate impact on women and girls. On the report of Mexico, the Committee referred to the low participation of women in leadership roles in relation to disaster risk reduction, despite the inclusion of a gender perspective in its 2012 national climate change policy, *Concluding observations on the ninth periodic report of Mexico*, CEDAW/C/MEX/CO/9 (6 July 2018), ¶ 43(d) ["Mexico observations"]. Referring to GR No 37, the Committee recommended accelerating the participation of women in decision-making processes and integrating a gender perspective in sustainable development policies, disaster risk reduction and post-disaster management.

88 *Concluding observations on the eighth periodic report of Mauritius*, CEDAW/C/MUS/CO/8 (30 October 2018) ["Mauritius observations"]. In the *Concluding observation on the combined second and third periodic reports of the Cook Islands*, CEDAW/C/COK/CO/2–3 (13 July 2018), ¶ 45, the Committee noted the adoption of a joint national action plan for disaster risk management and climate change adaptation 2016–2020, which included women in the drafting process. It, however, expressed concern about the disproportionate impact of climate change, droughts, rising sea levels and other weather-related disasters on women and the lack of a gender perspective in policies and programs on climate change, disaster response and risk reduction.

89 Mauritius observations, supra note 88, ¶ 36. See also *Concluding observations on the sixth periodic report of Nepal*, CEDAW/C/NPL/CO/6 (23 October 2018), ¶ 41, where the Committee referred to both GRs Nos 34 and 37 and recommended addressing the intersecting and multiple forms of discrimination facing women affected by conflict and natural disasters.

90 *Concluding observations on the eighth periodic report of Cyprus*, CEDAW/C/CYP/CO/8 (4 July 2018), ¶ 44.

91 *Concluding observations on the seventh periodic report of Chile*, CEDAW/C/CHL/CO/7 (21 February 2018) ["Chile observations"].

92 Ibid., ¶ 43.

93 Fiji observations, supra note 64.

94 Ibid., ¶ 53.

95 Ibid., ¶ 54.

96 *Concluding observations on the combined initial to third periodic reports of the Marshall Islands* CEDAW/C/MHL/CO/1–3 (2 March 2018) ["Marshall Islands observations"].

97 Ibid., ¶ 44.

98 Ibid., ¶ 45(a).

99 See *Concluding observations on the combined third and fourth periodic reports of Saudi Arabia*, CEDAW/C/SAU/CO/3–4 (27 February 2018), where the Committee noted that while the state party had adopted a climate action plan, no information was available on how women participated in the development of the plan and how a gender perspective was incorporated into mitigation and adaptation measures.

100 Ibid., ¶ 45.

101 *Concluding Observations on the eighth periodic report of the Republic of Korea*, CEDAW/C/KOR/CO/8 (22 February 2018) ["Korea observations"].

102 *Concluding observation on the combined fourth to sixth periodic reports of Suriname*, CEDAW/C/SUR/CO/4–6 (28 February 2018) ["Suriname observations"]. On the report of Jordan, the Committee commended it for its action plan and training manual on gender and climate change (2011) but noted that it required information on how women participated in its development and how a gender perspective was applied in the identification of adaptation and mitigation measures. It requested the state party to provide in its next periodic report information on the participation of women in the development of the plan; and best practices in adaptation and mitigation measures identified from a gender perspective. *Concluding observations on the sixth periodic report of Jordan*, CEDAW/C/JOR/CO/6 (16 February 2017), ¶¶ 51–52 ["Jordon observations"].

103 See *Concluding observations on the combined initial and second periodic reports of Nauru*, CEDAW/C/NRU/CO/1–2 (27 October 2017) ["Nauru observations"], where the Committee referred to the vulnerability of the state party to climate change which disproportionately affects women and girls and recommended ensuring their participation in planning and implementation of relevant policies and programs including decision-making bodies. On the report on Kenya, the Committee referred to the need to ensure the equal participation of rural women and girls in policy making on disaster mitigation and climate change and to implement the Climate Change Act of 2016 in a manner that prioritizes women's rights, *Concluding observations on the eighth periodic report of Kenya*, CEDAW/C/KEN/CO/8 (2 November 2017), ¶ 43.

104 CEDAW, *Concluding observations on the ninth periodic report of Norway*, CEDAW/C/NOR/CO/9 (7 November 2017) ["Norway observations"].

105 Ibid., ¶ 15.

106 See *Concluding observations on the combined eighth and ninth periodic reports of Haiti*, CEDAW/C/HTI/CO/8–9 (29 February and 1 March 2016) ["Haiti observations"]; *Concluding observations on the combined eighth and ninth periodic reports of Mongolia*, CEDAW/C/MNG/CO/8–9 (19 February 2016) ["Mongolia observations"]; *Concluding*

observations on the combined sixth and seventh periodic reports of Madagascar, CEDAW/C/MDG/CO/6–7 (24 November 2015) ["Madagascar observations"]. On the report of the Maldives, the Committee referred to the State party's vulnerability to climate change and the challenges it faces in disaster risk reduction and recommended ensuring women are represented and participate in decision-making and policymaking on plans and policies on disaster management and response to climate change. It recommended promoting *gender equality* as a component of such plans and policies. While reference has been made to incorporating a gender perspective in these policies and plans on many occasions, this is the first time the Committee has referred to the need to promote gender equality as a component. See *Concluding observations on the combined fourth and fifth periodic reports of Maldives*, CEDAW/C/MDV/CO/4–5 (11 March 2015), ¶¶ 42–43 ["Maldives observations"].

107 *Concluding observations on the combined third and fourth periodic reports of Tuvalu*, CEDAW/C/TUV/CO/3–4 (20 February 2015), ¶¶ 55–56 ["Tuvalu observations"]. See also comments on the report of Vanuatu where the Committee welcomed the adoption of the Climate Change and Disaster Risk Reduction Policy (2016–2030). Almost exact language as the Tuvalu report was used in relation to vulnerability of rural women. The Committee recommended seeking the active participation of women in decision-making and implementation processes on disaster risk reduction and climate policies. See *Concluding observations on the combined fourth and fifth periodic reports of Vanuatu*, CEDAW/C/VUT/CO/4–5 (9 March 2016), ¶¶ 36–37 ["Vanuatu observations"].

108 *Concluding observations on the combined initial and second report of Tuvalu*, CEDAW/C/TUV/CO/2 (29 July 2009) ["Second Tuvalu observations"].

109 Ibid., ¶ 31. It is noteworthy that the Committee did not confine this recommendation to women. The Committee noted the adoption of the 2007 Law on Natural Disaster Management by Indonesia and the need to include women, especially rural women, in the decision-making processes in policies and plans on poverty alleviation, sustainable development and natural disaster management. While the 2007 Law was aimed primarily at the victims of the 2005 Tsunami, it would be equally applicable to disasters caused by climate change. See *Concluding observations on the combined fourth and fifth periodic report of Indonesia*, CEDAW/C/IDN/CO/4–5 (27 July 2007), ¶¶ 38–39 ["Indonesia observations"].

110 Ibid., ¶ 32.

111 *Concluding observations on the combined eighth and ninth periodic reports of Bhutan*, CEDAW/C/BTN/CO/8–9 (25 November 2016), ¶ 34 ["Bhutan observations"]. The Committee acknowledged Sweden's leading role and commitment to the fight against climate change and achieving sustainable development and encouraged state party to continue to involve women in these efforts, *Concluding observations on the combined eighth and ninth periodic reports of Sweden*, CEDAW/C/SWE/CO/8–9 (10 March 2016), ¶ 5 ["Sweden observations"]. The Committee commended Honduras for including women as active participants in the formulation and implementation of national policies and action plans on climate change, disaster response and risk reduction, *Concluding observations on the combined seventh and eighth periodic reports of Honduras*, CEDAW/C/HND/CO/7–8 (25 November 2016), ¶ 43.

112 Bhutan observations, supra note 111, ¶ 35.

113 Korea observations, supra note 101, is an exception. While reference was made to 2030 Agenda, and the Committee called for the realization of substantive gender equality, the paragraph reproduced was missing. The concluding observations on the report of Saudi Arabia did not contain this paragraph either, although reference was made to the SDGs, *Concluding observations on the combined third and fourth periodic reports of Saudi Arabia*, CEDAW/C/SAU/CO/3–4 (27 February 2018).

114 For example, *Concluding observations on the combined initial, second and third periodic reports of Kiribati*, CEDAW/C/KIR/CO/1–3 (11 March 2020) ["Kiribati observations"];

Concluding observations on the combined fourth to seventh periodic reports of Latvia, CEDAW/C/LVA/CO/4–7 (11 February 2020); *Concluding observations on the sixth periodic report of Zimbabwe*, CEDAW/C/ZWE/CO/6 (13 February 2020). In addition, the Committee expressed satisfaction at the state party's commitment to the implementation of the SDGs. *Concluding observations on the fifth periodic report of Pakistan*, CEDAW/C/PAK/CO/5 (12 February 2020); and *Concluding observations on the third periodic report of Afghanistan*, CEDAW/C/AFG/CO/3 (18 February 2020). In its *Concluding observations on the combined sixth and seventh periodic reports of Luxembourg*, CEDAW/C/LUX/CO/6–7 (1 March 2018), the Committee called for the achievement of substantive equality when Implementing 2030 Agenda.

115 Kiribati observations, supra note 114, ¶ 14.

116 *Concluding observations on the seventh periodic report of Iraq*, CEDAW/C/IRQ/CO/7 (12 November 2019), ¶ 6.

117 *Concluding observations on the fourth periodic report of Andorra*, CEDAW/C/AND/CO/4 (23 October 2019).

118 Ibid., ¶ 45.

119 Ibid., ¶ 46.

120 For example, Suriname observations, supra note 102; *Concluding observations on the seventh periodic report of Burkina Faso*, CEDAW/C/BFA/CO/7 (24 October 2017) ["Burkina Faso observations"]; *Concluding observations on the combined second to fourth periodic reports of the Democratic People's Republic of Korea*, CEDAW/C/PRK/CO/2–4 (8 November 2017); *Concluding observations on the combined eighth and ninth periodic reports of Guatemala*, CEDAW/C/GTM/CO/8–9 (10 November 2017) ["Guatemala observations"]; *Concluding observations on the sixth periodic report of Israel*, CEDAW/C/ISR/CO/6 (31 October 2017); Nauru observations, supra note 103; *Concluding observations on the combined seventh and eighth periodic reports of Nigeria*, CEDAW/C/NGA/CO/7–8 (14 July 2017) ["Nigeria observations"]; Thailand observations, supra note 63; Jordan observations, supra note 102. The exact wording appears on the reports of Micronesia, supra note 84; *Concluding observations on the seventh periodic report of Turkey*, CEDAW/C/TUR/CO/7 (25 July 2016); *Concluding observations on the combined seventh and eighth periodic reports of Iceland*, CEDAW/C/ISL/CO/7–8 (10 March 2016); Madagascar, supra note 106; Vanuatu, supra note 107; Sweden, supra note 111; Bhutan, supra note 111; and *Concluding observations on the seventh periodic report of Argentina*, CEDAW/C/ARG/CO/7 (25 November 2016) ["Argentina observations"].

121 *Concluding observations on the combined initial, second and third periodic reports of Djibouti*, CEDAW/C/DJI/CO/1–3 (21 July 2011), ¶ 41 ["Djibouti observations"]; *Concluding observations on the combined initial, second and third report of Papua New Guinea*, CEDAW/C/PNG/CO/3 (22 July 2010) ["Papua New Guinea observations"]; Second Tuvalu observations, supra note 108, ¶ 60; Indonesia observations, supra note 109, ¶ 43.

122 *Concluding observations on the combined fourth to eighth periodic reports of Saint Vincent and the Grenadines*, CEDAW/C/VCT/CO/4–8 (20 July 2015), ¶ 50 ["Saint Vincent and the Grenadines observations"]; Tuvalu observations, supra note 107, ¶ 39; Maldives observations, supra note 106, ¶ 49; *Concluding observations on the combined initial to third periodic reports of Solomon Islands*, CEDAW/C/SLB/CO/1–3 (31 October 2014), ¶ 50 ["Solomon Islands observations"]; *Concluding observations on the combined fourth to sixth periodic reports of Iraq*, CEDAW/C/IRQ/CO/4–6 (18 February 2014), ¶ 56 ["Iraq observations"]; *Concluding observations on the combined fourth and fifth periodic reports of Cambodia*, CEDAW/C/KHM/CO/4–5 (8 October 2013), ¶ 49 ["Cambodia observations"].

123 Norway observations, supra note 104, ¶ 8.

124 Similar to the countries listed in note 120.

125 Norway observations, supra note 104, ¶ 8. Similar sentiments were expressed on the report of Oman, including commending measures taken to address climate change,

see *Concluding observations on the combined second and third periodic reports of Oman*, CEDAW/C/OMN/CO/2–3 (3 November 2017), ¶ 7; Barbados observations, supra note 83.

126 *Concluding observations on the eighth periodic report of the Democratic Republic of the Congo*, CEDAW/C/COD/CO/8 (6 August 2019), ¶ 46.

127 Ibid., ¶ 47.

128 Guyana observations, supra note 58, ¶ 44(c).

129 Ibid (emphasis added).

130 See UNDRIP (13 September 2007), available at: https://www.ohchr.org/en/indigenous-peoples/un-declaration-rights-indigenous-peoples.

131 Adopted in 1992 at the Rio Conference on Environment in Development, available at: https://www.cbd.int/doc/legal/cbd-en.pdf.

132 *Concluding observations on the seventh periodic report of Angola*, CEDAW/C/AGO/CO/7 (27 February 2019), ¶ 42(c).

133 Ibid. See also CEDAW, *Concluding observations on the eighth periodic report of Australia*, CEDAW/C/AUS/CO/8 (3 July 2018), ¶ 30(a), where the Committee recommended that the state party ensures that "all large-scale development and extractive industry projects are implemented with the prior informed consent of the local women affected, reflect adequate benefit-sharing arrangements and are preceded by a thorough gender impact assessment involving rural and indigenous women." In its Mexico observations, supra note 87, ¶ 46(c), the Committee recommended the establishment of a legal framework to ensure that development, agro-industrial and other business projects are implemented with the free, prior and informed consent of affected women and to ensure benefit sharing agreements concerning the use of natural resources and land in accordance with ILO Convention No 169; In its Chile observations, supra note 91, ¶¶ 46–47, the Committee recommended setting up a mandatory consultation mechanism to seek the free, prior and informed consent of indigenous women regarding the use of their natural resources and lands. In its Fiji observations, supra note 64, ¶ 50, the Committee called upon the state to "protect rural women from land grabbing and ensure that their prior informed consent and adequate benefit-sharing arrangements are required in any decisions regarding agreements on the lease or sale of land"; see Guatemala observations, supra note 120, ¶ 41, which refers to FPIC of indigenous women through consultations on any proposed intensive farming, development or extractive projects and activities on their traditional lands. See Thailand observations, supra note 63, where the Committee expressed concern at the restrictions women face regarding land and natural resources because of land acquisition for development projects, for use by mining and other extractive industries and zoning of national parks and referred to the need to obtain their free, prior and informed consent in such situations and provide compensation as necessary. On the report of Peru, the Committee recommended guaranteeing equal participation of women in rural and agricultural policies, to integrate a gender dimension into land acquisition and provide adequate compensation for large-scale land and other resource acquisitions, *Concluding observations on the combined seventh and eighth periodic reports of Peru*, CEDAW/C/PER/CO/7–8 (1 July 2014), ¶ 38.

134 United Kingdom observations, supra note 58, ¶ 53.

135 Ibid., ¶ 54(b).

136 Marshall Islands observations, supra note 96, ¶ 8.

137 *Report of the Special Rapporteur on the implications for human rights of the environmentally sound management and disposal of hazardous substances and wastes*, A/HRC/21/48/Add.1 (3 September 2012).

138 Marshall Islands observations, supra note 96, ¶ 8.

139 Ibid., ¶ 9.

140 Korea observations, supra note 101.

141 Suriname observations, supra note 102.
142 Ibid., ¶ 20.
143 Ibid., ¶ 21.
144 This cross-fertilization of law by referring to the IACtHR is significant.
145 Suriname observations, supra note 102, ¶ 45. See also Burkina Faso observations, supra note 120, ¶ 24, where the Committee noted that rural women have been disproportionately affected by the loss of fertile land and income and access to clean water, and recommended facilitating the acquisition and retention of land and natural resources by women, negotiating compensation for full recovery of losses, and establishing a legal framework to ensure that agro-industrial projects and extractive industries do not undermine rural women's right to land ownership and livelihoods. Such ventures should be permitted only after gender-impact assessments involving rural women.
146 Guatemala observations, supra note 120, ¶ 41.
147 Nauru observations, supra note 103, ¶¶ 34–35.
148 Micronesia observations, supra note 84, ¶ 42.
149 Mongolia observations, supra note 106, ¶ 32.
150 Ibid., 33(d).
151 Haiti observations, supra note 106 ¶ 37.
152 Ibid., ¶ 38. The Committee also made recommendations on climate change and natural disasters.
153 Solomon Islands observations, supra note 122, ¶ 39.
154 Iraq observations, supra note 122.
155 Cambodia observations, supra note 122, ¶ 42.
156 Ibid., ¶ 43.
157 Djibouti observations, supra note 121. On the report of Argentina, the Committee stressed the need to ensure that indigenous women have adequate access to safe and affordable water for personal and domestic use and for irrigation, Argentina observations, supra note 120.
158 Second Tuvalu observations, supra note 108. Same remarks were made in the Papua New Guinea observations, supra note 121.
159 *Concluding observations on combined initial and second periodic report of Mozambique, CEDAW/C/MOZ/1–2 (23 May 2007), and Concluding observations on the combined initial and second periodic report of Suriname, CEDAW/C/SUR/1–2 (2002),* where the Committee referred to lack of services and environmental pollution.
160 *Initial report of Kazakhstan*, CEDAW/C/KAZ/1 (18 and 23 January 2001). Similar comments were made on initial report of Uzbekistan, CEDAW/C/UZB/1 (25 and 30 January 2001).
161 Mauritius, CEDAW/C/MAR/1–2, in *Report of the Committee on the Elimination of Discrimination Against Women, Fourteenth Session* (31 May 1995).
162 Poland, CEDAW/C/5/Add.31, in *Report of the Committee on the Elimination of Discrimination Against Women, Fourteenth Session* (15 May 1987).
163 Cabo Verde observations, supra note 59.
164 Nigeria observations, supra note 120, ¶ 39(b).
165 SDG 5 refers to gender quality, SDG 7 refers to affordable and clean energy while SDG 13 refers to combating climate change.
166 Guatemala observations, supra note 120, ¶ 28. Similar concerns were expressed in Thailand observations, supra note 63, ¶ 30.
167 Cambodia observations, supra note 122, ¶ 42.
168 Thailand observations, supra note 63.
169 Ibid., ¶ 3.
170 See Chapter 6 where the ESCR Committee also extended this principle beyond indigenous groups.
171 ESCR Committee too made similar recommendations, see Chapter 6.

172 GR No 34, supra note 16. Here too we see the reference to *natural* disasters. See also GR No 36 (2017) which recognizes that women and children are the most vulnerable groups during disasters. *General Recommendation No 36 on the right of girls and women to education*, CEDAW/C/GC/36 (27 November 2017).

References

Concluding observations on combined eighth and ninth periodic reports of Bhutan (2016), CEDAW/C/BTN/CO/8–9.

Concluding observations on combined eighth and ninth periodic reports of Guatemala (2017), CEDAW/C/GTM/CO/8–9.

Concluding observations on combined eighth and ninth periodic reports of Haiti (2016), CEDAW/C/HTI/CO/8–9.

Concluding observations on combined eighth and ninth periodic reports of the Lao People's Democratic Republic (2018), CEDAW/C/LAO/CO/8–9.

Concluding observations on combined eighth and ninth periodic reports of Mongolia (2016), CEDAW/C/MNG/CO/8–9.

Concluding observations on combined eighth and ninth periodic reports of Sweden (2016), CEDAW/C/SWE/CO/8–9.

Concluding observations on combined fifth to eighth periodic reports of Barbados (2017), CEDAW/C/BRB/CO/5–8.

Concluding observations on combined fourth and fifth periodic reports of Cambodia (2013), CEDAW/C/KHM/CO/4–5.

Concluding observations on combined fourth and fifth periodic report of Indonesia (2007), CEDAW/C/IDN/CO/4–5.

Concluding observations on combined fourth and fifth periodic reports of Maldives (2015), CEDAW/C/MDV/CO/4–5.

Concluding observations on combined fourth and fifth periodic reports of Vanuatu (2016), CEDAW/C/VUT/CO/4–5.

Concluding observations on combined fourth to eighth periodic reports of Saint Vincent and the Grenadines (2015), CEDAW/C/VCT/CO/4–8.

Concluding observations on combined fourth to sixth periodic reports of Iraq (2014), CEDAW/C/IRQ/CO/4–6.

Concluding observations on combined fourth to sixth periodic reports of Suriname (2018), CEDAW/C/SUR/CO/4–6.

Concluding observations on combined fourth to seventh periodic reports of Antigua and Barbuda (2019), CEDAW/C/ATG/CO/4–7.

Concluding observations on the combined fourth to seventh periodic reports of Latvia (2020), CEDAW/C/LVA/CO/4–7.

Concluding observations on combined initial and second periodic report of Mozambique (2007), CEDAW/C/MOZ/1–2.

Concluding observations on combined initial and second periodic reports of Nauru (2017), CEDAW/C/NRU/CO/1–2.

Concluding observations on the combined initial and second periodic report of Suriname (2002), CEDAW/C/SUR/1–2.

Concluding observations on combined initial and second report of Tuvalu (2009), CEDAW/C/TUV/CO/2.

Concluding observations on combined initial to third periodic reports of Djibouti (2011), CEDAW/C/DJI/CO/1–3.

Concluding observations on combined initial to third periodic reports of the Federated States of Micronesia (2017), CEDAW/C/FSM/CO/1–3.

Concluding observations on combined initial to third periodic reports of Kiribati (2020), CEDAW/C/KIR/CO/1–3.

Concluding observations on combined initial to third periodic reports of the Marshall Islands (2018), CEDAW/C/MHL/CO/1–3.

Concluding observations on combined initial to third report of Papua New Guinea (2010), CEDAW/C/PNG/CO/3.

Concluding observations on combined initial to third periodic reports of Solomon Islands (2014), CEDAW/C/SLB/CO/1–3.

Concluding observation on combined second and third periodic reports of the Cook Islands (2018), CEDAW/C/COK/CO/2–3.

Concluding observations on combined second and third periodic reports of Oman (2017), CEDAW/C/OMN/CO/2–3.

Concluding observations on the combined second to fourth periodic reports of the Democratic People's Republic of Korea (2017), CEDAW/C/PRK/CO/2–4.

Concluding observations on combined seventh and eighth periodic reports of Honduras (2016), CEDAW/C/HND/CO/7–8.

Concluding observations on combined seventh and eighth periodic reports of Iceland (2016), CEDAW/C/ISL/CO/7–8.

Concluding observations on combined seventh and eighth periodic reports of Nigeria (2017), CEDAW/C/NGA/CO/7–8.

Concluding observations on combined seventh and eighth periodic reports of Peru (2014), CEDAW/C/PER/CO/7–8.

Concluding observations on combined sixth and seventh periodic reports of Luxembourg (2018), CEDAW/C/LUX/CO/6–7.

Concluding observations on combined sixth and seventh periodic reports of Madagascar (2015), CEDAW/C/MDG/CO/6–7.

Concluding observations on combined sixth and seventh periodic reports of Thailand (2017), CEDAW/C/THA/CO/6–7.

Concluding observations on combined third and fourth periodic reports of Saudi Arabia (2018), CEDAW/C/SAU/CO/3–4.

Concluding observations on combined third and fourth periodic reports of Tuvalu (2015), CEDAW/C/TUV/CO/3–4.

Concluding observations on combined third to fifth periodic reports of Mozambique (2019), CEDAW/C/MOZ/CO/3–5.

Concluding observations on eighth periodic report of Australia (2018), CEDAW/C/AUS/CO/8.

Concluding observations on eighth periodic report of Cyprus (2018), CEDAW/C/CYP/CO/8.

Concluding observations on eighth periodic report of the Democratic Republic of the Congo (2019), CEDAW/C/COD/CO/8.

Concluding observations on eighth periodic report of Kenya (2017), CEDAW/C/KEN/CO/8.

Concluding observations on eighth periodic report of Mauritius (2018), CEDAW/C/MUS/CO/8.

Concluding observations on eighth periodic report of New Zealand (2018), CEDAW/C/NZL/CO/8.

Concluding observations on eighth periodic report of the Republic of Korea (2018), CEDAW/C/KOR/CO/8.

Concluding observations on eighth periodic report of the United Kingdom of Great Britain and Northern Ireland (2019), CEDAW/C/GBR/CO/8.

Concluding observations on fifth periodic report of Fiji (2018), CEDAW/C/FJI/CO/5.
Concluding observations on fifth periodic report of Pakistan (2020), CEDAW/C/PAK/CO/5.
Concluding observations on fourth periodic report of Andorra (2019), CEDAW/C/AND/CO/4.
Concluding observations on fourth periodic report of Botswana (2019), CEDAW/C/BWA/CO/4.
Concluding observations on fourth periodic report of Côte d'Ivoire (2019), CEDAW/C/CIV/CO/4.
Concluding observations on ninth periodic report of Cabo Verde (2019), CEDAW/C/CPV/CO/9.
Concluding observations on ninth periodic report of Guyana (2019), CEDAW/C/GUY/CO/9.
Concluding observations on ninth periodic report of Mexico (2018), CEDAW/C/MEX/CO/9.
Concluding observations on ninth periodic report of Norway (2017), CEDAW/C/NOR/CO/9.
Concluding observations on second periodic report of Qatar (2019), CEDAW/C/QAT/CO/2.
Concluding observations on seventh periodic report of Angola (2019), CEDAW/C/AGO/CO/7.
Concluding observations on seventh periodic report of Argentina (2016) CEDAW/C/ARG/CO/7.
Concluding observations on seventh periodic report of Burkina Faso (2017), CEDAW/C/BFA/CO/7.
Concluding observations on seventh periodic report of Chile (2018), CEDAW/C/CHL/CO/7.
Concluding observations on seventh periodic report of Iraq (2019), CEDAW/C/IRQ/CO/7.
Concluding observations on seventh periodic report of Turkey (2016), CEDAW/C/TUR/CO/7.
Concluding observations on sixth periodic report of the Bahamas (2018), CEDAW/C/BHS/CO/6.
Concluding observations on sixth periodic report of Cambodia (2019), CEDAW/C/KHM/CO/6.
Concluding observations on sixth periodic report of Israel (2017), CEDAW/C/ISR/CO/6.
Concluding observations on sixth periodic report of Jordan (2017), CEDAW/C/JOR/CO/6.
Concluding observations on sixth periodic report of Nepal (2018), CEDAW/C/NPL/CO/6.
Concluding observations on sixth periodic report of the former Yugoslav Republic of Macedonia (2018), CEDAW/C/MKD/CO/6.
Concluding observations on sixth periodic report of Samoa (2018), CEDAW/C/WSM/CO/6.
Concluding observations on sixth periodic report of Seychelles (2019), CEDAW/C/SYC/CO/6.
Concluding observations on sixth periodic report of Zimbabwe (2020), CEDAW/C/ZWE/CO/6.
Concluding observations on third periodic report of Afghanistan (2020), CEDAW/C/AFG/CO/3.
Convention on Biological Diversity (1992), https://www.cbd.int/doc/legal/cbd-en.pdf.
General Recommendation No. 21 (1997), https://tbinternet.ohchr.org/Treaties/CEDAW/Shared%20Documents/1_Global/INT_CEDAW_GEC_4736_E.pdf.
General recommendation No. 27 (2010), CEDAW/C/GC/27.
General recommendation No. 34 (2016), CEDAW/C/GC/34.
General Recommendation No. 35 (2018), CEDAW/C/GC/35.
General recommendation No. 36 (2017), CEDAW/C/GC/36.
General Recommendation No. 37 (2018), CEDAW/C/GC/37.
Initial report of Kazakhstan (2001), CEDAW/C/KAZ/1.
Initial report of Uzbekistan (2001), CEDAW/C/UZB/1.
Mapping Human Rights Obligations Relating to the Enjoyment of a Safe, Clean, Healthy and Sustainable Environment, Individual Report on the UN Convention on the Elimination of All Forms of Discrimination against Women, Report No. 4 (2013), https://www.ohchr.org/en/special-procedures/sr-environment/mapping-report.

Optional Protocol to the Convention on the Elimination of All Forms of Discrimination against Women (2000), https://www.ohchr.org/EN/ProfessionalInterest/Pages/OPCEDAW.aspx.

Report of the Committee on the Elimination of Discrimination against Women (1987), https://undocs.org/A/42/38.

Report of the Committee on the Elimination of Discrimination against Women (1995), https://undocs.org/A/50/38.

Report of the Special Rapporteur on the implications for human rights of the environmentally sound management and disposal of hazardous substances and wastes, A/HRC/21/48/Add.1 (2012), https://undocs.org/A/HRC/21/48/Add.1.

Resolution 34/180 (1979), A/RES/34/180.

SDGs (2022), https://sdgs.un.org.

UN CEDAW, https://www.ohchr.org/en/instruments-mechanisms/instruments/convention-elimination-all-forms-discrimination-against-women.

UNDRIP (2007), https://www.ohchr.org/en/indigenous-peoples/un-declaration-rights-indigenous-peoples.

UNFCCC, Decision 18/CP.20 (2014), https://unfccc.int/resource/docs/2014/cop20/eng/10a03.pdf#page=35%22.

UNFCCC, Decision 3/CP.25 (2019), https://unfccc.int/sites/default/files/resource/cp2019_13a01E.pdf.

8 Committee on the Rights of the Child

8.1 Introduction and Mandate

It is well known that children are more susceptible to environmental degradation than adults. Climate change is a good example of a present-day challenge that has repercussions across generations. As the World Commission on Environment and Development cautioned as far back as 1987 our children will inherit the losses of our shortsighted decisions today:

> We borrow environmental capital from future generations with no intention or prospect of repaying. They may damn us for our spendthrift ways, but they can never collect on our debt to them. We act as we do because we can get away with it: future generations do not vote; they have no political or financial power; they cannot challenge our decision.[1]

It is against this backdrop that we must read the Convention on the Rights of the Child (CRC) which was opened for signature in 1989[2] and entered into force in 1990, in accordance with Article 49.[3] Except for the United States,[4] every state in the world has ratified the Convention, becoming the first and so far the only human rights treaty to have achieved near-universal ratification. Unlike the two Covenants adopted almost three decades prior to that, the CRC embodies both civil and political rights, and ESC rights in relation to the child, defined as a person under the age of 18 years.[5] Similar to CEDAW, the CRC is a specialized treaty applicable exclusively to children, in tandem with other human rights treaties.

A Committee is established under Article 43 of the Convention. The main purpose of the Committee is to examine the progress made by state parties in fulfilling the obligations under the Convention. The members of the Committee are elected by state parties for a term of four years among their nationals who serve in their personal capacity. The Committee consists of 18 experts of high moral standing and recognized competence in the field covered by the Convention.

State parties have undertaken to take "all appropriate legislative, administrative, and other measures"[6] to implement the rights recognized in the Convention.

DOI: 10.4324/9781003128847-11

Similar to the ICESCR, ESC rights are to be fulfilled to the maximum extent of available resources and through international cooperation.[7] State parties have undertaken to submit to the Committee reports on the measures they have adopted to give effect to the rights in the Convention within two years the Convention enters into force for the state party and every five years thereafter.[8] Moreover:

> Reports made under the present article shall indicate factors and difficulties, if any, affecting the degree of fulfilment of the obligations under the present Convention. Reports shall also contain sufficient information to provide the Committee with a comprehensive understanding of the implementation of the Convention in the country concerned.[9]

The Committee is required to submit to the UNGA, through ECOSOC, a report on its activities every two years.[10]

Three optional protocols to the CRC have been adopted. These are on children in armed conflict, child prostitution, and pornography, and the individual communications procedure.[11] Parties to the Protocol recognize the competence of the Committee as provided for by the Protocol. According to Article 5 "communications may be submitted by or on behalf of an individual or group of individuals, within the jurisdiction of a State party, claiming to be victims of a violation by that State party" of any of the rights in the Convention and the two Optional Protocols.[12] Under Article 6, the Committee may issue interim measures "in exceptional circumstances to avoid possible irreparable damage to the victim or victims of the alleged violations."[13] This is the only treaty body empowered to issue interim measures to avoid irreparable damage to the victim.[14] Similar to all other Committees, domestic remedies must be exhausted before a communication can be submitted.[15]

The CRC is the first (and so far, the only) global human rights treaty to embody a provision on environmental pollution. It calls upon states to take measures to:

> [c]ombat disease and malnutrition, including within the framework of primary health care, through, inter alia, the application of readily available technology and through the provision of adequate nutritious foods and clean drinking-water, taking into consideration the dangers and risks of *environmental pollution*.[16]

Moreover, states must ensure that people, in particular parents and children, are informed, and have access to education on, *inter alia*, hygiene and environmental sanitation.[17]

This chapter discusses the work of the CRC Committee in relation to the topics examined in this book: environmental protection, climate change and sustainable development, including SDGs. It will use the individual report prepared for the Independent Expert on Human Rights and Environment[18] as background for a discussion of environmental issues up to December 2013. This chapter will discuss developments since then drawing upon this excellent summary.

8.2 General Comments

Currently, there are no General Comments specifically on environmental degradation or climate change, although this is about to change. The GC No 15 on health[19] contains some relevant provisions. Recognizing that children's health is affected by many factors, the Committee referred to the impact of climate change and rapid urbanization on children's health. The GC elaborated on Article 24's reference to clean water and environmental pollution:

> States should take measures to address the dangers and risks that local environmental pollution poses to children's health in all settings. Adequate housing that includes non-dangerous cooking facilities, a smoke-free environment, appropriate ventilation, effective management of waste and the disposal of litter from living quarters and the immediate surroundings, the absence of mould and other toxic substances, and family hygiene are core requirements to a healthy upbringing and development. States should regulate and monitor the environmental impact of business activities that may compromise children's right to health, food security and access to safe drinking water and to sanitation.[20]

The GC referred to the need to recognize the challenges to children's health by humanitarian emergencies, including large-scale displacements due to natural or man-made disasters. While not referring specifically to climate change here, the reference to "man-made disasters" could include climate-induced disasters. It is noteworthy that the GC refers to the need to regulate and monitor the environmental impact of business activities. The GC recognized climate change as one of the biggest threats to children's health:

> The Committee draws attention to the relevance of the environment, beyond environmental pollution, to children's health. Environmental interventions should, inter alia, address climate change, as this is one of the biggest threats to children's health and exacerbates health disparities. States should, therefore, put children's health concerns at the centre of their climate change adaptation and mitigation strategies.[21]

The GC further elaborated on the nature of states' obligations in fulfilling their obligations by adopting the respect, protect, and fulfill typology applicable to human rights generally: "to *respect* freedoms and entitlements, to *protect* both freedoms and entitlements from third parties or from social or *environmental threats*, and to *fulfill* the entitlements through facilitation or direct provision."[22]

In addition, the provision on non-refoulement in the joint GC adopted with the Committee on Migrant Workers could be useful in the context of climate change although the GC itself does not refer to that. Noting that this principle has been interpreted "to be an implicit guarantee flowing from the obligations to respect, protect and fulfil human rights," the GC highlights that it prohibits

States from removing individuals from their jurisdiction "when they would be *at risk of irreparable harm* upon return, including persecution, torture, gross violations of human rights, or *other irreparable harm*."[23] In light of the decision in the *Teitiota Case*,[24] it could be argued that "irreparable harm" would include damage caused by climate change or serious risk to life. The Committees pointed out that some states interpreted the non-refoulement principle narrowly. However, as we discussed in Chapter 5, in the *Teitiota case*, the Human Rights Committee decided that the non-refoulement principle applies to serious threats to life caused by climate change. In light of that decision, this provision could be used by potential claimants affected by climate change even though the threshold laid down in that case was very high.[25]

GC No 16 that addresses the impact of the business sector on children's rights[26] is also relevant to our discussion. It recognizes that childhood is a unique period of development and exposure to violence, child labor, unsafe products, or *environmental hazards* "may have lifelong, irreversible and even transgenerational consequences."[27] In drawing up the GC, the Committee held consultations with many stakeholders including children and examined existing frameworks such as ILO Conventions, OECD guidelines, and the Guiding Principles on Business and Human Rights.

GC No 16 makes several references to environmental pollution, and hazardous conditions. Under general principles relating to the right to life, survival, and development (article 6), the GC recognizes that activities of business enterprises can impact the realization of article 6 in several ways. It notes that environmental degradation and contamination due to business activities can compromise children's rights to health, food security, and access to safe drinking water and sanitation. Moreover, alienating land to investors can deprive local populations of access to natural resources impacting their subsistence and cultural heritage, which may especially affect the rights of indigenous children.[28] The GC points out that states should adopt preventive measures such as effective regulation, monitoring of activities and evaluating the environmental impact of businesses.

Applying the typology of obligations of respect, protect and fulfill, the GC notes that the obligation to fulfill includes adopting laws and standards on, *inter alia*, environment, and the obligation to provide effective remedies and reparation for violations include environmental tribunals. Moreover, reparation should be timely to limit ongoing and future damage to the child. For example, if children are affected by environmental pollution, immediate steps should be taken by all relevant parties to prevent further damage to the health and development of the child and to repair any damage done. Additionally, to ensure that the best interests of the child are a primary consideration in legislation and policy development, states should adopt *continuous child-rights impact assessments* which should complement ongoing monitoring and evaluation of laws, policies, and programs on children's rights.[29] Continuous child-rights impact assessment is a novel tool advanced by the CRC Committee, similar to the notion of continuous environmental impact assessments proposed by Judge Weeramantry in the *Gabcikovo Nagymaros Case*.[30]

8.2.1 *Proposed General Comment on Environment and Climate Change*

In June 2021 the CRC Committee decided to draft a GC on children's rights and the environment with a special focus on climate change.[31] The Committee held consultations with children and young people around the world during February-May 2022. An online consultation and regional workshops on the draft GC were also held. Several states have commented on the concept note prepared by the Committee.

The concept note recognizes that environmental harm affects children more than adults and that every year over 1.5 million children under the age of five years die as a result of air pollution, water pollution, exposure to toxic substances and other types of environmental harm and contribute to increased incidence of asthma, diabetes, cancer, and early mortality. Moreover, the rise in zoonotic diseases caused by environmental degradation and biodiversity loss gives rise to viral epidemics. The climate crisis, described as the defining crisis of our time, continues to threaten both civil and political rights and ESC rights of present and future generations. Referring to the OHCHR report on climate change and children, the concept note stated: "climate change threatens children's rights to health, life, food, water and sanitation, education, housing, culture and development, among other rights."[32] Climate change will cause millions more deaths from malaria, diarrheal diseases, and undernutrition. Moreover, children with disability, children on the move, children in poverty, and those separated from their families will be at a greater risk. Due to the increase in frequency and severity of extreme weather events and disasters, children will be affected by heat waves, drought, wildfires, and floods. The concept note recognizes the disproportionate impact on certain communities including children who contributed little to the problem and that climate change will exacerbate inequality around the world. Moreover, children will experience the risks, burdens and impacts of climate change differently than adults.[33]

With regard to the scope of the General Comment, the Committee plans to highlight three issues – environmental pollution; loss of biodiversity; and climate change – focusing on the four general principles in the CRC: non-discrimination; best interests of the child; the right to life, survival and development; and obtaining the views of the child. Relying on the typology adopted by the UN Special Rapporteur on Human Rights and Environment, the Committee plans to cover all three types of obligations in the GC – substantive, procedural, and heightened obligations – that are owed to children in the context of the environment, with a special focus on climate change.[34]

The main objective of the GC is to provide authoritative guidance to state parties to undertake appropriate legislative, administrative, and other measures toward a child-rights approach to environmental issues, especially climate change. Toward fulfilling that objective, the GC seeks to: (a) clarify states' obligations relating to climate change and children's rights, including mitigation and adaptation; (b) emphasize the urgent need to address the adverse impacts of environmental harm and climate change on children; (c) clarify the relationship between children's rights and the protection of ecosystems, biodiversity and natural resources and state's obligations; (d) clarify how children could exercise their procedural rights[35] to protect against environmental harm; and (e) clarify the legal implications of concepts such as "international cooperation," extraterritorial obligations," "intergenerational equity," and "future generations."[36]

8.3 Concluding Observations

Similar to other treaty bodies, the CRC Committee has made comments on environmental protection/degradation and climate change in its concluding observations. An examination of concluding observations shows that the Committee has commented on three broad areas that overlap with environmental issues: climate change; environmental sanitation and health; and activities of businesses. In addition, almost all the concluding observations referred to SDGs as a yardstick to measure states' compliance, most of which are outside the issues discussed in this volume.

8.3.1 Climate Change

On the reports of the Cook Islands, the Committee made extensive comments on the impact of climate change on children's rights.[37] While noting the measures taken on adaptation and resilience, including disaster risk reduction, the Committee expressed concern about the adverse impact of climate change and natural disasters on children, including the rights to life, survival and development, education, health, adequate housing, and safe drinking water and sanitation. It expressed concern that the needs of children, including children with disabilities, have not been adequately included in policies and programs on climate action, including climate-related disasters and the lack of opportunities for children to participate in decision-making related to climate action.[38]

Drawing attention to SDG target 13.5[39] the Committee recommended taking into account the vulnerabilities and needs of children, and their views in developing policies and programs on climate change and disaster risk management; collecting disaggregated data identifying the different types of risk faced by children in relation to disasters that is necessary to formulate policies and frameworks; and strengthening measures to increase awareness and preparedness among children for climate change and natural disasters, and integrate these issues into the national curriculum. It further recommended the state party to seek bilateral, multilateral and regional cooperation in implementing the recommendations.[40]

The Committee welcomed the enactment of the 2013 Climate Change Law in the Federated States of Micronesia,[41] but expressed concern that the state party had not integrated vulnerabilities and needs of children into policies and programs addressing climate change and disaster risk management.[42] Referring to several SDG targets,[43] the Committee recommended that the state party: ensure that vulnerabilities and needs of children and their views are taken into account when developing policies and programs on climate change and disaster risk management;[44] increase awareness among children and their preparedness for climate change and natural disasters, by incorporating them into school curriculum and teacher training programs[45] and increasing the safety of school infrastructure; review emergency protocols and support for children including those with disabilities during emergencies and natural disasters; and improve data collection and assessment of risks and in particular the needs of children with disabilities.[46]

The Committee referred to mitigation commitments on the report of Austria[47] and recommended ensuring that its policies on climate mitigation and reduction of GHG emissions are compatible with the Convention, including the rights to health and an adequate standard of living, and that the vulnerabilities and needs of children, and their views, are considered when implementing these policies. It further recommended conducting an assessment of policies relating to the transport sector and the impacts of atmospheric pollution and GHG emissions on children's rights as a basis for designing a strategy, and eliminating any subsidies that undermine children's right to health.[48] The recommendations included: taking the views of children into account on policies and programs on climate change and disaster risk management; collecting disaggregated data identifying the risks faced by children,[49] and strengthening measures to increase children's awareness and preparedness for climate change and disasters; and providing climate education, and opportunities to participate in discussions and decision-making on climate action. In addition, the Committee recommended strengthening national policies for sustainable safe water supplies and sanitation to increase access to sufficient safe drinking water and providing adequate sanitation, including in the outer islands and seeking international cooperation in implementing these recommendations.[50] With regard to children and migration the Committee recommended developing legislation, policies, and programs governing the international migration of children.[51]

On the report of Antigua and Barbuda,[52] the Committee referred to the state party's vulnerability to natural disasters. Relying on target 1.5 of the SDGs[53] the Committee recommended identifying the types of risks children would face during natural disasters and ensuring that their needs and views are taken into account in developing policies and programs addressing climate, environmental change, and disaster risk management. It called upon the state party to seek international cooperation in relation to disaster risk reduction, and mitigation and adaptation measures with regard to climate change and environmental damage.[54]

On the report of Australia, the Committee expressed concern about the impact of climate change on children's rights and the state party's position that the Convention does not extend to protection from climate change.[55] Reminding the state party of the indivisibility and interdependence of rights in the Convention, the Committee stressed that climate change impacts affect children's rights, especially rights to life, survival and development, non-discrimination, health, and an adequate standard of living.[56] The Committee expressed concern about the insufficient progress made on the goals and targets in the Paris Agreement and continuing investment in extractive industries, in particular coal. The Committee expressed its:

> [d]isappointment that a protest led by children calling on government to protect the environment received a strongly worded negative response from those in authority, which demonstrates disrespect for the right of children to express their views on this important issue.[57]

Drawing attention to SDG target 13.5, the Committee urged the state party to ensure that children's views are considered in developing policies on climate change, the environment and disaster risk management; increase children's awareness and preparedness for climate change and natural disasters[58]; and take measures to reduce its GHG emissions by establishing targets and deadlines to phase out the domestic use and export of coal. It also urged the state party to accelerate the transition to renewable energy and commit to meet its electricity needs 100% with renewable energy.[59] Referring to GC No 12, the Committee recommended developing toolkits to hold public consultations with children on issues that affect them, including climate change and the environment.[60]

The Committee welcomed the adoption of the second NAPA by Bhutan to reduce climate-related risks and vulnerabilities.[61] Referring to SDG target 13.5, the Committee recommended that the special vulnerabilities of children, as well as their views, are taken into account when developing policies and programs addressing climate change and disaster risk management;[62] increase children's awareness and preparedness for climate change by incorporating these issues into the school curriculum and teacher training programs; and develop sustainable systems for water management and supply to address the drying up of spring water sources and prevent children from having to carry water.[63]

On the report of Cabo Verde, the Committee expressed concern about the lack of information on the impact of climate change on children's rights, as the state party is already experiencing consequences such as freshwater shortage, sea level rise, changes in rainfall patterns, desertification, and higher temperatures.[64] Drawing attention to SDG target 13.b, the Committee recommended that the State party take into account the vulnerabilities, needs, and views of children when developing policies and programs on climate change and disaster risk management;[65] collect disaggregated data identifying the types of risk faced by children to disasters; and increase children's awareness and preparedness for climate change and natural disasters by incorporating them into the school curriculum and teacher training programs.[66]

On the combined reports of Fiji, the Committee welcomed the adoption of the National Climate Change Policy and the establishment of the National Climate Change Coordinating Committee, and the efforts by the State party to teach children about climate change and measures to address it.[67] However, the Committee expressed concern about not providing for children's voices to be heard on decisions on climate change especially those living in coastal and low-lying areas where climate change is causing salinization and impacting agricultural land.

Drawing attention to reports that children face higher risks from disasters and are more vulnerable to climate change than adults, the Committee recommended considering the special needs and vulnerabilities of children and their views when developing policies and programs to address climate change and disaster risk management. The Committee further recommended strengthening social protection systems for children and families affected by climate change; adopting national legislation and policies to increase access to safe drinking water, and providing adequate sanitation in particular on the outer islands; increasing children's awareness and preparedness for climate change and disasters by incorporating it to school curriculum and teacher training programs[68] and seek international cooperation in implementing these recommendations.[69]

On the report of Jamaica, the Committee welcomed the guidelines on child-friendly disaster management and response through its Office of Disaster Preparedness and Emergency Management[70] but expressed concern at the impact of climate change and disasters on children, including the rights to education, health, housing, safe and drinkable water and sanitation, and the potential to undermine social safety nets, especially for those in poverty. The Committee recommended developing strategies to reduce the vulnerabilities of children exacerbated by climate change by including vulnerability reduction strategies into its national plan on climate change and disaster preparedness, and strengthening its social safety nets to mitigate social, economic, and environmental impacts of climate change.[71]

Further to similar recommendations made on several reports, the Committee called upon Japan to ensure that climate mitigation policies are compatible with the Convention, and to reduce GHG emissions in line with its international commitments to avoid threatening children's rights, particularly the rights to health, food, and an adequate standard of living.[72] It called upon the state party to reconsider funding coal-fired power plants in other countries and replace them with sustainable sources energy, and seek international cooperation in implementing these recommendations.[73]

The Committee, relying on SDG target 13.b, referred to the need to provide sufficient funds to implement the disaster risk management policy in Malawi.[74] It recommended integrating disaster risk reduction into development planning; registering persons affected, particularly vulnerable groups such as children, as part of disaster response to ensure that they receive appropriate services; and increasing children's preparedness for climate change and disasters by incorporating it into the school curriculum and teacher training programs.[75]

Similar to other treaty bodies, the CRC Committee also made extensive comments on the report of the Marshall Islands on the impact of climate change on children's rights.[76] While welcoming the adoption of the Joint National Action Plan for Climate Change Adaptation and Disaster Risk Management 2014–2018 and the National Climate Change Policy Framework, the Committee expressed concern that climate adaptation and disaster risk reduction is not sufficiently incorporated into the school curriculum; the social protection system is not disaster-sensitive; special needs of children, including those with disabilities, are not included in disaster risk planning and evacuation centers, particularly in the outer islands.[77] Referring to SDG target 13.b, the Committee recommended implementing its National Action Plan; increasing children's preparedness for climate change and disasters by including it in the school curriculum and establish early warning systems in schools; developing a disaster-sensitive social protection system taking into account the views of children; reviewing emergency protocols to assist children with disabilities during emergencies and disasters; increasing the number of evacuation centers; and improving data to establish an evidence base for risk reduction.[78]

The Committee welcomed the adoption of Palau's Climate Change Policy for Climate and Disaster Resilient Low Emissions Development[79] but expressed concern about the lack of information on whether climate adaptation and disaster risk reduction are part of the school curriculum, whether there is a disaster-sensitive social protection system in place, and the measures regarding children in vulnerable situations, including children with disabilities, when planning for disasters.[80] Referring again to SDG target 13.b, the Committee recommended including climate adaptation and disaster risk reduction in the school curriculum and establishing early warning systems and trainings on disasters; developing a social protection system taking into account vulnerabilities of children and their views; reviewing emergency protocols to include support for children with disabilities; improving data and assessments to have an evidence base for risk reduction and preparedness taking into account the needs of preschool children and children with disabilities; and seeking international cooperation in implementing these recommendations.[81]

On the report of Switzerland the Committee voiced its concern about the high carbon footprint through investments made in fossil fuels by its financial institutions, and the negative impact of climate change and air pollution on children's health.[82] Referring to SDG targets 3.9[83] and 13.3,[84] the Committee recommended: reducing GHG emissions in line with its international commitments including achieving net zero emissions by 2050; assessing policies and practices relating to aviation and transportation, and the impact of atmospheric pollution and GHG emissions on children's rights and design a strategy to remedy the situation, including investments in carbon-neutral technologies; ensuring that financial institutions consider the implications of their investments with regard to climate change and the harmful impacts on children, and *adopting binding rules*[85]; strengthening awareness-raising among children on climate change and environmental health, relevant air quality and climate legislation and their right to health; ensuring that children's needs and views are taken into account when developing policies and programs addressing climate change; and collecting data on the impact of climate change on children and provide information in its next report.[86]

The Committee referred to the increased frequency and intensity of natural disasters experienced by Tajikistan partly due to climate change and human and property losses as well as damage to the socio-economic and cultural infrastructure.[87] It urged the state party to adopt a child-focused approach to climate adaptation and include children in the development of the National Climate Change Adaptation Strategy (2016–2030), with special attention to children with disabilities and girls taking note of SDG target 13.5.[88]

On the report of Tonga, the Committee noted that it is among the most vulnerable countries to natural hazards and welcomed the revised 2018 Joint National Action Plan on Climate Change and Disaster Risk Management but pointed out that more could be done to include special needs of children and those with disabilities and improve school infrastructure especially in rural areas.[89] Drawing attention to SDG targets 13.1, 13.3 and 13.b, the Committee recommended: implementing the National Action Plan addressing gaps in the previous plan; building children's awareness and preparedness for climate change and disasters; reviewing emergency protocols to include assistance for children, particularly those with disabilities; improving data to have an evidence base for risk reduction and preparedness, including children with disabilities; and investing resources in health care, given the major public health burden of climate change.[90]

On the initial report of Tuvalu,[91] the Committee welcomed the adoption of the National Climate Change Policy and the National Strategic Action Plan for Climate Change and Disaster Risk Management but expressed deep concern at the adverse impact of climate change and natural disasters on children's rights to education, health, housing, and safe drinking water and sanitation. Referring to the concerns raised by the Special Rapporteur on Right to Water and Sanitation[92] the Committee noted the lack of access to safe drinking water and sanitation in schools but the affected populations, mainly women and children, were not involved in discussions on climate change and policymaking. The Committee recommended ensuring that the vulnerabilities of children and their views are considered in developing policies or programs on climate change and disaster risk management; and collecting disaggregated data identifying the types of risk children face in order to formulate policies, frameworks and agreements. Moreover, the state party should implement national legislation and policies, including the Water and Sanitation Policy 2013 to increase access to safe drinking water and sanitation, in particular in the outer islands and in schools, especially for adolescent girls; increase children's awareness of and preparedness for climate change and natural disasters by incorporating them into the school curriculum and teacher training programs; and seek international cooperation in implementing these recommendations.[93]

On a subsequent report of Tuvalu, the Committee noted the measures taken by the state party to promote climate change resilience in schools, including disaster risk reduction but expressed concern about the adverse impact of climate change on children, including the rights to life, survival and development; nondiscrimination; education; health; housing; and safe drinking water and sanitation.[94] The Committee voiced concern about the contamination of underground water supplies due to rising sea levels, insufficient inclusion of children's rights, including children with disabilities, in policies and programs on climate action, disaster risk reduction and preparedness; and insufficient opportunities for children to participate in decision-making related to climate action.

Similarly, on the report of Vanuatu,[95] the Committee welcomed the mainstreaming of Disaster Risk Management, Climate Change Adaptation and Energy in the National Reviewed Curriculum but noted that the State party has not included climate adaptation and disaster risk reduction in the school curriculum. Moreover, the lack of a comprehensive social protection system, insufficient action to include the needs of children including children with disabilities in planning disaster risk reduction and response, and lack of resilience of school infrastructure to natural disaster, particularly in remote areas were identified as areas of concern.[96]

The Committee noted the establishment of the Youth Climate Council and the Youth Ecological Council on the report of Poland.[97] Referring to SDG targets 3.9 and 13.3, the Committee urged the state party to: adopt mitigation measures in line with GHG emission targets and deadlines compliant with the commitments in the Paris Agreement; phase out the funding of coal power plants and accelerate the transition to renewable energy; expedite the implementation of the national air protection program; place the rights and participation of children at the center of adaptation and mitigation strategies; assess the impact of air pollution from coal power plants and transport on children's health, design a strategy to address it, and regulate the maximum allowed air pollutant emissions, including emissions by private businesses; and increase children's awareness and preparedness for climate change and disasters by incorporating it into school curricula and teacher training programs.[98]

8.3.2 *Environmental Health/Pollution/Degradation*

The Committee referred to environmental health and/or sanitation on 29 reports during the period 2013–2021, sometimes in combination with pollution and occasionally in combination with climate change. On the report of Argentina,[99] the Committee expressed concerns about the harmful effects of open-pit mining activities and the use of agrochemicals, especially by private and transnational corporations, for the environment and the health of children living in areas where mining and soya bean production are carried out. It recommended strengthening the implementation of legislation and other measures to protect the health of children, particularly indigenous children, from environmental harm. It further recommended ensuring that the impact of mining and agrochemicals on underlying determinants of health, such as food, safe drinking water and sanitation, is minimized, those responsible are held accountable, and victims are afforded effective remedies.[100]

On the report of Belgium, the Committee noted a high level of air pollution, particularly from road transport and its negative impact on the climate and children's health, and an increase in asthma and respiratory diseases.[101] Referring to SDG targets 3.9 and 13.5, the Committee recommended assessing the impact of air pollution on children's health and designing a strategy to remedy the situation, and regulating the air pollutant emissions, including from road transport; developing a national plan to reduce GHG emissions taking into account the vulnerabilities, needs and views of children; and raising awareness of environmental health and climate change among children.[102]

The Committee expressed concern about the negative impact of groundwater contamination by uranium and of air pollution on children's health on the report of Bosnia and Herzegovina.[103] Referring to SDG target 3.9,[104] the Committee recommended developing comprehensive plans to monitor children's environmental health at all levels of government; conducting an assessment of air pollution impacts and uranium contamination on children; and regulating the maximum concentration of air pollution; and cleaning up all sites contaminated by uranium.[105]

Similar concerns were expressed on the report of Brazil regarding the negative effects of polluted air, water and soil, and of food contamination, on children's health, especially the excessive use of agrochemicals, and using crop dusters to spray pesticides and other toxic chemicals close to villages and schools, resulting in the poisoning of children.[106] Moreover, the Committee expressed concern about the contamination of water resources in the state of Pará caused by mining activities and industrial projects; and the decreased availability of drinking water, its poor quality and the increased incidence of water-related diseases such as malaria caused by the construction of the Belo Monte dam and similar projects, which affects indigenous children's health.[107] The Committee recommended: enforcing laws and regulations on the use of agrochemicals, particularly using crop dusters in proximity to villages and schools, and ban agrochemicals that have been banned in other countries; improving water supply infrastructure and guaranteeing access to safe drinking water, especially for those living near contaminated water canals; ending illegal mining activities, and mitigating the negative effects of the construction of the Belo Monte dam on the rights of indigenous children and their families; undertaking awareness-raising programs for communities living in affected areas to minimize the impact of contaminated water and food, and for users of agrochemicals; and conducting an assessment of the effects of polluted air, water, and soil on children's health, implement a strategy to remedy the situation, and monitor the levels of air, water, and soil pollutants and of pesticide residues in the food chain.[108]

While noting the initiatives by Colombia to promote a healthy environment, the Committee expressed concern over the negative effects of polluted air, water and soil and electromagnetic pollution on children's health and the insufficiency of measures taken to address them.[109] It recommended conducting an assessment of pollution on children's health, designing a strategy to remedy the situation, and regulating the maximum concentrations of air and water pollutants. It called upon the state party to comply with the orders of its Constitutional Court on electromagnetic pollution, including regulating the positioning of mobile-telephone towers at a safe distance from educational institutions, homes, neighborhoods, and hospitals.[110]

On the report of Czechia, the Committee noted a high level of air pollution due to the carbon-intensive economy and recalling SDG targets 3.9[111] and 13.3[112] recommended that the state party assess the impact of air pollution on children's health; regulate the maximum concentrations of air pollutant emissions, establish monitoring mechanisms and introduce deterrent sanctions for non-compliance; and strengthen awareness-raising of environmental health and climate change among children.[113] It is noteworthy that the Committee used the SDG targets as the benchmark to achieve in relation to air pollution, hazardous activities, and climate change, including education on these issues.

Similar concerns were expressed on the report of Eswatini where the Committee noted that industrial pollution is threatening access to safe drinking water from rivers and, urban waste and agricultural chemicals, and poses a serious risk to both the environment and children's health, and the high vulnerability to recurring droughts and related food insecurity has adversely affected the implementation of the Convention.[114] Referring to SDG target 13.b,[115] the Committee recommended: promoting the sustainable management of natural resources, such as land and water, and evaluating the consequences for the present and future generations; identifying the impacts of hazardous chemicals and waste on children and formulating a national policy on chemicals and waste management, and reviewing regulations on chemicals such as mercury; increasing efficiency in biomass usage to protect the forest ecosystems and accelerating the transition to renewable energy; reducing the vulnerability of children to the impacts of climate change by building their adaptive capacity and resilience; ensuring that children's vulnerabilities and views are taken into account in preparing early warning systems and disaster risk management plans; and increasing children's awareness of climate change and environmental degradation by incorporating environmental education into the school curriculum.[116]

The Committee expressed appreciation for the information provided by Congo regarding oil extraction and forest concession contract clauses that provide for the adoption of measures to protect the rights to health and education of children living in the areas of industrial activity.[117] However, the Committee was concerned that it is *not* mandatory to carry out environmental and social impact assessments prior to the approval of investment projects, particularly as a consequence of forced displacement and expropriation, pollution, and damage to cultural assets and traditions.[118]

On the report of Cote d'Ivoire, the Committee expressed concern about the negative impact of dumping toxic waste in 18 localities of Abidjan on children's health and well-being, and the delay in providing compensation to the victims.[119] Relying on GC No 16 and the Guiding Principles on Business and Human Rights, the Committee recommended adopting regulations to hold businesses accountable for complying with international standards relating to labor and the environment that are relevant to children's rights, and ensuring that the victims affected by the dumping of toxic waste in Abidjan in 2016, including children, have access to health care and receive compensation, investigations are undertaken, and those responsible held accountable.[120]

The Committee noted the information provided by Ecuador about the Environmental Management Act and the duty of State-owned companies to provide information about the environmental impacts of their activities.[121] Referring to

GC No 16, the Committee recommended that the state party establish a regulatory framework for companies in the oil and mineral sectors to ensure that their activities do not impact human rights or endanger environmental or other standards, especially those relating to children's rights.[122] It further recommended to ensure that these companies implement international and national environment and health standards, monitor their implementation, provide remedies for violations, and seek appropriate international certification.[123] Although reference is made to obtaining "appropriate" international certification, more specific language would have been helpful as to what kind of certification should be obtained.

The Committee welcomed the positive steps taken to address deforestation in Gabon but expressed concern about the policy to increase mono-cropping and noted that land laws do not reflect the nomadic lifestyle of pygmy communities including children who rely on the forests for their livelihood.[124] The Committee recommended reviewing polices on mono-cropping with the participation of pygmy communities, including children, and ensuring a transparent human rights due diligence process with the participation of these communities before demarcating land for commercial purposes or turning them into national parks.[125]

The Committee expressed concern about the lack of a comprehensive strategy on health, including environmental and water and sanitation issues in Gambia.[126] Referring to CG No 15, the Committee recommended strengthening educational programs to inform parents about child health and nutrition, environmental sanitation, increasing access to safe drinking water and sanitation, and developing a strategy on health including environmental, water, and sanitation issues.[127]

On the report of Iran, the Committee referred to the adverse environmental effects of river diversion programs, sugar-cane farming and industrial pollution in Khuzestan province and recommended taking urgent steps to address the impact on agriculture and human health, which includes environmental pollution and water shortages.[128] Similarly, on the report of Iraq, the Committee expressed concern that many regions are affected by a high toxic level of lead and mercury contamination, and depleted uranium pollution, leading to a high infant mortality rate and an increase in cancer rates and birth defects among children.[129] The Committee recommended prioritizing the provision of drinking water and environmental sanitation, and access to food, and requesting UNICEF and WHO for assistance.

The Committee made extensive comments on environmental health on the report of Japan, noting the adoption of the Nuclear Accident Victimized Children Support Act, the Fukushima Citizen Health Management Fund and the Comprehensive Support Projects for Health and Life of Afflicted Children.[130] Recalling SDG target 3.9, the Committee recommended: reaffirming that radiation exposure in evacuation zones is consistent with internationally accepted levels for children; continuing to provide support to evacuees from areas not designated for return; increasing medical services to children affected by radiation in Fukushima prefecture; conducting health check-ups for children in areas with radiation doses exceeding 1 millisievert per year;[131] and providing, in school materials, accurate information about the risk of radiation exposure and children's vulnerability.[132]

The Committee welcomed the recognition of rights to housing, sanitation, food, water, and social security in the 2010 Constitution of Kenya, but noted that forced eviction and displacement of people, including children has taken place due to development projects and environmental conservation.[133] Furthermore, climate change, population growth, and unsustainable development projects are adding further pressure on children's access to safe drinking water, sanitation, and food in arid and semi-arid lands, negatively impacting children's health. The Committee recommended integrating measures to protect children's rights to housing, sanitation, food, water, and health in policies and programs on climate change and disaster risk management, National Adaptation Plan, and ensuring the participation of communities at risk, including children.[134]

On the report of Mongolia, the Committee expressed serious concern over the impact of air pollution on children including reduced fetal growth, and respiratory diseases and pneumonia, one of the leading causes of child mortality there.[135] Moreover, mining activities and rapid urbanization have caused water and soil contamination affecting children's access to safe drinking water. Referring to its previous recommendations, the Committee urged the state party to: adopt measures to mitigate the impact of air pollution on children, including providing high-efficiency particulate air (HEPA) filters in prenatal care packages and expanding the availability of alternatives to coal for heating in winter; declare access to clean water and sanitation and protection of children from air pollution as priorities and allocate resources to mitigate the impacts of environmental pollution on children; and provide technical knowledge and expertise to officials to monitor air and water pollutants.[136] On the reports of Spain, the Committee recommended carrying out an assessment of air pollution from coal-fired power plants on children's health and on the climate, and design a strategy to remedy the situation, and regulate air-pollutants, including emissions by private businesses.[137]

The Committee voiced its concern about the environmental impact of weapons and armed conflict on the report of Syrian Arab Republic, the first (and the only) time that the Committee referred to the link between armed conflict, weapons, environmental health, and children's rights:

> [It is] concerned about the damage to the environment generated and exacerbated by weapons-related contamination, damage to critical infrastructure, including water treatment facilities and sewage systems, and the breakdown of the environmental services in the context of the armed conflict and its immediate and long-term risks to children's health, and the long-term environmental consequences, the Committee recommends that the State party devise and implement a system of response and assistance to enhance protection of the population, especially children, and the environment.[138]

8.3.3 *Children's Rights and the Business Sector*

An important feature of the Concluding Observations of the CRC Committee is a reference to the impact of the business sector on children's rights due to environmental damage with the Committee often devoting an entire section to this topic and making extensive recommendations. A search of the Concluding Observations during 2013–2021 revealed that the Committee referred to the impact of the business sector on children's rights due to environmental damage on 25 reports.

For example, on the report of Argentina,[139] the Committee reiterated its concern about the lack of clear guidelines and regulations for corporations, both national and transnational, on children's rights. Referring to GC No 16, the Committee recommended establishing *a legally binding framework* for domestic and foreign businesses, ensuring that their activities do not negatively affect human rights or violate environmental and other standards. It further recommended requiring all companies to undertake due diligence with regard to children's rights, and carry out periodic child-rights impact assessments, consultations, and full public disclosure of the environmental, health-related, and human rights impact of business activities, and plans to address these impacts.[140]

Similar comments were made on the report of Australia. Referring to GC No 16 and recalling its previous recommendations, the Committee called on the state party to ensure that Australian companies *and their subsidiaries* are held legally accountable for violations of children's rights including environmental health and establish mechanisms to investigate and redress; and recommended in almost identical language as the comments on Argentina, the need to carry out impact assessments and consultations, and plans to address the impacts.[141] Similar observations were on the reports of Bahrain where the Committee recommended establishing regulations to ensure that the business sector including multinational corporations, complies with international and national human rights, labor, and environmental standards in relation to children's rights.[142]

The Committee welcomed the establishment of the national forum in Belgium on the Children's Rights and Business Principles and the state party's efforts to develop a national plan for business and human rights. Relying on GC No 16 and the Guiding Principles on Business and Human Rights, the Committee recommended establishing regulations to ensure that the business sector complies with international and national human rights, labor, environmental and other standards; and establish a regulatory framework for industries to ensure that their activities do not affect human rights or endanger environmental and other standards, especially those relating to children's rights.[143]

The Committee expressed concern about the lack of information on measures to address social and environmental responsibility of business corporations and extractive industries on the report of Cabo Verde.[144] Relying on GC No 16 and the Guiding Principles on Business and Human Rights, the Committee recommended adopting regulations to ensure that the business sector, especially tourism and extractive industries, complies with international human rights and environmental standards, particularly with regard to children's rights.[145] Similar concerns were expressed on the report of Cameroon regarding the activities of transnational corporations in extractive and agribusiness sectors resulting in forced displacement of indigenous and other minority groups, the contamination of water resources and food, and environmental degradation, to the detriment of children, and the lack of regulatory frameworks to address them.[146] Referring to GC No 16 and the Guiding Principles, the Committee recommended adopting a regulatory framework to ensure that activities of the business sector do not affect human rights or environmental or other standards, especially those relating to children's rights; ensuring implementation by companies of international and national environmental and health standards,[147] monitoring and imposing appropriate sanctions and/or remedies when violations occur; and requiring companies to undertake assessments of, consultations on and public disclosure of the environmental, health-related and human rights impacts of their business activities and plans to address such impacts.[148] Similar recommendations were made on the reports of Malawi[149] and the Committee, referring to GC No 16, requested that the state party include in its draft mines and minerals bill to require environmental and health monitoring at all stages of the mining process and assessment of the impacts on children's rights to water, food, health and a clean environment.[150]

On the report of Guinea, the Committee noted the reforms undertaken to adapt its legal and institutional framework in the mining sector but expressed concern about the negative impact of mining operations on the living conditions of children including exposure to harmful substances and other health hazards, environmental degradation and deforestation, as well as children having to walk long distances to find clean drinking water.[151] The Committee also referred to the insufficient enforcement of the Mining Code (2011), and the lack of a legal obligation to consult and share revenues with local communities in contracts with international mining companies.[152]

The Committee made extensive comments on children's rights and the business sector on the reports of Peru, expressing concern at the impact of mining and hydroelectric projects on the living conditions of children and their families in regions such as La Oroya, Cerro de Pasco, and Cajamarca.[153] It referred to health hazards and environmental degradation, especially the contamination of drinking water caused by these projects and noted that environmental impact assessments are not always carried out prior to granting licenses to companies. The Committee expressed concern that the law on the right of indigenous peoples to prior consultation does not apply to all self-identified groups of indigenous peoples affected by such projects and that some projects relating to the mining sector are excluded from the consultation process. The Committee, referring to GC No 16 and the "Protect, Respect and Remedy" Framework accepted by the HRC made recommendations very similar to those discussed above (conducting assessments, holding consultations, complying with standards, providing remedies, and seeking international certification).[154]

On the report of the Russian Federation, the Committee noted that while the legislation provides for compensation for indigenous persons for environmental damage by businesses, it was not clear whether indigenous persons' claims for compensation have been met.[155] The Committee expressed concern about the negative impact of oil- and gas-extracting businesses on the traditional lifestyle of indigenous peoples, including children, through deforestation and pollution and by endangering the species that are crucial to their livelihoods. The Committee was further concerned about the negative health impact of coal extraction and the production of asbestos, especially on children living in the Kemerovo and Ural regions.[156] Drawing attention to GC No 16 and HRC resolution 8/7,[157] the Committee recommended establishing regulations to ensure that the business sector complies with international and national human rights, labor, environmental and other standards, especially with regard to children's rights, and: establish a regulatory framework for the oil and gas industries and coal extraction businesses to ensure that their activities do not affect human rights or endanger environmental and other standards, especially those relating to indigenous children's rights; curtail the production of asbestos and raise awareness of the toxicity of asbestos and its health impacts; and provide reparation for ongoing and future damage to the health and development of the children affected and repair any damage done. The rest of the recommendations are very similar to those made on reports discussed above (conducting assessments, holding consultations, complying with standards, providing remedies, and seeking international certification).[158]

On the report of South Africa[159] the Committee expressed concern that activities of business enterprises, in particular extractive industries, have a negative impact on children's rights, through environmental pollution and the exploitation of child labor.[160] Similar concerns were voiced on the report of Zambia where the Committee noted the negative impact of the mining sector, especially lead mining in Kabwe, on children's rights, such as the rights to health, development and play, and their standard of living.[161] Referring to GC No 16 and the UN "Protect, Respect and Remedy" Framework, the Committee recommended that the state party establish a regulatory framework for the mining industries to ensure that their activities do not affect environmental and other standards, particularly those relating to children's rights; and take measures to protect the rights of children in Kabwe, in all future lead mining activities, such as the new World Bank-funded environmental project. The rest of the recommendations are very similar to those made on reports discussed above (conducting assessments, holding consultations, complying with standards, providing remedies and seeking international certification).[162]

8.3.4 *Sustainable Development/SDGs*

The Committee referred to SDGs on almost all the reports and the following paragraph appeared in many of them:

> The Committee recommends that the State party ensure the realization of children's rights in accordance with the Convention [and the optional protocols] throughout the process of implementing the 2030 Agenda for Sustainable Development. It also urges the State party to ensure the meaningful participation of children in the design and implementation of policies and programmes aimed at achieving all 17 Sustainable Development Goals as far as they concern children.[163]

SDGs are often used as a yardstick to measure compliance with state parties' human rights and environmental obligations. As the above discussion revealed, SDG targets are often mentioned in recommendations on climate change and in relation to activities of the business sector. Two targets that were repeatedly mentioned were targets 13.5 and 13.b followed by target 3.5.

Sustainable development was mentioned in the context of either climate change or environmental health and occasionally in relation to the activities of the business sector. Many of these references appeared in the preceding sections. A few other references are discussed here. For example, on the report of Niger, the Committee recommended the establishment of a body responsible for long-term sustainable development.[164] While the Committee referred to the three pillars of sustainable development (economic, environmental, and social pillars) in the context of the Qatar National Vision 2030, the phrase "sustainable development" does not appear in the concluding observations.[165]

The Committee referred to sustainable land management on the reports of Suriname and recommended ensuring that Amerindian and Maroon communities are protected from illegal and uncontrolled logging and mining, which have negative environmental impacts by adopting legislation on sustainable land management in consultation with local communities and promote corporate social responsibility.[166]

On the report of Belgium, the Committee recommended that the state party ensure the realization of children's rights in accordance with the Convention and its Optional Protocols throughout the process of implementing the 2030 Agenda. It also urged the state party to ensure the meaningful participation of children in the design and implementation of policies and programs aimed at achieving the 17 Goals to the extent they concern children.[167]

8.3.5 *Other Issues Discussed/Recommendations Made*

On several occasions, the Committee stressed the need to provide opportunities to get the views of children in relation to environmental matters that affect them[168] and to ensure their participation in the decision-making process. The Committee also referred the need to include environment and climate change and rights-based environmental education[169] in school curricula and teacher training materials. On the report of Norway, the Committee recommended increasing alternative sources of energy and establishing safeguards to protect children from the negative impacts of fossil fuels, both internally and abroad.[170] The Committee referred to climate change in the context of nutrition on the report of South Africa and expressed concern about child food insecurity caused by, *inter alia*, poverty, energy costs, and climate change.[171]

8.4 Individual Complaints

While the individual complaints procedure under the CRC is rather new, the Committee issued a groundbreaking decision on climate change in 2021. In 2019, a group of children filed a communication against their states for their failure to address climate change. There were five complaints altogether filed against Argentina, Brazil, France, Germany, and Turkey. While not successful, *Sacchi and others v. Argentina, Brazil, France, Germany, and Turkey* was a significant complaint submitted to the CRC Committee. The communications were identical and were taken together. Therefore, reference will be made only to *Sacchi v. Argentina.* The communication noted that the climate crisis is not an abstract future threat and petitioners and children around the world already carry the burden of climate change. Because this is the first communication brought before the Committee involving climate change, it will be discussed in some detail here.

The authors argued that climate crisis is a children's right crisis, mitigating climate change is a human rights imperative, and that human rights obligations are informed by principles of international environment law. They argued that state parties have failed to: prevent foreseeable domestic and extraterritorial human rights violations from climate change; cooperate internationally; apply the precautionary principle; and ensure intergenerational justice for children and posterity.[172] They further claimed that the acts and omissions of states have exposed them to "foreseeable, life threatening risks" of climate change in the form of floods, heat waves, storms, droughts and disease;[173] caused injuries to their mental and physical health from asthma to emotional trauma which will worsen as the world continues to warm; and jeopardized subsistence practices of indigenous authors from Alaska to the Marshall Islands affecting not just their livelihoods but also their specific way of being, seeing, and acting that are essential to their cultural identity. Invoking articles 6 (right to life), 24 (right to health), 30 (minority rights and the right to enjoy culture, language, and religion) as well as article 3 which embodies the best interests of the child principle, the authors argued:

> By supporting climate policies that delay decarbonization, the State party is shifting the enormous burden and costs of climate change onto children and future generations. In doing so, it has breached its duty to ensure the enjoyment of children's rights for posterity, and failed to act in accordance with the principle of intergenerational equity... Ultimately, at stake are the rights of every child, everywhere. If the State party, acting alone and in concert with other states, does not immediately take available measures to stop the climate crisis, the devastating effects of climate change will nullify the ability of the Convention to protect the rights of any child, anywhere. No state acting rationally in the best interests of the child would ever impose this burden by choosing such delay.[174]

The authors requested the Committee to find that: (1) climate change is a children's rights crisis; (2) state parties are perpetuating the climate crisis by knowingly acting in disregard of the available scientific evidence on mitigation; and (3) by perpetuating life-threatening climate change, the state parties are acting in violation of authors' rights to life, health, cultural rights, and their best interests. They further requested the Committee to recommend that (i) the state parties review and amend laws and policies to ensure that mitigation and adaptation measures are accelerated to the maximum available resources and on the basis of best available scientific evidence; (ii) protect the authors' rights and make best interests of the child a primary consideration, especially in allocating costs and burdens of climate action; (iii) engage in international cooperation; and (iv) pursuant to Article 12, ensure that the children's right to be heard and express their views be enforced.[175]

8.4.1 *Amici Curiae Brief of Special Rapporteurs*

The UN Special Rapporteur on Human Rights and the Environment, David Boyd, and his predecessor John Knox, filed an amici curiae brief ("the Brief") in support of this communication.[176] The Brief, noting that the world is facing a climate crisis and relying on the IPCC 1.5 degree report, pointed out that rising GHG emissions have already caused global average temperatures to increase by 1.1 degree Celsius over pre-industrial levels causing severe effects on human lives and well-being and, therefore, human rights. Unless states take urgent action to reduce emissions, the future impacts will be even worse. The Brief pointed out that despite commitments dating back to the UNFCCC, "the world is neither addressing the crisis at an adequate pace nor headed in the right direction."[177]

Noting that children are particularly at risk from the climate crisis, the Brief stressed that children are more vulnerable than adults to environmental harm which interferes with a range of rights protected under the CRC from the right to life and health to play and recreation. They are especially vulnerable to health problems exacerbated by climate change which, in turn, worsens existing social and economic inequalities and increases poverty. Furthermore, climate crisis will worsen throughout their lives and today's actions or omissions will have greater consequences for them than for adults. Finally, their voices are usually not heard or included in the decision-making process. Moreover, these vulnerabilities intersect with others, such as poverty and marginalized populations and climate change will harm the poorest and most vulnerable children.[178]

Noting that this is "one of the most important communications ever received by this Committee," the Brief pointed out that the Committee's decision could provide vital guidance to other human rights bodies, tribunals, international organizations, and individuals all over the world. Responding to the objections raised by three respondent states, the Brief stressed that the Committee has jurisdiction over the Communication. While a communication is inadmissible if it is "manifestly ill-founded or not sufficiently substantiated,"[179] authors meet the criteria necessary to meet the standard there: (a) allege harm to authors, not an *actio popularis* kind of action; (b) allege a violation of a protected right; and (c) sufficiently substantiate the allegations by presenting facts and arguments. The Brief pointed out that the authors are personally suffering the harms that are detailed in the communication and the harms they are foreseeably likely to suffer in the future as a result of climate change. These harms interfere with their rights protected under the Convention, including rights to life, health, culture and to have their best interests treated as a primary consideration in all actions concerning them. The fact that other children are suffering similar harm does not bar their claim or make it an *actio popularis* claim. Moreover, they need not be uniquely situated or harmed to a greater degree than others for their claim to be admissible.[180] Similarly, the fact that some of the harms have not yet occurred does not prevent the authors from bringing claims based on the likelihood of those harms materializing as the:

> [h]arms threatened are far more than a theoretical possibility; they are virtually certain to occur if the international community… does not take action. Therefore, it is incumbent upon States to take preventive action to avoid foreseeable risks to human rights.[181]

Noting that none of the respondent states had mentioned children in their INDCs, the Brief stressed that "the climate crisis demands urgent, effective and rights-based action, because the failure to act now ensures that children will suffer greater harms in the future."[182]

With regard to the requirement that domestic remedies must be exhausted unless remedies are unduly prolonged or unlikely to bring effective relief, the Brief pointed out that the authors' application fell within these exceptions as the longer actions to reduce emissions are deferred, the more substantial the adverse impacts will become: "climate justice delayed is climate justice denied."[183] However,

> [r]ejecting this communication at the admissibility stage and not even considering the merits would send a terrible message, contradicting everything the Committee has said about the importance of listening to children's voices, empowering children and placing children's rights and the best interests of the child at the center of climate action.[184]

In addition to detailing again the impacts of the climate emergency on children's rights, their amici curiae brief on the merits pointed out that respondents are failing to comply with their obligations under the CRC in relation to climate change. Just because it is a global problem does not mean that their responsibility to protect against harm to human rights goes away. Each of the respondent states has duties to take effective action to protect human rights from climate change which they are not fulfilling. Quoting Greta Thunberg, one of the sixteen authors, the Brief noted: "I want you to act as if our house is on fire. Because it is."[185]

8.4.2 CRC Committee Decision

In October 2021, the CRC Committee issued its decision in all five cases.[186] The Committee noted that the authors claim that:

> [b]y recklessly causing and perpetuating life-threatening climate change, the State party has failed to take necessary preventive and precautionary

> measures to respect, protect, and fulfil their rights to life, health, and culture. They claim that the climate crisis is not an abstract future threat. The 1.1°C rise in global average temperature is presently causing devastating heat waves, fostering the spread of infectious diseases, forest fires, extreme weather patterns, floods, and sea level rise. Because children are among the most vulnerable to these life-threatening impacts, physiologically and mentally, they will bear the burden of these harms far more and far longer than adults.[187]

One of the main objections of the state party (Argentina in this case) was that the communication is "absolutely generic and legally indeterminate"[188] and that it is inadmissible *ratione loci* in relation to authors who are not nationals of the state party. While the state party acknowledged the existence of international obligations of an extraterritorial nature, and the possibility of transboundary environmental damage, it is not the case in this communication. Relying on the jurisprudence of the Human Rights Committee and the European and Inter-American human rights systems according to which "jurisdiction" is not limited to territory but to the relationship of power, authority, or effective control between an individual and a state, the state party argued that "there must always be a causal link between the damage caused and the act or omission of the State of origin in relation to activities in its territory or under its jurisdiction or control."[189]

The authors responded that the Committee is competent to examine the communication as the state party has effective control over economic activities in its territory that emit GHGs which contribute to violations of the authors' rights caused by climate change. The authors further claimed that the constitutional remedy *amparo* is ill-suited to complex cases like climate change. While they do not dispute the existence of a right to a healthy environment under Article 41 of the Constitution, the remedy provided under the General Law on the Environment is designed to deal with less complex cases. Remediation of environmental harm which, while broader than *amparo* action, can only address past or existing localized harms: "It is not a vehicle for transforming the State party's national and international climate policies with the aim of preventing harm that would materialize in the future."[190]

The Committee thus had to decide whether the claim is admissible under the Optional Protocol. On the one hand, the state party asserted that the Committee lacks jurisdiction and, therefore, the communication is inadmissible. The authors argued, on the other hand, that they are within the state party's jurisdiction as victims of foreseeable consequences of state party's contributions to climate change knowingly emitted, permitted, or promoted by the state party from within its territory.

The Committee noted that state parties have the obligation, under Article 2(1) of the Convention, to respect and ensure the rights of "each child within their jurisdiction." Under Article 5(1) of the Optional Protocol, the Committee can receive communications submitted by individuals within the jurisdiction of a state party claiming to be a victim of the rights violated by the state party. The Committee pointed out that "while neither the Convention nor the Optional Protocol make any reference to 'territory' in its application of jurisdiction, extraterritorial jurisdiction should be interpreted restrictively."[191]

The Committee noted the relevant jurisprudence of the Human Rights Committee and the European Court of Human Rights on extraterritorial jurisdiction which was, however, developed to factual situations very different from the circumstances in the present communication which raises "novel jurisdictional issues of transboundary harm related to climate change."[192] Relying on the *Advisory Opinion of the Inter-American Court of Human Rights on Environment and Human Rights* the Committee noted that it clarified the scope of extraterritorial jurisdiction in relation to environmental protection:

> The Court noted that when transboundary damage occurs that affects treaty-based rights, it is understood that the persons whose rights have been violated are under the jurisdiction of the State of origin, if there is a causal link between the act that originated in its territory and the infringement of the human rights of persons outside its territory. The exercise of jurisdiction arises when the State of origin exercises effective control over the activities that caused the damage and consequent human rights violation. In cases of transboundary damage, the exercise of jurisdiction by a State of origin is based on the understanding that it is the State in whose territory or under whose jurisdiction the activities were carried out that has the effective control over them and *is in a position to prevent them from causing transboundary harm* that impacts the enjoyment of human rights of persons outside its territory.[193]

Thus, the Court noted that:

> [t]he obligation to prevent transboundary environmental damage is an obligation recognized by international environmental law, under which States may be held responsible for any significant damage caused to persons outside their borders by activities originating in their territory or under their effective control or authority.[194]

The Committee also referred to its Joint Statement on Human Rights and Climate change[195] where it expressed the view that climate change posed a significant risk to the enjoyment of rights protected under the Convention, such as rights to life, adequate food, housing, health, water, and culture. The failure to take measures to prevent foreseeable harm to human rights caused by climate change or regulate activities contributing to such harm, could constitute a violation of human rights obligations.[196]

In light of the above, the Committee pointed out that the appropriate test for jurisdiction is that adopted by the Inter-American Court in its advisory opinion:

> This implies that when transboundary harm occurs, children are under the jurisdiction of the State on whose territory the emissions originated for the

> purposes of article 5(1) of the Optional Protocol if there is a causal link between the acts or omissions of the State in question and the negative impact on the rights of children located outside its territory, when the State of origin exercises effective control over the sources of the emissions in question.[197]

The Committee further pointed out that the harm suffered by the victims needs to have been reasonably foreseeable to the State party *at the time of its acts or omissions* even for the purpose of establishing jurisdiction. With regard to climate as a global problem, the Committee's observations are noteworthy:

> The Committee notes the authors' claims that, while climate change and the subsequent environmental damage and impact on human rights it causes is a global collective issue that requires a global response, *States parties still carry individual responsibility for their own acts or omissions* in relation to climate change and their contribution to it.[198]

The Committee further noted that there is scientific evidence that carbon emissions from the state party contribute to worsening climate change which has an adverse effect over the enjoyment of rights by individuals both within and beyond the territory of the state party. Thus, through its ability to regulate activities that are the source of these emissions and to enforce such regulations, the state party has effective control over the emissions. Significantly, the Committee referred to the principle of common but differentiated responsibility in recognizing the individual responsibility of the state party:

> In accordance with the principle of common but differentiated responsibility, as reflected in the Paris Agreement, the Committee finds that *the collective nature of the causation of climate change does not absolve the State party of its individual responsibility* that may derive from the harm that the emissions originating within its territory may cause to children, whatever their location.[199]

With regard to the foreseeability element, the Committee noted the author's uncontested argument that the state party has known about the harmful effects of its contribution to climate change for decades and that it signed the UNFCCC in 1992 and the Paris Agreement in 2016. Relying on existing scientific evidence, the Committee considered that the potential harm of the state party's acts or omissions regarding carbon emissions was reasonably foreseeable to the state party.

The third element that the Committee must consider for the purpose of establishing jurisdiction[200] is whether there is a sufficient causal link between the harm alleged and the state party's actions or omissions. Relying again on the Advisory Opinion of the Inter-American Court, the Committee pointed out that not every negative transboundary impact gives rise to the responsibility of the State in whose territory the activities took place. The possible grounds for jurisdiction must be justified based on the circumstances of the specific case, and that the harm needs to be "significant." Relying on the ILC draft articles on prevention of transboundary harm from hazardous activities,[201] the Committee stated that such harm should be more than "detectable" but need not be serious or substantial. The Committee further noted that the harm must lead to a real detrimental effect on matters such as human health, industry, property, or the environment and it must be capable of being measured by "factual and objective standards."[202]

The Committee's discussion of the authors' victim status is particularly noteworthy. It noted that the authors have been personally affected by the consequences of climate change such as smoke from wildfires, heat-related pollution, worsening asthma, the spread of vector-borne diseases resulting in some authors contracting malaria, dengue, and chikungunya multiple times, and exposure to heat waves threatening their health. They were further affected by drought impacting water security, subsistence way of life threatened for indigenous authors, and Marshall Islands and Palau are at the risk of becoming uninhabitable within decades due to rising sea levels; and the mental health impacts resulting in some authors suffering from climate anxiety:

> The Committee considers that, as children, the authors are particularly impacted by the effects of climate change, both in terms of the manner in which they experience such effects as well as the potential of climate change to affect them throughout their lifetime, in particular if immediate action is not taken. Due to the particular impact on children, and the recognition by States parties to the Convention that children are entitled to special safeguards, including appropriate legal protection *states have heightened obligations to protect children from foreseeable harm.*[203]

In light of these factors, the Committee concluded that the authors have sufficiently justified, for the purposes of establishing jurisdiction, that the impairments of the Convention rights as a result of the state party's acts or omissions regarding carbon emissions was reasonably foreseeable, and that the authors have established *prima facie* that they have personally experienced a real and significant harm to justify their victim status. Thus, the Committee finds that it is not precluded by Article 5(1) of the Optional Protocol from considering the communication.

Finally, the Committee turned to the issue of domestic remedies. Noting that authors must make use of all judicial and administrative avenues available, unless they objectively have no prospect of success, the Committee stressed that "mere doubts or assumptions about the success or effectiveness of remedies do not absolve the authors from exhausting them."[204] In the present case the Committee noted that the authors had not attempted to initiate any domestic proceedings in the state party and, therefore, the communication is inadmissible under article 7(e) of the Optional Protocol.[205]

Thus, while the communication (together with the other four communications) was dismissed for the failure to exhaust domestic remedies, the decision of the Committee is significant for several reasons. *First,* it recognized that climate change posed a significant risk to the enjoyment of rights protected under the

Convention, and the failure to take measures to prevent foreseeable harm could constitute a violation of human rights obligations. *Second*, it noted, relying on the principles of international environmental law and the 2017 Advisory Opinion of the IACtHR, that states may be held responsible for any significant damage caused to persons *outside* their borders by activities originating in their territory or under their effective control or authority. *Third*, in deciding whether activities are under the effective control of states, the Committee relied upon the test adopted in the Advisory Opinion – when transboundary harm occurs, children are under the jurisdiction of the State on whose territory the emissions originated if there is *a causal link* between the acts or omissions of the State in question and the negative impact on the rights of children located outside its territory, *when the State of origin exercises effective control over the sources of the emissions in question*. *Fourth*, the collective nature of climate change does not absolve the State party of its individual responsibility to limit its emissions. *Fifth*, scientific evidence showed that the potential harm of the state party's acts or omissions regarding carbon emissions was reasonably foreseeable to the state party. *Sixth*, in deciding whether there is a sufficient causal link between the harm alleged and the state party's actions or omissions, the Committee noted that the harm must lead to a real detrimental effect on matters such as human health, industry, property, or the environment which could be measured by objective standards. *Finally*, because children are especially impacted by climate change and states have agreed that they are entitled to special safeguards, states have heightened responsibility to protect children from foreseeable harm.[206]

8.5 Conclusion

As the discussion in this chapter showed, the CRC Committee has made extensive and elaborate comments and recommendations with regard to environmental degradation, climate change and sustainable development, including SDGs. While it is impossible to summarize all the important principles, we will refer to some of the salient points here.

In many of the recommendations the Committee referred to the need to take the views of children into account on policies and programs on climate change and disaster risk management, the need to collect disaggregated data identifying the risks faced by children. The Committee stressed the need to strengthen measures to increase children's awareness and preparedness for climate change and disasters by incorporating them into school curricula and teacher training programs, and providing children opportunities to participate in discussions and decision-making on climate action.

The Committee also made very specific recommendations on climate change, almost to the point of transgressing into the authority of other international bodies like the Conference of Parties under the UNFCCC. The Committee's recommendations to Australia to reduce its GHG emissions by establishing targets and deadlines to phase out coal, accelerate the transition to renewable energy, and commit to meeting 100% of its electricity needs with renewable energy, provide a good example.[207]

On several reports the Committee addressed the impact of businesses on children's rights and stressed the need to require companies to undertake assessments, consultations, and public disclosure of the environmental, health-related, and human rights impacts of their activities, and to establish regulations to ensure that the business sector complies with human rights, labor, and environmental standards, especially with regard to children's rights. On several reports the Committee referred the need to ensure that businesses obtain "appropriate international certification" but it is not clear what this means.

The Committee also referred, where appropriate, to the reports of other special mandate holders and the Guiding Principles on Business and Human Rights as well as to the jurisprudence of other human rights courts, especially, the Inter-American Court of Human Rights. In relation to standards, the Committee referred extensively to international and national human rights, environmental, and labor standards as well as WHO and health standards.

Undoubtedly the most important contribution of the Committee in this context is the decision in *Sacchi v. Argentina*. In rejecting the argument made by the state party that it cannot be held responsible for a global environmental problem, the Committee stressed that the global nature of climate change does not absolve states of their individual responsibility to address it. It endorsed the "effective control" test adopted by the IACtHR in relation to extraterritorial environmental damage. The Committee's commitment to environmental issues, including climate change is clear by its decision to adopt a new General Comment which, no doubt, will clarify states' obligations vis-à-vis children's rights and climate change.[208]

Notes

1 World Commission on Environment and Development, (October 1987) *Our Common Future*, Oxford: Oxford University Press, p. 8.
2 UNGA resolution 44/25 A/RES/44/25 (20 November 1989), https://undocs.org/A/RES/44/25.
3 Convention on the Rights of the Child (20 November 1989), https://www.ohchr.org/en/instruments-mechanisms/instruments/convention-rights-child.
4 US signed it in 1995 but never submitted to Senate for ratification.
5 CRC, supra note 3, art. 1.
6 Ibid., art. 4.
7 Ibid.
8 Ibid., art. 44(1).
9 Ibid., art. 44(2).
10 Ibid., art. 44(5).
11 Third Optional Protocol adopted and opened for signature by UNGA resolution A/RES/66/138 (19 December 2011), and entered into force on 14 April 2014, https://www.ohchr.org/en/instruments-mechanisms/instruments/optional-protocol-convention-rights-child-communications.
12 Ibid., art. 5.
13 Ibid., art. 6.

14 Only courts are typically empowered to issue provisional measures to avoid irreparable damage to victims or to prevent a situation from escalating. The Inter-American Commission of Human Rights and the African Commission of Human and Peoples' Rights area also empowered to issue them. See Weissbrodt, D. Martin, J. & Newman F (late) (2021) *International Human Rights: Law, Policy, and Process* (5th ed.), Durham, NC: Carolina Academic Press, Chapter 8–9.

15 Ibid., art. 7, where grounds of inadmissibility are enumerated.

16 Ibid., art. 24(2)(c) (emphasis added).

17 Ibid., art. 24(2)(e).

18 *Mapping Human Rights Obligations Relating to the Enjoyment of a Safe, Clean, Healthy and Sustainable Environment, Individual Report on the United Nations Convention on the Rights of the Child* (December 2013) [Mapping Report No. 5], available at: https://www.ohchr.org/en/special-procedures/sr-environment/mapping-report.

19 *General comment No. 15 (2013) on the right of the child to the enjoyment of the highest attainable standard of health (art. 24)*, CRC/C/GC/15 (17 April 2013), https://undocs.org/CRC/C/GC/15.

20 Ibid., ¶ 49.

21 Ibid., ¶ 50.

22 Ibid., ¶ 71 (emphasis added).

23 *Joint General Comment No. 3 (2017) of the Committee on the Protection of the Rights of All Migrant Workers and Members of Their Families and No. 22 (2017) of the Committee on the Rights of the Child on the general principles regarding the human rights of children in the context of international migration*, CMW/C/GC/3-CRC/C/GC/22 (16 November 2017), ¶ 45, https://undocs.org/CMW/C/GC/3.

24 Discussed in Chapter 5.

25 See Atapattu, S. (2022) "Migrating with Dignity: Protecting the Rights of 'Climate Refugees' with the *non-refoulement* Principle," in *Climate Refugees: Global, Local and Critical Approaches*, Behrman, S. & Kent, A., eds., Cambridge: Cambridge University Press, p. 127.

26 *General comment No. 16 (2013) on State obligations regarding the impact of the business sector on children's rights*, CRC/C/GC/16 (17 April 2013), https://tbinternet.ohchr.org/_layouts/15/treatybodyexternal/Download.aspx?symbolno=CRC%2fC%2fGC%2f15&Lang=e^{n}

27 Ibid., ¶ 4(a) (emphasis added).

28 Ibid., ¶ 19.

29 Ibid., ¶ 21 (emphasis added). *See also* GC *No. 11 on Indigenous children and their rights under the Convention*, CRC/C/GC/11 (12 February 2009), ¶ 35, https://undocs.org/CRC/C/GC/11, which states: "States parties should closely consider the cultural significance of traditional land and the quality of the natural environment while ensuring the children's right to life, survival and development to the maximum extent possible." It also urges states to take all reasonable measures to ensure that indigenous children and their communities receive information and education on issues relating to, inter alia, environmental sanitation and the dangers of pesticides and herbicides, ¶ 53. It refers to the importance of MDGs, ¶ 36, the predecessor to the SDGs. GC *No. 9, The rights of children with disabilities*, CRC/C/GC/9 (27 February 2007), https://undocs.org/CRC/C/GC/9, which also refers to MDGs in the context of universal education. It notes that certain disabilities result directly from human-caused or natural disasters. While it does not refer to climate change, the reference to "human-caused disasters" can include extreme weather events and disasters caused by climate change.

30 *Case Concerning the Gabcikovo Nagymaros Project (Hungary v. Slovakia), ICJ Rep (1996)*.

31 *Concept note: General comment on children's rights and the environment with a special focus on climate change*, https://www.ohchr.org/EN/HRBodies/CRC/Pages/CRC_GC26_concept_note.aspx.

32 Ibid. OHCHR report is discussed in Chapter 3.

33 Ibid. (footnotes omitted). It is noteworthy that the CRC Committee is referring to reports of OHCHR and other special rapporteurs, especially the Special Rapporteurs on Poverty and Human Rights, and Human Rights and the Environment.

34 Here the concept note refers to the reports of the IPCC.

35 These are access to information, participation and access to justice. See Chapter 2 for an overview.

36 The Committee plans to adopt the GC in late 2022. It will contribute to clarifying states' obligations in relation to environmental issues and climate change under the CRC, especially if the GC elaborates on concepts such as extraterritorial obligations and intergenerational equity.

37 *Concluding observations on the combined second to fifth periodic reports of the Cook Islands*, CRC/C/COK/CO/2–5 (2 April 2020), ["Cook Islands observations"].

38 Ibid., ¶ 44.

39 Target 13.5 of SDGs refers to raising capacity for climate change-related planning, https://sdgs.un.org/goals/goal13.

40 Cook Islands observations, supra note 37, ¶ 45.

41 *Concluding observations on the second periodic report of the Federated States of Micronesia*, CRC/C/FSM/CO/2, (3 April 2020) ["Micronesia observations"].

42 Ibid., ¶ 56.

43 Targets 13.2, 13.3 and 13.b. For targets under Goal 13 on climate action, see https://sdgs.un.org/goals/goal13.

44 A similar recommendation was made on the *Concluding observations on the combined fourth to sixth periodic reports of Tunisia*, CRC/C/TUN/CO/4–6 (7 June 2021), ¶ 34 ["Tunisia observations"].

45 Micronesia observations, supra note 41, ¶ 57. Similar recommendations were made on the report of Tunisia.

46 Micronesia observations, supra note 41, ¶ 57.

47 *Concluding observations on the combined fifth and sixth periodic reports of Austria*, CRC/C/AUT/CO/5–6 (6 March 2020).

48 Ibid., ¶ 35.

49 Ibid., ¶ 42.

50 Ibid., ¶ 43.

51 Ibid., ¶ 47.

52 *Concluding observations on the combined second to fourth periodic reports of Antigua and Barbuda*, CRC/C/ATG/CO/2–4 (30 June 2017) ["Antigua and Barbuda observations"].

53 Target 1.5 relates to building resilience of the poor and those in vulnerable situations by 2030, https://sdgs.un.org/goals/goal1.

54 Antigua and Barbuda observations, supra note 52, ¶ 46.

55 *Concluding observations on the combined fifth and sixth periodic reports of Australia*, CRC/C/AUS/CO/5–6 (1 November 2019), ¶ 4 ["Australia observations"].

56 Ibid., ¶ 40.

57 Ibid.

58 Similar recommendations were made on the reports of Serbia and the need to collect disaggregated data identifying the types of risks faced by children regarding: disasters, *Concluding observations on the combined second and third periodic reports of Serbia*, CRC/C/SRB/CO2–3 (4 March 2017), ¶ 53.

59 Australia observations, supra note 55, ¶ 41. This is a very specific recommendation, something that one would expect from COP to the UNFCCC.

60 Ibid., ¶ 22. Similar recommendations were made on the report of Hungary to develop toolkits for consulting children on issues that affect them, including climate change. See *Concluding observations on the sixth periodic report of Hungary*, CRC/C/HUN/CO/6 (3 March 2020), ¶ 18.

61 *Concluding observations on the combined third to fifth periodic reports of Bhutan*, CRC/C/BTN/CO/3–5 (5 July 2017), ¶ 36.

62 Ibid. See also the *Concluding observations on the fifth periodic report of Mongolia*, CRC/C/MNG/CO/5 (12 July 2017), ¶ 36 ["Mongolia observations"], where the Committee urged the state party to include children in the development of its climate policies and plans and adopt a child-focused approach to coping and adapting to climate change. It referred to target 13.5 of the SDGs on raising capacity for climate change-related planning and to the impact of climate change on fragile ecosystems and the direct impact on children due to extreme winters leading to significant losses in livestock. Similar recommendations were made on the report of New Zealand where the Committee referred to the need to pay attention to children especially affected by climate change, including Maori and Pasifika children and those living in poverty, and to undertake health impact assessments of children to inform climate legislation and policies, see *Concluding observations on the fifth periodic report of New Zealand*, CRC/C/NZL/CO/5 (21 Oct 2016), ¶ 34.

63 Mongolia observations, supra note 62, ¶ 36.

64 *Concluding observations on the second periodic report of Cabo Verde*, CRC/C/CPV/CO/2 (27 June 2019) ["Cabo Verde observations"].

65 Similar recommendations were made on the report of Guinea and the need to strengthen reforestation measures, *Concluding observations on the combined third to sixth periodic reports of Guinea*, CRC/C/GIN/CO/3–6 (28 February 2019), ¶ 36 ["Guinea observations"].

66 Cabo Verde observations, supra note 64, ¶¶ 72–73. Similar recommendations regarding school curriculum and teacher training programs were made on the report of Chile, and the need to develop and implement a national plan for disaster prevention and risk management, *Concluding observations on the combined fourth and fifth periodic reports of Chile*, CRC/C/CHL/CO/4–5 (30 October 2015) ["Chile observations"].

67 *Concluding observations on the combined second to fourth periodic reports of Fiji*, CRC/C/FJI/CO2–4 (13 October 2014), ¶ 55 ["Fiji observations"].

68 Similar recommendations were made on *Concluding observations on the second periodic report of Lesotho*, CRC/C/LSO/CO/2 (25 June 2018); and *Concluding observations on the fifth periodic report of the Democratic People's Republic of Korea*, CRC/C/PRK/CO/5 (23 October 2017). In addition, the Committee recommended providing children with immediate access to treatment for malnutrition during climate-related emergencies, cooperating with UNICEF and the World Food Program and paying attention to child rights in initiatives on disaster risk reduction, response and preparedness.

69 Fiji observations, supra note 67, ¶ 56.

70 *Concluding observations on the combined third and fourth periodic reports of Jamaica*, CRC/C/JAM/CO/3–4 (10 March 2015) ["Jamaica observations"].

71 Ibid., ¶ 51. Almost identical language was used on the report of Saint Lucia, *Concluding observations on the combined second to fourth periodic reports of Saint Lucia*, CRC/C/LCA/CO/2–4 (4 July 2014), ¶¶ 52–53. Similar recommendations were made on the reports of Solomon Islands where the Committee noted the particular vulnerability of the state party to climate change but expressed concern that it has not included climate adaptation or disaster risk reduction in school curriculum and that early warning systems are not in place in schools: *Concluding observations on the second and third periodic reports of the Solomon Islands*, CRC/C/SLB/CO/2–3 (28 February 2018), ¶¶ 42–43; *Concluding observations on the combined second and third periodic reports of Saint Vincent and the Grenadines*, CRC/C/VCT/CO/2–3 (13 March 2017); and *Concluding observations on the combined third and fourth periodic reports of Suriname*, CRC/C/SUR/CO/3–4 (4 November 2016) ["Suriname observations"].

72 *Concluding observations on the combined fourth and fifth periodic reports of Japan*, CRC/C/JPN/CO/4–5 (4 March 2019), ¶ 37.

73 Ibid.
74 *Concluding observations on the combined third to fifth periodic reports of Malawi*, CRC/C/MWI/CO/3–5 (4 March 2017), ¶ 36 ["Malawi observations"]. The Committee referred to target 13.b in the context of natural resource use, chemical pollution etc. in *Concluding observations on the combined second to fourth periodic reports of Eswatini*, CRC/C/SWZ/CO/2–4 (22 October 2021), ¶ 59 ["Eswatini observations"].
75 Malawi observations, supra note 74, ¶ 36. The Committee made similar recommendations on the reports of Niger and noted the significant impacts of climate change in the form of deforestation, desertification, and water and food shortages. It recommended strengthening policies and programs to address climate change and disaster risk management, through replanting trees, regenerating land and increasing solar energy and measures to protect rights to housing, sanitation, food, water and health with the participation of children, *Concluding observations on the combined third to fifth periodic reports of the Niger*, CRC/C/NER/CO/3–5 (21 November 2018), ¶ 36 ["Niger observations"].
76 *Concluding observations on the combined third and fourth periodic reports of the Marshall Islands*, CRC/C/MHL/CO/3–4 (27 February 2018), ¶ 33, ["Marshall Islands observations"]. See also Chapters 6 and 7.
77 Ibid.
78 Ibid., ¶ 34. Similar recommendations were made on the reports of Samoa with respect to children with disabilities, raising capacity, taking the vulnerabilities, needs and views of children into account, and collecting disaggregated data on children. The Committee urged the state party to provide information on whether child protection issues have been included in the Strategy for the Development of Samoa 2016–2020, see *Concluding observations on the combined second to fourth periodic reports of Samoa*, CRC/C/WSM/CO/2–4 (12 July 2016), ¶¶ 48–49.
79 *Concluding observations on the second periodic report of Palau*, CRC/C/PLW/CO/2 (28 Feb 2018).
80 Ibid., ¶ 48.
81 Ibid., ¶ 49.
82 *Concluding observations on the fifth and sixth reports of Switzerland*, CRC/C/CHE/CO/5–6 (22 October 2021) ["Switzerland observations"].
83 Refers to substantially reducing by 2030 the number of deaths and illnesses from hazardous chemicals and air, water and soil pollution and contamination, see https://sdg.humanrights.dk/en/goals-and-targets.
84 Refers to improving education, awareness-raising and human and institutional capacity on climate change mitigation, adaptation, impact reduction and early warning.
85 (Emphasis added). This may be the first time that a treaty body is recommending adopting binding rules to govern financial institutions.
86 Switzerland observations, supra note 82, ¶ 37. These are very specific and extensive recommendations, especially those relating to the financial sector's impact on climate change.
87 *Concluding observations on the combined third to fifth periodic reports of Tajikistan*, CRC/C/TJK/CO/3–5 (29 September 2017), ¶ 38.
88 Ibid.
89 *Concluding observations on the initial report of Tonga*, CRC/C/TON/CO/1 (2 July 2019), ¶ 53.
90 Ibid., ¶ 54.
91 *Concluding observations on the initial report of Tuvalu*, CRC/C/TUV/CO/1 (30 October 2013), ¶ 55 ["Tuvalu observations"].
92 *Report of the Special Rapporteur on the human right to safe drinking water and sanitation*, A/HRC/24/44/Add.2 (1 July 2013), https://undocs.org/A/HRC/24/44/Add.2.
93 Tuvalu observations, supra note 91, ¶ 56.

94 *Concluding observations on the combined second to fifth periodic reports of Tuvalu*, CRC/C/TUV/CO/2–5 (31 March 2020), ¶ 42(a).

95 *Concluding observations on the second to fourth periodic reports of Vanuatu*, CRC/C/VUT/CO/2–4 (29 September 2017), ¶ 42.

96 Ibid., ¶ 43. Similar recommendations were made: *Concluding observations on the combined third to fifth periodic reports of the Democratic Republic of the Congo*, CRC/C/COD/CO/3–5 (28 February 2017), ¶ 38; Fiji observations, supra note 67; and Jamaica observations, supra note 70.

97 *Concluding observations on the combined fifth and sixth periodic reports of Poland*, CRC/C/POL/CO/5–6 (6 December 2021), ¶ 37 ["Poland observations"].

98 Ibid.

99 *Concluding observations on the combined fifth and sixth periodic reports of Argentina*, CRC/C/ARG/CO/5–6 (1 October 2018), ¶ 34.

100 Ibid.

101 *Concluding observations on the combined fifth and sixth periodic reports of Belgium*, CRC/C/BEL/CO/5–6 (28 February 2019) ["Belgium observations"].

102 Ibid., ¶ 35. On the report of Venezuela, the Committee referred to the impact of excessive air pollution resulting in pneumonia in children under five years in light of the report by Ministry of Environment that air pollution exceeds the permitted standards in some cities. The Committee referred to the need to provide access to safe drinking water resulting in numerous cases of infant deaths associated with infections and diarrhea, *Concluding observations on the combined third to fifth periodic reports of the Bolivarian Republic of Venezuela*, CRC/C/VEN/CO/3–5 (13 October 2014), ¶ 62. It is noteworthy that these comments were made under "standard of living."

103 *Concluding observations on the combined fifth and sixth periodic reports of Bosnia and Herzegovina*, CRC/C/BBIH/CO/5–6 (5 December 2019), ¶ 36 ["Bosnia and Herzegovina observations"].

104 SDG 3.9 refers to substantially reducing the number of deaths and illnesses from hazardous chemicals and air, water and soil pollution and contamination by 2030, https://sdgs.un.org/goals/goal3.

105 Bosnia and Herzegovina observations, supra note 103, ¶ 36.

106 *Concluding observations on the combined second to fourth periodic reports of Brazil*, CRC/C/BRA/CO/2–4 (10 October 2015).

107 Ibid., ¶ 65.

108 Ibid., ¶ 66.

109 *Concluding observations on the combined fourth and fifth periodic reports of Colombia*, CRC/C/COL/CO/4–5 (6 March 2015), ¶ 49 ["Colombia observations"].

110 Ibid., ¶ 50.

111 See supra note 104 for Target 3.9.

112 Target 13.3 refers to Improving education, awareness-raising and human and institutional capacity on climate change mitigation, adaptation, impact reduction and early warning. See https://sdg.humanrights.dk/en/goals-and-targets.

113 *Concluding observations on the combined fifth and sixth periodic reports of Czechia*, CRC/C/CZE/CO/5–6 (22 October 2021), ¶ 39. A similar observation was made on the report of the United Arab Emirates where the Committee voiced concern about the consequences of the high consumption of petroleum on children's health, in particular about air pollution and lead poisoning and recommended that the state party take steps to address them, *Concluding observations on the second periodic report of the United Arab Emirates*, CRC/C/ARE/CO/2, 30 October 2015, ¶¶ 55–56. Likewise, the Committee expressed concern at the high level of air pollution and its impact on climate change and children's health and rights in both the state party and other countries, on the report of UK. It recommended the adoption of a legal commitment to expedite the implementation of plans to reduce air pollution levels, especially near

schools and residential areas and place children's rights at the center of adaptation and mitigation strategies, including through its new domestic climate strategy, its international climate change program and provide financial support, *Concluding observations on the fifth periodic report of the United Kingdom of Great Britain and Northern Ireland*, CRC/C/GBR/CO/5 (4 July 2016), ¶¶ 68–69.

114 Eswatini observations, supra note 74, ¶ 58.

115 Target 13.b refers to promoting mechanisms for raising capacity for climate change-related planning and management in least developed countries, including focusing on women, youth and local and marginalized communities, see https://sdg.humanrights.dk/en/goals-and-targets?page=3.

116 Eswatini observations, supra note 74, ¶ 59.

117 *Concluding observations on the combined second to fourth periodic report of the Congo*, CRC/C/COG/CO/2–4, (25 February 2014).

118 Ibid., ¶ 26.

119 *Concluding observations on the second periodic report of Côte d'Ivoire*, CRC/C/CIV/CO/2 (12 July 2019).

120 Ibid., ¶ 16.

121 *Concluding observations on the combined fifth and sixth periodic reports of Ecuador*, CRC/C/ECU/CO/5–6 (26 October 2017).

122 Ibid., ¶ 15.

123 Ibid.

124 *Concluding observations on the second periodic report of Gabon*, CRC/C/GAB/CO/2 (8 July 2016), ¶ 51.

125 Ibid., ¶ 52.

126 *Concluding observations on the combined second and third periodic reports of the Gambia*, CRC/C/GMB/CO/2–3 (20 February 2015).

127 Ibid., ¶ 61.

128 *Concluding observations on the combined third and fourth periodic reports of the Islamic Republic of Iran*, CRC/C/IRN/CO/3–4 (14 March 2016), ¶¶ 73–74.

129 *Concluding observations on the combined second to fourth periodic reports of Iraq*, CRC/C/IRQ/CO/2–4 (3 March 2015), ¶ 61.

130 *Concluding observations on the combined fourth and fifth periodic reports of Japan*, CRC/C/JPN/CO/4–5 (4 March 2019), ¶ 36.

131 It is not clear where this standard came from.

132 In addition, the Committee recommended that the state party implement the recommendations of the Special Rapporteur on the right health, A/HRC/23/41/Add.3 (31 July 2013), https://undocs.org/A/HRC/23/41/Add.3. On the report of the Marshall Islands the Committee recommended the implementation of the recommendations of the Special Rapporteur on hazardous waste, especially those relating to the impacts of nuclear tests by the US on children's health, the right to live in a healthy environment and indefinite displacement. See Marshall Islands observations, supra note 76. See Chapters 6 and 7 for comments by the ESCR Committee and CEDAW Committee.

133 *Concluding observations on the combined third to fifth periodic reports of Kenya*, CRC/C/KEN/CO/3–5 (21 March 2016), ¶ 55 ["Kenya observations"].

134 Ibid., ¶ 56.

135 *Concluding observations on the fifth periodic report of Mongolia*, CRC/C/MNG/CO/5 (12 July 2017), ¶ 34.

136 Ibid., ¶ 35. These are very specific recommendations.

137 *Concluding observations on the combined fifth and sixth periodic reports of Spain*, CRC/C/ESP/CO/5–6 (5 March 2018).

138 *Concluding observations on the fifth periodic report of the Syrian Arab Republic*, CRC/C/SYR/CO/5 (6 March 2019), ¶ 41 ["Syrian Arab Republic observations"]. In *Concluding*

observations on the combined fifth and sixth periodic reports of Portugal, CRC/C/PRT/CO/5–6 (9 December 2019), ¶ 39, the Committee, drawing attention to SDG target 1.3, recommended ensuring that families with children, including families of the Roma community and of African descent, are protected against pollution and have access to safe drinking water and sanitation.

139 *Concluding observations on the combined fifth and sixth periodic reports of Argentina*, CRC/C/ARG/CO/5–6 (1 October 2018), ¶ 13.

140 Ibid.

141 Ibid., ¶ 17.

142 *Concluding observations on the combined fourth to sixth periodic reports of Bahrain*, CRC/C/BHR/CO/4–6 (27 February 2019), ¶ 14 ["Bahrain observations"]. See also, *Concluding observations on the combined fifth and sixth periodic reports of Belarus*, CRC/C/BLR/CO/5–6, (28 February 2020). See *Concluding observations on the combined third and fourth periodic reports of Ireland*, CRC/C/IRL/CO/3–4 (1 March 2016) ["Ireland observations"], where similar recommendations were made as well as the need to ensure compliance with public procurement. The Committee, while welcoming the state party's national action plan on business and human rights, noted that there was no firm commitment to children's rights; nor has it taken serious note of GC No. 16.

143 Bahrain observations, supra note 142, ¶ 14.

144 Cabo Verde observations, supra note 64.

145 Ibid., ¶¶ 21–22.

146 *Concluding observations on the combined third to fifth periodic reports of Cameroon*, CRC/C/CMR/CO/3–5 (6 July 2017), ¶ 12 ["Cameroon observations"].

147 See *Concluding observations on the combined third and fourth periodic reports of Mozambique*, CRC/C/MOZ/CO/3–4 (27 November 2019) ¶ 15, where similar recommendations were made, referring specifically to extractive, agricultural, forestry and fishery industries and to the "protect, respect and remedy" framework of the Guiding Principles of Business and Human Rights.

148 Cameroon observations, supra note 146, ¶ 12. Almost identical recommendations were made on the report of Chile, supra note 66, ¶ 21. See also Colombia observations, supra note 109, ¶ 18; and Ireland observations, supra note 142, ¶ 23.

149 Malawi observations, supra note 74, ¶ 12.

150 Ibid.

151 Guinea observations, supra note 65, ¶ 14. Similar recommendations made on the report of Kenya, supra note 133, ¶ 19.

152 Ibid.

153 *Concluding observations on the combined fourth and fifth periodic reports of Peru*, CRC/C/PER/CO/4–5 (2 March 2016) ["Peru observations"].

154 It is not clear what "appropriate international certification" here is. Similar recommendation was made on the report of Chile observations, supra note 66, ¶ 21(b).

155 *Concluding observations on the combined fourth and fifth periodic reports of the Russian Federation*, CRC/C/RUS/CO/4–5 (25 Feb 2014), ¶ 20 ["Russian Federation observations"].

156 Ibid.

157 Resolution 8/7 *Mandate of the Special Representative of the Secretary-General on the issue of human rights and transnational corporations and other business enterprises*, (18 June 2008), https://ap.ohchr.org/documents/E/HRC/resolutions/A_HRC_RES_8_7.pdf.

158 Russian Federation observations, supra note 155, ¶ 21. These are very elaborate recommendations. Very similar recommendations were made on the reports of Singapore, *Concluding observations on the combined fourth and fifth periodic reports of Singapore*, CRC/C/SGP/CO/4–5, 28 June 2019, ¶ 16 ["Singapore observations"]. In addition, the

Committee recommended that the state party "require companies to undertake due diligence in their operations and *throughout the supply chain* with regard to the harmful effects of environmental degradation on children's rights" (emphasis added).

159 *Concluding observations on the second periodic report of South Africa*, CRC/C/ZAF/CO/2 (27 Oct 2016) ["South Africa observations"].

160 Ibid., ¶ 19.

161 *Concluding observations on the combined second to fourth periodic reports of Zambia*, CRC/C/ZMB/CO/2–4 (14 March 2016), ¶ 21.

162 Ibid., ¶ 22.

163 Singapore observations, supra note 158, ¶ 6. Identical language appears in the Switzerland observations, supra note 82, ¶ 5; Syrian Arab Republic observations, supra note 138, ¶ 6; Tunisia observations, supra note 44, ¶ 5, with the addition of reference to the national development plan that succeeded the plan on achieving the Sustainable Development Agenda, 2016–2020; Poland observations, supra note 97, ¶ 5.

164 Niger observations, supra note 75, ¶ 37.

165 *Concluding observations on the combined third and fourth periodic reports of Qatar*, CRC/C/QAT/CO/3–4 (22 June 2017), ¶ 7.

166 Suriname observations, supra note 71, ¶ 36.

167 Belgium observations, supra note 101, ¶ 5.

168 Australia observations, supra note 55, ¶ 22; Cabo Verde observations, supra note 64, ¶ 30.

169 Cabo Verde observations, supra note 64, ¶ 79.

170 *Concluding observations on the combined fifth and sixth periodic reports of Norway*, CRC/C/NOR/CO/5–6 (4 July 2018), ¶ 27. This recommendation was included under the heading "environmental health," rather than climate change.

171 South Africa observations, supra note 159, ¶ 53.

172 *Decision adopted by the Committee on the Rights of the Child under the Optional Protocol to the Convention on the Rights of the Child on a communications procedure in respect of Communication No. 104/2019*, CRC/c/88/D/104/2019 (8 October 2021), ¶ 3.3 ["*Sacchi v. Argentina*"], https://undocs.org/CRC/C/88/D/104/2019.

173 Ibid., ¶ 3.1.

174 Ibid., ¶ 3.7.

175 Ibid., ¶ 3.8.

176 *Amici Curiae Brief of Special Rapporteurs on Human Rights and the Environment in Support of Admissibility, Before the Committee on the Rights of the Child, C.S. et al. v. Argentina, Brazil, France, German and Turkey*, N.104/2019, N.105/2019, N.106/2019, N.107/2019, and N.108/2019 ["Brief"], https://www.hausfeld.com/uploads/documents/crc_admissibility_brief_boyd_knox_final_-_1_may_2020.pdf.

177 Ibid., ¶ 5.

178 Ibid., ¶ 11 (footnotes omitted).

179 Ibid., ¶ 25.

180 *Cf Teitiota v New Zealand*, CCPR/C/127/D/2728/2016 (7 January 2020), https://undocs.org/CCPR/C/127/D/2728/2016, where the UN Human Rights Committee stated that the author must show harm above and beyond that suffered by others. See Chapter 5.

181 Brief, ¶ 29 (footnotes omitted).

182 Ibid., ¶ 40.

183 Ibid., ¶ 49.

184 Ibid., ¶ 53.

185 *Amici Curiae Brief of Special Rapporteurs on Human Rights and the Environment on the Merits, Before the Committee on the Rights of the Child, C.S. et al. v. Argentina, Brazil, France, German and Turkey*, N.104/2019, N.105/2019, N.106/2019, N.107/2019, and N.108/2019, ¶ 53, http://climatecasechart.com/climate-change-litigation/wp-content/uploads/sites/16/non-us-case-documents/2021/20210901_Communication-

No.-1042019-Argentina-Communication-No.-1052019-Brazil-Communication-No.-1062019-France-Communication-No.-1072019-Germany-Communication-No.-1082019-Turkey_na.pdf.

186 *Decision adopted by the Committee under the Optional Protocol to the Convention on the Rights of the Child on a communications procedure, concerning Communication No. 104/2019*, CRC/C/88/D/104/2019 (8 October 2021), https://undocs.org/CRC/C/88/D/104/2019. All five cases were taken together as they were identical.

187 Ibid., ¶ 3.1

188 Ibid., ¶ 4.2

189 Ibid., ¶4.3 relying on IACtHR, *The Environment and Human Rights*, Advisory Opinion, OC-23/17 (15 Nov 2017), No 23, ¶¶ 102 and 103, https://www.corteidh.or.cr/docs/opiniones/seriea_23_ing.pdf.

190 CRC Decision, supra note 186, ¶ 8.3.

191 Ibid., ¶ 10.3, relying on the IACtHR Advisory Opinion, supra note 189, ¶ 8; and European Court of Human Rights, *Catan and others v. the Republic of Moldova and Russia*, Application Nos. 43370/04, 8252/05 and 18454/06, https://hudoc.echr.coe.int/fre#{%22itemid%22:[%22002-7212%22]}.

192 CRC Decision, supra note 186, ¶ 10.4.

193 Ibid., ¶ 10.5 (footnotes omitted, emphasis added).

194 Ibid. (footnotes omitted).

195 *Joint Statement on Human Rights and Climate Change*, (16 September 2019), https://www.ohchr.org/en/statements/2019/09/five-un-human-rights-treaty-bodies-issue-joint-statement-human-rights-and.

196 CRC Decision, supra note 186, ¶ 10.6.

197 Ibid., ¶ 10.7.

198 Ibid., ¶ 10.8 (emphasis added).

199 Ibid., ¶ 10.10.

200 The other two being effective control over sources of emissions, and reasonable foreseeability of harm.

201 ILC, *Draft Articles on Prevention of Transboundary Harm from Hazardous Activities* (2001), https://legal.un.org/ilc/texts/instruments/english/commentaries/9_7_2001.pdf.

202 CRC Decision, supra note 186, ¶ 10.12.

203 Ibid., ¶ 10.13 (emphasis added).

204 Ibid., ¶ 10.17.

205 See Tigre, M. & Lichet, V., "The CRC Decision in Sacchi v. Argentina," *Insights, American Society of International Law*, vol. 25, no. 26 (13 December 2021), https://www.asil.org/sites/default/files/ASIL_Insights_2021_V25_I26.pdf. They believe that the decision whether to waive the requirement to exhaust domestic remedies was crucial in this case; had the CRC accepted the cases, it would have become a tribunal of first instance. Moreover, it would have undermined cases at the national level to hold states accountable.

206 Tigre & Lichet, supra note 205, believe that the decision:

> [c]ontributes to a broader toolbox of responses at local, national, regional, and international levels to the climate emergency. Overall, the CRC highlighted that climate change is indeed a child rights crisis and that states are responsible for their emissions… Children willing to sue states for failing to address climate change now have a direct legal framework to bring a case before the CRC if they exhaust local remedies.

They believe that the Committee's decision to draft a General Comment on children's rights and the environment signaled "the potential of human rights litigation to contribute to normative development beyond a specific case."

207 Australia observations, supra note 55, ¶ 41.

208 The Committee held a day of general discussion (DGD) on children's rights and the environment in 2016. The overall objective was to promote understanding of the relationship between children's rights and the environment; identify what needs to be done for child rights-related laws, policies and practices to take account of environmental issues and for environment-related laws, policies and practices to be child-sensitive.

References

Amici Curiae Brief of Special Rapporteurs on Human Rights and the Environment in Support of Admissibility, Before the Committee on the Rights of the Child, C.S. et al. v. Argentina, Brazil, France, German and Turkey, N.104/2019, N.105/2019, N.106/2019, N.107/2019, and N.108/2019, available at: https://www.hausfeld.com/uploads/documents/crc_admissibility_brief_boyd_knox_final_-_1_may_2020.pdf.

Amici Curiae Brief of Special Rapporteurs on Human Rights and the Environment on the Merits, Before the Committee on the Rights of the Child, C.S. et al. v. Argentina, Brazil, France, German and Turkey, N.104/2019, N.105/2019, N.106/2019, N.107/2019, and N.108/2019, http://climatecasechart.com/climate-change-litigation/wp-content/uploads/sites/16/non-us-case-documents/2021/20210901_Communication-No.-1042019-Argentina-Communication-No.-1052019-Brazil-Communication-No.-1062019-France-Communication-No.-1072019-Germany-Communication-No.-1082019-Turkey_na.pdf.

Atapattu, S. (2022) "Migrating with Dignity: Protecting the Rights of 'Climate Refugees' with the non-refoulement Principle," in *Climate Refugees: Global, Local and Critical Approaches*, Behrman, S. & Kent, A., eds., 127–145, Cambridge: Cambridge University Press.

Case Concerning the Gabcikovo Nagymaros Project (Hungary v. Slovakia), ICJ Rep (1996)

Catan and others v. the Republic of Moldova and Russia (2012), Application Nos. 43370/04, 8252/05 and 18454/06, https://hudoc.echr.coe.int/fre#{%22itemid%22:[%22002-7212%22]}.

Concept note: General comment on children's rights and the environment with a special focus on climate change (2022), https://www.ohchr.org/EN/HRBodies/CRC/Pages/CRC_GC26_concept_note.aspx.

Concluding observations on the combined fifth and sixth periodic reports of Argentina (2018), CRC/C/ARG/CO/5–6.

Concluding observations on the combined fifth and sixth periodic reports of Australia (2019), CRC/C/AUS/CO/5–6.

Concluding observations on the combined fifth and sixth periodic reports of Austria (2020), CRC/C/AUT/CO/5–6.

Concluding observations on the combined fifth and sixth periodic reports of Belarus (2020), CRC/C/BLR/CO/5–6.

Concluding observations on the combined fifth and sixth periodic reports of Belgium (2019), CRC/C/BEL/CO/5–6.

Concluding observations on the combined fifth and sixth periodic reports of Bosnia and Herzegovina (2019), CRC/C/BBIH/CO/5–6 (5 December 2019).

Concluding observations on the combined fifth and sixth periodic reports of Czechia (2021), CRC/C/CZE/CO/5–6.

Concluding observations on the combined fifth and sixth periodic reports of Ecuador (2017), CRC/C/ECU/CO/5–6.

Concluding observations on the combined fifth and sixth periodic reports of Norway (2018), CRC/C/NOR/CO/5–6.

Concluding observations on the combined fifth and sixth periodic reports of Poland (2021), CRC/C/POL/CO/5–6.

Concluding observations on the combined fifth and sixth periodic reports of Portugal (2019), CRC/C/PRT/CO/5–6.

Concluding observations on the combined fifth and sixth periodic reports of Spain (2018), CRC/C/ESP/CO/5–6.

Concluding observations on the combined fifth and sixth reports of Switzerland (2021), CRC/C/CHE/CO/5–6.

Concluding observations on the combined fourth and fifth periodic reports of Chile (2015), CRC/C/CHL/CO/4–5.

Concluding observations on the combined fourth and fifth periodic reports of Colombia (2015), CRC/C/COL/CO/4–5.

Concluding observations on the combined fourth and fifth periodic reports of Japan (2019), CRC/C/JPN/CO/4–5.

Concluding observations on the combined fourth and fifth periodic reports of Peru (2016), CRC/C/PER/CO/4–5.

Concluding observations on the combined fourth and fifth periodic reports of the Russian Federation (2014), CRC/C/RUS/CO/4–5.

Concluding observations on the combined fourth and fifth periodic reports of Singapore (2019), CRC/C/SGP/CO/4–5.

Concluding observations on the combined fourth to sixth periodic reports of Bahrain (2019), CRC/C/BHR/CO/4–6.

Concluding observations on the combined fourth to sixth periodic reports of Tunisia (2021), CRC/C/TUN/CO/4–6.

Concluding observations on the combined second and third periodic reports of the Gambia (2015), CRC/C/GMB/CO/2–3.

Concluding observations on the combined second and third periodic reports of Saint Vincent and the Grenadines (2017), CRC/C/VCT/CO/2–3.

Concluding observations on the combined second and third periodic reports of Serbia (2017), CRC/C/SRB/CO2–3.

Concluding observations on the second and third periodic reports of the Solomon Islands (2018), CRC/C/SLB/CO/2–3.

Concluding observations on the combined second to fifth periodic reports of the Cook Islands (2020), CRC/C/COK/CO/2–5.

Concluding observations on the combined second to fifth periodic reports of Tuvalu (2020), CRC/C/TUV/CO/2–5.

Concluding observations on the combined second to fourth periodic reports of Antigua and Barbuda (2017), CRC/C/ATG/CO/2–4.

Concluding observations on the combined second to fourth periodic reports of Brazil (2015), CRC/C/BRA/CO/2–4.

Concluding observations on the combined second to fourth periodic report of the Congo (2014), CRC/C/COG/CO/2–4.

Concluding observations on the combined second to fourth periodic reports of Eswatini (2021), CRC/C/SWZ/CO/2–4.

Concluding observations on the combined second to fourth periodic reports of Fiji (2014), CRC/C/FJI/CO2–4.

Concluding observations on the combined second to fourth periodic reports of Iraq (2015), CRC/C/IRQ/CO/2–4.

Concluding observations on the combined second to fourth periodic reports of Saint Lucia (2014), CRC/C/LCA/CO/2–4.
Concluding observations on the combined second to fourth periodic reports of Samoa (2016), CRC/C/WSM/CO/2–4.
Concluding observations on the combined second to fourth periodic reports of Vanuatu (2017), CRC/C/VUT/CO/2–4.
Concluding observations on the combined second to fourth periodic reports of Zambia (2016), CRC/C/ZMB/CO/2–4.
Concluding observations on the combined third and fourth periodic reports of Ireland (2016), CRC/C/IRL/CO/3–4.
Concluding observations on the combined third and fourth periodic reports of the Islamic Republic of Iran (2016), CRC/C/IRN/CO/3–4.
Concluding observations on the combined third and fourth periodic reports of Jamaica (2015), CRC/C/JAM/CO/3–4.
Concluding observations on the combined third and fourth periodic reports of the Marshall Islands (2018), CRC/C/MHL/CO/3–4.
Concluding observations on the combined third and fourth periodic reports of Mozambique (2019), CRC/C/MOZ/CO/3–4.
Concluding observations on the combined third and fourth periodic reports of Qatar (2017), CRC/C/QAT/CO/3–4.
Concluding observations on the combined third and fourth periodic reports of Suriname (2016), CRC/C/SUR/CO/3–4.
Concluding observations on the combined third to fifth periodic reports of Bhutan (2017), CRC/C/BTN/CO/3–5.
Concluding observations on the combined third to fifth periodic reports of the Bolivarian Republic of Venezuela (2014), CRC/C/VEN/CO/3–5.
Concluding observations on the combined third to fifth periodic reports of Cameroon (2017), CRC/C/CMR/CO/3–5.
Concluding observations on the combined third to fifth periodic reports of the Democratic Republic of the Congo (2017), CRC/C/COD/CO/3–5.
Concluding observations on the combined third to fifth periodic reports of Kenya (2016), CRC/C/KEN/CO/3–5.
Concluding observations on the combined third to fifth periodic reports of Malawi (2017), CRC/C/MWI/CO/3–5.
Concluding observations on the combined third to fifth periodic reports of the Niger (2018), CRC/C/NER/CO/3–5.
Concluding observations on the combined third to fifth periodic reports of Tajikistan (2017), CRC/C/TJK/CO/3–5.
Concluding observations on the combined third to sixth periodic reports of Guinea (2019), CRC/C/GIN/CO/3–6.
Concluding observations on the fifth periodic report of the Democratic People's Republic of Korea (2017), CRC/C/PRK/CO/5.
Concluding observations on the fifth periodic report of Mongolia (2017), CRC/C/MNG/CO/5.
Concluding observations on the fifth periodic report of New Zealand (2016), CRC/C/NZL/CO/5.
Concluding observations on the fifth periodic report of the Syrian Arab Republic (2019), CRC/C/SYR/CO/5.

Concluding observations on the fifth periodic report of the United Kingdom of Great Britain and Northern Ireland (2016), CRC/C/GBR/CO/5.
Concluding observations on the initial report of Tonga (2019), CRC/C/TON/CO/1.
Concluding observations on the initial report of Tuvalu (2013), CRC/C/TUV/CO/1.
Concluding observations on the second periodic report of Cabo Verde (2019), CRC/C/CPV/CO/2.
Concluding observations on the second periodic report of Côte d'Ivoire (2019), CRC/C/CIV/CO/2.
Concluding observations on the second periodic report of the Federated States of Micronesia (2020), CRC/C/FSM/CO/2.
Concluding observations on the second periodic report of Gabon (2016), CRC/C/GAB/CO/2.
Concluding observations on the second periodic report of Lesotho (2018), CRC/C/LSO/CO/2.
Concluding observations on the second periodic report of Palau (2018), CRC/C/PLW/CO/2.
Concluding observations on the second periodic report of South Africa (2016), CRC/C/ZAF/CO/2.
Concluding observations on the second periodic report of the United Arab Emirates (2015), CRC/C/ARE/CO/2.
Concluding observations on the sixth periodic report of Hungary (2020), CRC/C/HUN/CO/6.
Convention on the Rights of the Child (1989).
The Danish Institute for Human Rights (2022), *The Human Rights Guide to the Sustainable Development Goals*, https://sdg.humanrights.dk/en/goals-and-targets.
Decision adopted by the Committee under the Optional Protocol to the Convention on the Rights of the Child on a communications procedure, concerning Communication No. 104/2019 (2021), CRC/C/88/D/104/2019.
General Comment No. 9 (2007), CRC/C/GC/9.
General Comment No. 11 (2009), CRC/C/GC/11.
General Comment No. 15 (2013), CRC/C/GC/15.
General Comment No. 16 (2013), CRC/C/GC/16.
IACtHR, *The Environment and Human Rights*, Advisory Opinion, OC-23/17 (2017), https://www.corteidh.or.cr/docs/opiniones/seriea_23_ing.pdf.
ILC (2001) *Draft Articles on Prevention of Transboundary Harm from Hazardous Activities*, https://legal.un.org/ilc/texts/instruments/english/commentaries/9_7_2001.pdf.
Joint General Comment No. 3 (2017), CMW/C/GC/3-CRC/C/GC/22.
Joint Statement on Human Rights and Climate Change (2019), https://www.ohchr.org/en/statements/2019/09/five-un-human-rights-treaty-bodies-issue-joint-statement-human-rights-and.
Mapping Human Rights Obligations Relating to the Enjoyment of a Safe, Clean, Healthy and Sustainable Environment, Individual Report on the United Nations Convention on the Rights of the Child (2013), https://www.ohchr.org/en/special-procedures/sr-environment/mapping-report.
Report of the Special Rapporteur on the human right to safe drinking water and sanitation (2013), A/HRC/24/44/Add.2.
Report of the Special Rapporteur on the right of everyone to the enjoyment of the highest attainable standard of physical and mental health (2013), A/HRC/23/41/Add.3.
Resolution 8/7 (2008), A/HRC/RES/8/7.
Resolution 44/25 (1989), A/RES/44/25.
Resolution 66/128 (2011), A/RES/66/138.
Sacchi v. Argentina, Communication No. 104/2019 (2021), CRC/C/88/D/104/2019.
Teitiota v New Zealand (2020), CCPR/C/127/D/2728/2016.

Tigre, M., & Lichet, V. (2021) "The CRC Decision in Sacchi v. Argentina," *Insights, American Society of International Law*, vol. 25, no. 26, https://www.asil.org/sites/default/files/ASIL_Insights_2021_V25_I26.pdf.

Weissbrodt, D., Martin, J. & Newman, F. (late) (2021). *International Human Rights: Law, Policy, and Process* (5th edn.), Durham, NC: Carolina Academic Press, Chapter 8–9.

World Commission on Environment and Development (1987) *Our Common Future*, Oxford: Oxford University Press.

9 Other Treaty Bodies

In addition to the human rights bodies discussed in the previous chapters, several other treaty bodies have addressed environmental issues within their mandates and this chapter seeks to discuss the most pertinent ones. Thus, this chapter is a "catch all" chapter. The treaty bodies discussed in this chapter are: the Committee on the Rights of Persons with Disabilities ("Disabilities Committee"), the Committee on the Protection of Rights of All Migrant Workers and Members of their Families ("Committee on Migrant Workers"), and the Committee on the Elimination of Racial Discrimination ("Committee on Racial Discrimination").

9.1 The Committee on the Rights of Persons with Disabilities

The Disabilities Committee oversees the implementation of the Convention on the Rights of Persons with Disabilities (2006).[1] Established under Article 34 of the Convention, the mandate of the Committee includes the consideration of country reports submitted pursuant to Article 35,[2] adoption of general recommendations, and submitting reports to UN bodies together with recommendations where technical advice is needed. State parties have undertaken to cooperate with the Committee and assist its members in fulfilling their mandate.[3] The Committee, in turn, is required to cooperate with other treaty bodies and specialized agencies to ensure consistency and avoid duplication. The Committee is required to report every two years to the UNGA and the ECOSOC and may make recommendations based on the reports of the state parties.[4]

9.1.1 General Comments

Several General Comments adopted by the Committee are relevant to our discussion. Thus, GC No 2 which elaborates on Article 9 of the Convention on accessibility refers to sustainable development and the built environment.[5] Noting that accessibility is a key underlying principle of the Convention without which persons with disability cannot lead independent lives and fully participate in society, the GC provided that "Accessibility should be viewed not only in the context of equality and non-discrimination, but also as a way of investing in society and as an integral part of the sustainable development agenda."[6] The General Comment

DOI: 10.4324/9781003128847-12

again referred to sustainable development in the context of international cooperation and stressed that "Accessibility must be an integral part of any sustainable development effort, especially in the context of the post-2015 development agenda."[7]

GC No 3 on women and girls with disability refers to the need to protect women and girls with disabilities from sexual exploitation during natural disasters and humanitarian emergencies, as they are at an increased risk and are less likely to have access to recovery and rehabilitation services or access to justice.[8] It further refers to the need to combat multiple discrimination by ensuring that all international cooperation is disability- and gender-sensitive and inclusive in the implementation of the 2030 Agenda, including SDGs and indicators.[9]

GC No 5 makes several references to sustainable development. It provides that policies and plans for social inclusion of persons with disabilities, including promoting the right to independent living as a cost-effective mechanism "to ensure the enjoyment of rights, sustainable development and a reduction in poverty."[10] Referring to the New Urban Agenda adopted by the UN Conference on Housing and Sustainable Urban Development (Habitat III) as an integral part of 2030 Agenda, the General Comment notes that it advocates for a vision of cities where all persons can enjoy equal rights and opportunities, by promoting inclusive, healthy, accessible, resilient, and sustainable cities and human settlements. In this regard, GC refers to SDG targets 10.2[11] and 11.1.[12]

Likewise, GC No 6 refers to sustainable development and Agenda 2030.[13] It provides that disability-inclusive indicators must be developed and used in a manner consistent with the 2030 Agenda[14] and notes that all international cooperation efforts, including the 2030 Agenda, must be accessible to persons with disabilities and be guided by the Convention. It calls upon states to develop monitoring frameworks with human rights indicators, benchmarks, and targets for each indicator, consistent with SDG 10.[15]

GC No 7 on the participation of persons with disabilities including children with disabilities through their representative organizations has several relevant provisions.[16] It calls upon state parties to consider the general principles in the Convention when implementing, monitoring, and advancing the 2030 Agenda and SDGs.[17] It distinguishes between organizations of persons with disabilities and "civil society organizations" and notes that state parties should give priority to the views of organizations of persons with disabilities when addressing issues relating to disability. They should further develop frameworks to request civil society organizations to consult organizations of persons with disabilities with regard to the rights enshrined in the Convention and other topics, such as non-discrimination, peace, and *environmental rights*.[18] The GC calls upon state parties to strengthen the participation of organizations of persons with disabilities at the international level, including the high level Political Forum on Sustainable Development, and regional and universal human rights mechanisms.[19] It further provides that organizations of persons with disabilities should be consulted and involved in developing and implementing international cooperation plans and programs, including the 2030 Agenda and the Sendai Framework for Disaster Risk Reduction ("Sendai Framework").[20]

9.1.2 Concluding Observations

The pertinent references in the Concluding Observations of the Committee fall into two main categories: those referring to sustainable development and SDGs (and MDGs prior to that) and those referring to humanitarian situations and disasters and the Sendai Framework with an occasional reference to climate change. We will discuss the salient references in this section.

In the majority of the Concluding Observations since 2015, the Committee referred to the 2030 Agenda or post-2015 agenda and SDGs, mainly in the context of international cooperation. In the Concluding Observations issued in 2015, reference is made to the MDGs. Thus, for example, on the report of the Dominican Republic, the Committee expressed concern about the lack of information on the mainstreaming of the rights of persons with disabilities in plans and programs relating to the MDGs.[21] It recommended monitoring the impact of projects supported by international cooperation and mainstreaming the perspective of persons with disabilities in the implementation of the MDGs and the SDGs in the post-2015 UN sustainable development agenda.[22]

The recommendations made on the report of the European Union in relation to international cooperation are more detailed. Expressing concern about the lack of a systematic and institutionalized approach to mainstreaming rights of persons with disabilities across EU's international cooperation policies and programs, as well as lack of coordination among EU institutions, the Committee recommended that the EU adopt a harmonized policy on disability-inclusive development, "appoint disability focal points in related institutions and *take the lead in the implementation of disability-inclusive Sustainable Development Goals.*"[23]

The Committee made more specific recommendations on the report of Luxembourg: adopt a development policy in line with the Convention's principles and values in relation to development cooperation policies and programs; mainstream disability in the implementation of the 2030 Agenda and the SDGs; and ensure meaningful participation of persons with disabilities through their organizations in designing and implementing programs developed in international cooperation efforts.[24] On the report of Chile, the Committee referred to the linkages between Article 9 of the Convention[25] and SDG 11 on sustainable cities and communities, especially targets 11.2 and 11.7[26] as well as GC No 2 on accessibility.[27]

On the report of Slovakia, the Committee referred to the link between Article 28 of the Convention and SDG target 10.2[28] in the context of poverty alleviation and recommended that the state party provide an adequate standard of living to persons with disabilities and to ensure that social protection schemes are regularly monitored to track poverty alleviation.[29] It referred to Article 28 of the Convention and SDG targets 1.3 and 1.4[30] on the reports of Guatemala, Honduras (in the context of indigenous peoples with disabilities), Armenia, Moldova, Panama, Latvia, Slovenia, and Niger.

The second category of Concluding Observations related to disaster risk reduction and climate change and the Committee referred to the Sendai Framework on several occasions. On the report of Chile, the Committee expressed concern about the limited support for persons with disabilities in disaster risk situations, inadequate accessibility of evacuation routes, and the inaccessibility of services and information on disaster risk reduction for persons with disabilities.[31] The Committee recommended ensuring training for civil protection personnel on the inclusion of persons with disabilities in disaster risk reduction strategies, provision for accessibility in the design of infrastructure and evacuation routes, and providing information on disaster risk reduction. The Committee stressed that states should pay particular attention to the accessibility of information and take into account the Sendai Framework.[32]

On the report of UK and Northern Ireland, the Committee expressed concern about the impact on persons with disabilities in emergencies, and the absence of policies on disaster risk reduction that include persons with disabilities in the planning, implementation, and monitoring.[33] The Committee recommended adopting, in consultation with organizations of persons with disabilities, a comprehensive disaster risk reduction plan that provide for accessibility in situations of risk in line with the Sendai Framework. Moreover, the state party should ensure that organizations of persons with disabilities participate in resilience teams and play a role in formulating policies on disaster preparedness and planning.[34]

On a few occasions, the Committee referred to climate change in the context of disaster risk reduction. Thus, on the report of Guatemala, the Committee recommended that the state party:

> [mainstream] disability in its climate change policies and programmes, taking into consideration the outcomes of the Sendai Framework for Disaster Risk Reduction 2015–2030 with regard to persons with disabilities, the outcome document of the Climate Summit and the Charter on Inclusion of Persons with Disabilities in Humanitarian Action.[35]

Likewise, on the report of Colombia, the Committee recommended that the state party include, in line with the Sendai Framework, persons with disabilities in strategies for climate change adaptation and disaster risk reduction; include accessibility in infrastructure and evacuation routes; and provide information on disaster risk reduction in an appropriate mode.[36] On the report of Honduras, the Committee recommended that the National Risk Management system include a protocol to mitigate the risks to persons of disabilities with accessible early warning systems, focusing particularly on persons in rural and remote areas. It further recommended mainstreaming disability in its climate change policies and programs.[37]

On the report of Vanuatu, the Committee noted its vulnerability to natural disasters and expressed concern that persons with disabilities are not sufficiently involved in disaster risk reduction plans. It noted that some persons with disabilities were left behind during the evacuation from Ambae Island in 2018 and there have been no unified measures taken to establish an emergency notification system that is accessible for all persons with disabilities.[38] The Committee recommended consulting persons with disabilities through their representative organizations on the design and implementation of all disaster risk reduction plans and adopting a comprehensive strategy which incorporates recommendations made in the joint study on disability inclusion after Hurricane Pam[39] in accordance with the Sendai Framework and SDGs 11 and 13.[40]

The Committee referred to intersectional discrimination on the report of Panama in the context of an adequate standard of living and social protection. Noting with concern the number of persons with disabilities living in (extreme) poverty, the Committee recommended taking measures to enable persons with disabilities to enjoy a decent standard of living, with access to drinking water, electricity and sanitation, and to mitigate the impact of disability-related poverty, especially the impact on groups *subject to intersectional discrimination*, such as women, children, Afrodescendants and indigenous persons with disabilities.[41]

Similar concerns were expressed on the report of Guatemala where the Committee, noting that disability is not properly taken into account in policies on indigenous peoples, referred to their exclusion, lack of access to drinking water, sanitation, and decent housing, and overall conditions of poverty.[42] The Committee recommended enhancing efforts to mainstream disability in its programs and policies on indigenous peoples, adopting a rural and community-based approach, and taking measures to eliminate the disadvantages faced by indigenous women, children and older persons with disabilities who live in extreme poverty.[43]

9.2 Committee on the Protection of Rights of All Migrant Workers and Members of Their Families

The Committee on Migrant Workers was established under the International Convention on the Protection of Rights of All Migrant Workers and Members of their Families (1990).[44] Under the Convention, a Committee is established to review the application of the Convention.[45] State parties have undertaken to submit reports to the UN Secretary General for consideration by the Committee on the legislative, judicial, administrative and other measures they have taken to give effect to the provisions of the Convention within one year after the Convention enters into force for the state party and every five years thereafter.[46] After examining the state reports, the Committee is required to transmit its comments to the state party. In addition, the Committee is required to submit an annual report to the UNGA.[47] A state party may declare that it recognizes the competence of the Committee to consider communications of non-compliance under the Convention by another party.[48]

9.2.1 *General Comments*

So far, the Committee has adopted five General Comments, a couple of them jointly with the CRC Committee. None of these refer directly to the issues under discussion in this volume. The Committee is currently working on a General

Comment on the Convergence of the Convention and the Global Compact for Migration adopted by the UNGA in 2018[49] and is soliciting submissions. It held a day of general discussion in September 2022.[50] The Concept Note prepared for the draft General Comment refers to the need to put human rights back at the heart of discussions and actions in the context of international migration and take account of positive effects of migration on countries, in connection with the SDGs.[51] The outline prepared for the draft General Comment refers to SDGs and several targets.[52]

9.2.2 *Concluding Observations*

While the Committee has referred to the SDGs and Agenda 2030 on a few occasions, it referred to climate change only once even though the role of climate change as a driver of migration is well-recognized.[53] On the report of Guatemala, the Committee noted the link between climate change and migration:

> The Committee notes that a large part of the population in Guatemala suffers from adverse living conditions as a result of high rates of poverty, corruption, violence and crime, social exclusion of certain groups and underemployment, and *natural disasters and climate change*, in particular in areas of the so-called "dry corridor." These are also the structural causes that lead to forced migration. The Committee notes that the foregoing has a higher impact on the most vulnerable sectors of the population, such as indigenous peoples, peasant farmers and poor people. The Committee takes note of the State party's efforts to reverse this situation and uphold their people's right to migrate or not to be forced to migrate.[54]

On several reports the Committee recommended that "the State party avail itself of technical cooperation from the international community in order to follow up on the recommendations contained in the present Concluding Observations, in line with the 2030 Agenda for Sustainable Development."[55] The Committee referred to SDG target 17.18 on several reports.[56]

9.3 Committee on the Elimination of Racial Discrimination

One of the oldest human rights treaties, the International Convention on the Elimination of All Forms of Racial Discrimination (CERD) was adopted by the international community in 1965.[57] Part II of the Convention embodies provisions on the Committee on Racial Discrimination established under the Convention.[58] State parties have undertaken to submit a report to the Committee on the legislative, judicial, administrative, or other measures which they have adopted to give effect to the provisions of the Convention within one year after the Convention enters into force for the state concerned and every two years thereafter.[59] The Committee reports annually to the UNGA of its activities and recommendations to states. Under Article 14, a state party may declare that it recognizes the competence of the Committee to receive communications from individuals or groups of individuals within its jurisdiction claiming to be victims of a violation by that state party of any of the rights in the Convention.[60]

The report prepared for the Independent Expert on Human Rights and Environment, John Knox, in preparation for the mapping report will be used as background for this chapter.[61] As is the case with all human rights instruments, CERD does not include a provision on environmental protection. However, the Committee has addressed issues relating to environmental degradation, pollution, and sustainable development, mainly in the context of indigenous peoples.[62] As noted in Mapping Report No 3, human rights have been addressed by the Committee when considering the impact of environmental harm to include the right not to be discriminated against in the enjoyment of the right to property and the right to health of indigenous peoples.[63]

Mapping report No 3 notes that the CERD Committee has identified substantive and procedural obligations of state parties with respect to indigenous peoples. Substantive obligations include the duties to recognize and protect their property rights; ensure healthy conditions for them; and to develop specific policies to address environmental impacts affecting them.[64] Procedural obligations include the duties to conduct EIAs in relation to development activities on indigenous peoples' lands; ensure their participation in decision-making concerning their lands; ensure reasonable benefit or compensation to indigenous peoples impacted by the exploitation of their natural resources, and those who are affected by such exploitation has access to effective remedies.[65]

9.3.1 General Recommendations

Committee has adopted several General Recommendations (GR) on issues ranging from gender, indigenous peoples, refugees, and displaced persons. GR 23 on rights of indigenous peoples affirms that discrimination against indigenous peoples falls under CERD.[66] It recognized that indigenous peoples lost their lands and resources to colonists, commercial companies, and state enterprises which jeopardized their culture and identity. The Committee called upon state parties to "provide indigenous peoples with conditions allowing for a sustainable economic and social development compatible with their cultural characteristics."[67] Moreover, states should ensure that indigenous peoples have equal rights in respect of participation in public life and that no decisions relating to their rights and interests are taken without their informed consent.[68] The Recommendation calls upon state parties to:

> [r]ecognize and protect the rights of indigenous peoples to own, develop, control and use their communal lands, territories and resources and, where they have been deprived of their lands and territories traditionally owned or otherwise inhabited or used without their free and informed consent, to take steps to return those lands and territories. Only when this is for factual reasons not possible, the right to restitution should be substituted by the right

to just, fair and prompt compensation. Such compensation should as far as possible take the form of lands and territories.[69]

9.3.2 Concluding Observations

The Committee has pronounced on environmental pollution and degradation on a number of occasions. For example, on the report of Slovakia, the Committee expressed concern that a disproportionately large number of Roma suffer higher mortality rates, have poorer nutrition levels, and poor access to clean drinking water, inadequate sanitation, and high exposure to environmental pollution in Roma settlements.[70] Likewise, the Committee referred to the Serb and Roma minorities on the report of Croatia where it recommended that the state party continue its efforts for sustainable development of "areas of special State concern," which are inhabited by minorities, including Serb and Roma minorities, by eliminating economic and social disparities between regions.[71]

On the reports of Australia, the Committee reiterated its recommendation that the state party should ensure effective participation by indigenous communities in decisions affecting their land rights as required under Article 5(c) of the Convention and GR 23 which stresses the importance of securing their "informed consent."[72] The Committee requested the state party to provide full information on this issue in its next periodic report[73] and expressed serious concern about the continuing discrimination faced by indigenous peoples in relation to their ESC rights and the dramatic inequality still experienced by them who represents only 2.1% of the total population.[74]

On the report of Finland, the Committee noted that the issue of land ownership of the Sami has not been resolved and that Finland had not acceded to ILO Convention No 169.[75] The Committee also expressed concern about activities authorized in Sami reindeer-breeding areas which may threaten Sami culture and their traditional way of life. The Committee urged the state party to resolve land disputes together with the Sami people, in line with GR 23, and requested the state party to provide full information in the next periodic report.[76] Similar recommendations were made on the report of Sweden where the Committee expressed concern over land rights of Sami people, in particular their hunting and fishing rights that are being threatened by privatization of traditional Sami lands.[77] The Committee recommended that the state party introduce legislation recognizing Sami land rights reflecting the importance of reindeer husbandry to their way of life, and ratify ILO Convention No 169.[78]

The Committee, while noting the adoption of the Restitution of Land Rights Amendment Act of 2004 by South Africa, expressed concern about the extent of restitution, sustainable development of resettled communities and the enjoyment of their rights to housing, health, and access to water and education.[79] It expressed concern about the situation of indigenous peoples and in particular, hunter-gatherer, pastoralist, and nomadic groups and noted the absence of information on them. It requested the state party to provide in its next periodic report, information on any special measures taken in line with GR XXIII ensuring the enjoyment of their rights under the Convention including freedom of movement and their right to participate in decisions that affect them.[80]

On the report on Ethiopia the Committee expressed concern about the establishment of national parks and indigenous peoples' ability to pursue their traditional way of life and recommended that the state party provide information on the effective participation of these communities in the decisions that affect their rights and interests, including their informed consent in the establishment of the parks.[81] The Committee recommended, in light of GR No 23, adopting "measures to guarantee that national parks established on ancestral lands of indigenous communities allow for *sustainable economic and social development* compatible with the cultural characteristics and living conditions of those indigenous communities."[82] The Committee referred to the extreme poverty of people of minority ethnic groups in remote areas, including their access to food and water, despite the adoption of a Sustainable Development and Poverty Reduction Program.[83]

National parks were again the subject of comments on the report of Congo where the Committee expressed concern about the reports of violence, and cruel, inhuman, and degrading treatment by the "eco-guards" for the management of the ecosystems adjacent to the Ndoki National Park directed at the indigenous peoples in the northern region. It urged the state party to conduct a thorough investigation and provide information in its next periodic report.[84]

On the report of India, the Committee recommended, referring to GR No 23, that the state party formally recognize its tribal people as distinct groups entitled to special protection under national and international law.[85] It noted that the state party had not implemented the right of ownership of tribal communities, individual or collective, over lands traditionally occupied by them. Large scale projects such as construction of dams have been carried out without seeking the prior informed consent of tribal people which has resulted in the forced resettlement and affected the traditional lifestyle of these communities. The Committee urged the state party to implement the right of ownership of tribal communities over their traditional lands in accordance with ILO Convention No 107 and seek the prior, informed consent of communities affected by the construction of dams or similar projects on their traditional lands and provide adequate compensation and alternative land and housing to those communities. It should further protect tribes such as the Jarawa against encroachments on their lands and resources by settlers, poachers, and private companies and implement the Indian Supreme Court order to close the sections of the road that run through the Jarawa reserve.[86]

The Committee made extensive comments on environmental contamination on the report of Guyana where it expressed concern that indigenous peoples are disproportionately affected by malaria and environmental pollution, especially mercury and bacterial contamination of rivers caused by mining activities in indigenous areas.[87] In addition to providing access to trained health personnel and allocating sufficient funds, the Committee recommended that the state party undertake environmental impact assessments and "seek the informed consent of concerned indigenous communities prior to authorizing any mining or similar operations which may threaten the environment in areas inhabited by these communities."[88]

On the report of the Democratic Republic of the Congo, the Committee expressed concern that the rights of the Pygmies to own, control and use their lands and resources and communal territories are not guaranteed on which concessions are granted without prior consultation.[89] The Committee recommended taking measures to protect their land and forest rights of indigenous peoples in domestic legislation, declare a new moratorium on forest lands, and take the interests of Pygmies and environment conservation needs into account in relation to land use. The Committee also encouraged the state party to improve the indigenous peoples' enjoyment of ESC rights.[90]

On the report of Argentina, the Committee observed that the state party's national plan against discrimination seeks to ensure that indigenous peoples have access to justice and that INADI[91] is backing indigenous groups' complaints regarding their ancestral lands, their opposition to logging operations, and pollution of rivers but expressed concern about the failure to prosecute violent acts during forced evictions.[92] It expressed deep concern over reports that indigenous communities have been expelled from their ancestral lands. Despite ratification of ILO Convention No 169, the state party has not set up any consultative mechanisms to obtain the free, prior, and informed consent of communities that might be affected by development projects or exploitation of natural resources and recommended establishing such mechanisms. Where eviction is deemed necessary, the state party should ensure that those evicted receive adequate compensation and are provided with relocation sites with basic services such as drinking water, electricity, washing facilities, sanitation, and adequate social services.[93]

On the report of Guatemala, the Committee expressed concern that 90% of its 38 hydrographic basins are polluted which affects indigenous communities' access to safe drinking water and is causing the spread of diseases associated with lack of sanitation.[94] It urged the state party to take steps to ensure access to safe drinking water for indigenous communities, develop tools to prevent and monitor water pollution, treat the hydrographic basins that are already polluted, and adopt legislation guaranteeing all communities have access to safe drinking water.[95]

The Committee reiterated its concern over the awarding of licenses for hydroelectric projects, exploitation of natural resources and single-cropping on lands belonging to indigenous peoples on the 2015 report of Guatemala[96] which were granted without consulting indigenous peoples. It noted that Article 46 of the Constitution gives ILO Convention No 169 precedence over internal law and consultations should be respected *even in the absence of any national legislation*.[97] Stressing that such consultations are mandatory when natural resources belong to indigenous peoples and when the state has reserved to itself the ownership of underground resources, the Committee referred to the decisions of the Constitutional Court which ordered such consultations to be conducted. It noted that a legal framework governing land, territories, and natural resources had not been adopted, despite UN Special Rapporteur on the Rights of Indigenous Peoples' observation that the current legal protection does not meet international standards. Referring to the scant protection provided to indigenous peoples, the Committee pointed out that the state party continues to allow indigenous lands to be taken from them without consultation. Noting the impact of not observing the right to consultation and the right to land in the conflicts triggered by the concessions of the projects, the Committee reiterated, in line with GR No 23, its recommendation on the right to consultation and urged the state party to adopt ways of consulting with indigenous populations through institutions representing them. It stressed that indigenous peoples can be moved and resettled only in exceptional circumstances in which case *their consent is necessary.*[98] The Committee further recommended the adoption, in consultation with indigenous peoples, of a legal framework on the right to consultation and updating other laws, including the Mining Act, the Environmental Protection and Enhancement Act and the Regulations on Environmental Impact Assessments. It recommended a temporary moratorium on new licenses until a consultation mechanism is in place and urged the state party to recognize the right of indigenous peoples to their lands and territories.[99]

The Committee made extensive comments on the report of Colombia concerning indigenous peoples and regretted that the state party had not implemented the decisions of the Constitutional Court which warned that some indigenous peoples are at risk of physical and cultural extinction[100] and urged the preparation of ethnic protection plans with the effective participation of affected peoples. Expressing satisfaction at the progress made to recognize rights of indigenous and Afro-Colombian peoples to their territories and efforts made to return lands from which they were forcibly displaced, the Committee expressed concern about the lack of effective implementation of the law that recognizes the collective ownership of Afro-Colombians over their territories, and the limited right to restitution to victims displaced from their lands. The Committee recommended that the state party guarantee the right of indigenous and Afro-Colombian peoples to use, develop, and control their lands, territories, and natural resources freely and with full security, and provide legal recognition and protection.[101]

Despite the recognition of the right to consultations as a fundamental right by Colombia, the Committee expressed concern about its lack of implementation, where consultations are carried out without proper information, or hastily without due consideration for the concerns of indigenous peoples.[102] Nor is the bill on prior consultations fully in line with international standards and Afro-Colombian peoples were not consulted about the National Development Plan. Moreover, high level government officials' statements that prior consultation is an obstacle to development and infrastructure projects are concerning. The Committee recommended, in line with GR No 23, ensuring consultations, *in order to obtain the free, prior, and informed consent* of indigenous and Afro-Colombian peoples, as a tool of effective participation in relation to legislative or administrative provisions that could affect their rights, especially with regard to their land and natural resources; adopting procedures for prior consultations ensuring respect for traditions and customs; and avoiding statements criticizing the exercise of the right to free, prior, and informed consent and their right to sustainable development.[103]

The Committee further discussed projects involving exploitation of natural resources and their negative impacts, including illegal mining, on the territories of indigenous and Afro-Colombian peoples which cause irreparable damage to the environment and affect their traditional forms of subsistence and create tensions between outsiders and indigenous communities.[104] In an effort to protect human rights and eliminate racial discrimination which are essential for sustainable economic development,[105] the Committee urged the state party to guarantee the enjoyment of lands and natural resources by indigenous and Afro-Colombian peoples; implement protection measures against negative environmental impacts and the traditional ways of life of indigenous and Afro-Colombian peoples; provide compensation to them for damage or loss suffered; and ensure that they benefit from natural resource activities.[106]

In addition, the Committee expressed concern over the lack of access to safe drinking water and sanitation by indigenous and Afro-Colombian people and the impact of disproportionate use of water and pollution of rivers due to mining activities. It expressed concern over the El Cercado dam and El Cerrejon mine and the diversion of the Rancheria River had affected the Wajuu people's access to water in La Guajira. The Committee recommended guaranteeing access to safe drinking water and sanitation services for indigenous and Afro-Colombian peoples, and that water use by the mining industry does not affect access to water by these communities including the adoption of water processing and desalination standards.[107]

The Committee, while welcoming the adoption of a Concept Paper in 2009 by the Russian Federation on the sustainable development of indigenous peoples, expressed concern over the slow implementation of the objectives and that changes to federal legislation regulating the use of land, forests, and water bodies have diminished the rights of indigenous peoples to land, wildlife, and other natural resources by granting licenses to private businesses.[108] A new draft federal law could diminish the status of protected territories and would allow them to be exploited by third parties, including extractive industries. Moreover, the obligation to consult with indigenous peoples prior to any industrial development as required under law is often disregarded and indigenous peoples rarely receive compensation for the destruction of their lands and resources by private companies, and face obstacles to engaging in economic activities beyond their "traditional activities." The Committee recommended including in its next report measures taken to implement the 2009 Concept Paper; ensuring that any legislative changes enhance, *not diminish*, the rights of indigenous peoples as enshrined in UNDRIP; ensuring that indigenous communities are consulted through their freely elected representatives for any decisions that may impact them; providing compensation to communities that have been adversely affected by activities of private companies in line with GR No 23; and implementing recommendations of the UN Special Rapporteur on Indigenous Peoples after his visit in 2009.[109]

On the report of Niger, the Committee referred to the negative impact of the exploitation of natural resources, especially uranium, on the environment, health, and safety and to the absence of consultations with communities about the impact of extractive industries on water pollution and restricted access to water. It expressed concern at the poor management of these resources and that the communities do not benefit from such exploitation despite levying a mining royalty of 15% to be paid to the communities. The Committee recommended strengthening measures for the exploitation of natural resources to protect the environment and the health of local residents; declaring a moratorium on projects until human rights impact studies are done and until consultations with local communities *to secure their consent* are carried out; guaranteeing access to water for populations living on their traditional lands; engaging in consultations with the population to ensure that their rights are not violated and if expropriated, they receive redress; and conducting uranium mining activities so that people's health and the environment are not harmed and regularly inspecting such sites.[110]

On the report of Argentina, the Committee pointed out that structural discrimination prevents indigenous people and people of African descent from enjoying minimum international standards for development, including those in the SDGs.[111] Referring to the difficulties of indigenous communities to access water which is compounded by lack of titles to their lands and activities of companies that exploit natural resources, the Committee recommended that the state party take measures or affirmative action to eliminate structural discrimination against indigenous peoples and to meet the SDGs.[112]

The Committee expressed concern that the public tenders used by Chile for the recovery of land by the National Indigenous Development Corporation could prevent many indigenous communities from gaining access to their ancestral lands.[113] The tracts of land given to them in exchange for their ancestral lands have proven to be unproductive and do not form part of an overall strategy for the restitution of indigenous peoples' rights. While regulations on EIAs were about to enter into force, the Committee reiterated its concern that indigenous peoples' territories "continue to be negatively affected by the development of natural resources, the establishment of waste disposal sites and the pollution of water and other subsoil resources located in or on those lands"[114] but the plans to halt some production activities did not provide for redress. The Committee encouraged the state party to expedite the restitution of ancestral lands and protect indigenous peoples' rights to them; undertake environmental impact assessments and *hold free, prior, and informed consultations with a view to obtaining free and fully informed consent* before authorizing any investment projects; and provide redress for the damage suffered and resolve the environmental problems caused by such activities, which are affecting the lives and livelihoods of indigenous peoples.[115]

We have the rare occasion to read recommendations made on the report of the USA as CERD is one of the few human rights treaties the country has ratified.[116] In 2014, the Committee made extensive comments in relation to environmental justice concerns involving racial and ethnic minorities, indigenous peoples, and low-income communities, as well as activities of transnational corporations abroad. While welcoming the acknowledgment that low-income and minority communities are exposed to an unacceptable amount of pollution,[117] the Committee expressed concern that racial and ethnic minorities and indigenous peoples "continue to be disproportionately affected by the negative health impact of pollution caused by the extractive and manufacturing industries."[118] The Committee called upon the state party to enforce federal legislation prohibiting environmental pollution at state and local levels; undertake an independent investigation into environmentally polluting activities and their impact on communities and bring those responsible to account; ensure that victims have access to remedies; and clean up any radioactive and toxic waste as a matter of urgency, paying attention to areas inhabited by racial and ethnic minorities and indigenous peoples.[119]

With regard to indigenous people, while acknowledging the measures taken to recognize their culture and traditions and supporting the UNDRIP as announced by President Obama in 2012, the Committee expressed concern about not guaranteeing their free, prior, and informed consent and not protecting their sacred sites against polluting activities due to resource extraction, industrial development, construction of border walls, tourism, and urbanization. Referring to GR No 23, the Committee called upon the state party to guarantee the right of indigenous peoples to effective participation in public life and in decisions that affect them, based on free, prior, and informed consent; eliminate obstacles to the recognition of tribes; adopt measures to protect the sacred sites of indigenous peoples in the context of development, national security projects, and exploitation of natural resources; and hold those responsible for damage accountable.[120]

In its comments on the 2022 report of USA, the Committee noted the measures adopted to address the longstanding effects of pollution and climate change on disadvantaged communities[121] but expressed concern:

> [a]t the disproportionate health, socio-economic and cultural impact of pollution, climate change and natural disasters on racial and ethnic minorities and Indigenous Peoples, caused by extractive and manufacturing industries, such as petrochemical facilities and methanol complexes, as for instance in the case of "Cancer Alley" in Louisiana, and by radioactive and toxic waste.[122]

Reiterating its recommendation that the state party guarantee that federal legislation prohibiting environmental pollution is effectively enforced, the Committee recommended cleaning up radioactive and toxic waste, especially areas inhabited by racial and ethnic minorities and indigenous peoples; adopting moratoriums on new heavy industry facilities and expanding existing ones, such as petrochemical plants; and investigating cases of polluting activities affecting racial and ethnic minorities and indigenous peoples, holding those responsible accountable and providing effective remedies for the victims. It further recommended protecting important cultural sites from harm by extractive and manufacturing industries.[123]

The Committee while noting Costa Rica's difficulty of accessing indigenous territories, expressed concern that only 7.6% of indigenous peoples have their basic needs met[124] and about the situation in Talamanca and in the banana plantations. It recommended removing economic, social and geographical barriers to access basic services in indigenous territories, especially in Talamanca and banana plantations, so that they are not compelled to leave their ancestral lands. While domestic legislation protects indigenous peoples' right to land tenure, the situation is different in practice and the Committee noted the trend of indigenous lands falling into the hands of non-indigenous settlers.[125] It urged the state party to guarantee their right to land tenure and to take steps to implement the Constitutional Chamber decision on the delimitation of lands of communities, to recover lands lost through improper transfer, and to protect indigenous peoples' cultural heritage including their languages.[126] Noting the lack of a gender policy on indigenous women, the Committee recommended taking steps to combat double discrimination on the basis of gender and ethnicity[127] and adopt one in line with GR No 25.[128]

On the report of Indonesia, the Committee noted with concern the plan to establish oil palm plantations along the Indonesia-Malaysia border as part of the Kalimantan Border Oil Palm Mega-project, and the threat this poses to indigenous peoples and their culture and the ensuing conflicts between local communities and palm oil companies.[129] The Committee recommended undertaking consultations with the communities to obtain their consent whilst this project is being studied.[130] While under Indonesian law land, water, and natural resources are controlled by the state and exploited for the benefit of the people, the Committee recommended that it must be done in line with the rights of indigenous peoples, and that the state party should review its laws, especially the 2004 Law on Plantations, to ensure that they respect the rights of indigenous peoples to their communal lands.[131]

On the report of Peru, the Committee noted that it guarantees the protection of individual and collective rights of indigenous peoples established as campesino communities in the Andes and in the Amazon region who are considered as "indigenous peoples" for the purposes of ILO Convention No 169 and UNDRIP.[132] However, with regard to indigenous peoples and Afro-Peruvian communities not yet so recognized the Committee recommended adopting a law that covers all Indigenous and Native communities.[133] It referred to the tension and even violence generated by the exploitation of subsoil resources of indigenous territories and noted that sometimes indigenous peoples are not consulted and their prior informed consent obtained prior to allowing such exploitation. It further expressed concern on the negative health and environmental impact of extractive activities which are conducted at the expense of cultural rights of indigenous peoples. The Committee urged the state party to adopt the bill on the consultation and participation of indigenous peoples in environmental matters, in line with GR No 23 and to consult indigenous communities *at each step of the process* and obtain their consent before plans to extract natural resources are implemented.[134] It expressed concern at the violence surrounding such exploitation and noted the steps taken to address the situation including launching an investigation. The Committee welcomed the visit by the UN Special Rapporteur on Indigenous Peoples and urged the state party to implement his recommendations and to establish an independent commission that includes indigenous representatives to carry out an impartial investigation.[135] It also recommended that the Commission's findings should feed into the discussions on the bill on consultations and the regulations for mining and oil sectors.

Noting the complexity of the sustainable management of natural resources, the Committee noted the state party's efforts to improve legislation, particularly in relation to water resources but expressed concern that the catchment basins may have a negative impact on the wetland areas of Peru and the way of life of indigenous peoples.[136] The Committee recommended that the water management policy consider the needs of the communities likely to be affected, to guarantee the use of water by the community of Ancomarca, and provide compensation for the damage suffered.[137] The Committee expressed concern that conflicts may arise due to the lack of consensus on national policy, in particular in the areas of education, development projects, and environmental protection and recommended conducting a participatory and inclusive process to determine the vision of the nation.[138]

The Committee made extensive comments again on the report of Peru and expressed concern that "concessions for the extraction of natural resources continue to infringe the rights of indigenous peoples over their lands, traditional and ancestral territories and natural resources, including waters, and generate environmental problems, such as the pollution of aquifers"[139] and at the lack of implementation of measures to mitigate environmental impacts. Referring to GR No 23 and the recommendations made by the Special Rapporteur on Indigenous Peoples following his visit to Peru, the Committee urged strengthening legislation to protect indigenous peoples with regard to the exploitation of natural resources; guaranteeing the enjoyment by indigenous peoples of their lands and natural resources; implementing safeguards against environmental impacts; and providing compensation to indigenous peoples for damage suffered by natural resource activities and ensuring that they benefit from such activities.[140]

The Committee referred again to the absence of mechanisms to protect indigenous peoples' lands and resources on a subsequent report of Peru, in part due to lack of land titling and concentration of land and exploitation of natural resources by private entities, leading to serious conflicts.[141] It expressed concern that development of natural resources has a negative impact on their territories, especially water pollution which affects their traditional means of subsistence. The Committee recommended establishing a mechanism for filing land claims and restitution of ancestral territories, and protecting their right to own, use, and exercise full control over their lands and resources by providing legal recognition and safeguards in accordance with international standards.[142] It also recommended conducting social and environmental impact assessments of natural resource development projects on indigenous peoples' territories and provide compensation to affected communities for damage suffered and ensure they share in the benefits derived from those activities. It expressed concern about the lack of prior consultations on natural resource development projects including mining projects on indigenous lands and recommended that sufficient information is provided, and prior consultations are conducted in a timely and reasonable manner.[143]

The Committee commended Kenya for its engagement with the 2030 Agenda and for including in its development blueprint – Vision 2030 – strategies to achieve substantive equality.[144] However, it expressed alarm at reports that the Sengwer people are being forcibly evicted from their traditional forest lands in violation of a court injunction and that activities affecting ancestral lands occupied by indigenous peoples have been undertaken without their free, prior, and informed consent.[145] The Committee, in line with GR No 23, urged the state party to ensure legal acknowledgment of the collective rights of indigenous peoples to own and use their lands, resources, and communal territories, and to participate in the exploitation, management, and conservation of natural resources.[146] It also urged the state party to carry out consultations with affected communities to develop, conserve, or exploit indigenous ancestral lands or its natural resources and obtain their free, prior, and informed consent.

On the report of New Zealand, the Committee expressed concern about the application of the Marine and Coastal Area Act of 2011 on Māori land and resource rights which the state party had not reviewed to align with the recommendations made in the previous concluding observations. Nor has the state party applied the free, prior, and informed consent principle consistently in relation to Māori customary marine interests.[147] The Committee recommended reviewing the Act to protect land and resource rights of Māori communities and their access to places of cultural significance, and provide in its next report information on the recognition of their customary interests. The Committee also expressed concern over reports of granting private companies the right to use freshwater resources located on Māori land despite their opposition. While welcoming the adoption of the Te Awa Tupua (Whanganui River Claims Settlement) Act of 2017[148] and the state party's efforts to consult with iwi and hapu concerning their freshwater rights and advising Māori organizations on geothermal assets, the Committee urged the state party to respect the rights of Māori communities to freshwater and geothermal resources as protected by the Act and the Convention.[149]

While welcoming the adoption of the National Sustainable Development Strategy in Kyrgyzstan, the Committee requested the state to ensure the participation of Mugat and Uzbek ethnic minorities in its development and implementation.[150] On the report of Bolivia, the Committee requested information on how demands for sustainable development, promotion of agriculture, and protection of rights of indigenous and peasant communities will be included in the draft bill on land reform.[151]

The Committee referred to the impact of activities of transnational corporations on a few reports. For example, it noted the adverse extraterritorial impacts of economic activities of transnational corporations registered in Canada in the exploitation of natural resources and their impact on the rights to land, health, living environment, and the way of life of indigenous people living in these regions.[152] The Committee encouraged the state party to take legislative or administrative measures to prevent transnational corporations negatively impacting indigenous peoples' rights in territories outside Canada and explore ways to hold them accountable. The Committee requested the state party to include in its next periodic report on the measures taken in this regard.[153]

Similar concerns were raised on the report of USA regarding the adverse effects of activities related to the exploitation of natural resources by transnational corporations registered in the US and operating abroad on the rights to land, health, environment, and the way of life of indigenous peoples and minority groups living there.[154] The Committee recommended taking measures to prevent activities of transnational corporations infringing the rights of local populations, especially indigenous peoples and minorities, in other countries.[155]

9.4 Conclusion

This chapter discussed the work of three treaty bodies that are not usually associated with environmental protection or climate change: Committee on the Rights of Persons with Disabilities, Committee on the Protection of Rights of Migrant Workers, and the Committee on the Elimination of Racial Discrimination. However, as the foregoing discussion showed, these bodies have addressed environmental protection and/or pollution, climate change, and sustainable development including SDGs within their mandates. SDGs are mainly used as a yardstick to measure states' progress with their human rights obligations.

While the Committee on Migrant Workers had the least number of references to environmental issues, all three Committees have acknowledged the link between environmental issues and the enjoyment of rights embodied in their respective treaties. The Committee on Racial Discrimination has addressed environmental protection almost exclusively in relation to indigenous peoples although a few broader recommendations have been made, including a comment about adopting a national policy that is accepted by society as a whole, not just indigenous peoples. Most recommendations of the Committee on Racial Discrimination endorsed the free, prior, and informed consent principle, consultations, and participation of indigenous peoples and on one occasion said it applies at every step of the process with regard to projects involving natural resource extraction. The Committee's comments on the report of Guatemala are significant – because the Constitution gives ILO Convention No 169 precedence over internal law, consultations should be respected even in the absence of any national legislation. It further stressed that indigenous peoples' consent is necessary to move them from ancestral lands which can be done only in exceptional circumstances. The recommendations made with regard to controlling the activities of transnational corporations abroad are also significant. The requirement to hold free, prior, and informed consultations with a view to obtaining free and fully informed consent seems to go even beyond the requirements of UNDRIP.

The recommendations made by the Disabilities Committee recognize the intersectional nature of vulnerability as well as marginalization that can lead to even greater injustices facing people with disabilities. Many of the environment/climate-related comments of the Disabilities Committee were made in the context of disasters, even though yet again we see references to "natural disasters."

Notes

1 As of May 2022, 185 states have ratified it. https://www.un.org/development/desa/disabilities/convention-on-the-rights-of-persons-with-disabilities.html.
2 State parties are required to submit, within two years of the Convention's entry into force for that party, a comprehensive report on how it is implementing the obligations under the Convention. Thereafter, reports are to be submitted at least every four years. UN Convention on the Rights of Persons with Disabilities, art. 35: https://www.un.org/disabilities/documents/convention/convoptprot-e.pdf.
3 Ibid., art. 36.
4 Ibid., art. 39.
5 *General Comment No. 2 (2014) Article 9: Accessibility*, CRPD/C/GC/2 (22 May 2014).
6 Ibid., ¶ 4.
7 Ibid., ¶ 47.
8 *General comment No. 3 (2016) on women and girls with disabilities*, CRPD/C/GC/3 (25 November 2016), ¶ 49.
9 Ibid., ¶ 63(e). In addition, GC No. 4 makes several references to SDG 4 on education. It provides that "Disability-inclusive indicators must be developed and used in a manner consistent with the 2030 Agenda for Sustainable Development." *General Comment No. 4 (2016) on the right to inclusive education*, CRPD/C/GC/4 (25 November 2016), ¶12(i). It refers to 2030 Agenda in the context of providing compulsory primary education to all children.
10 *General Comment No. 5 (2017) on living independently and being included in the community*, CRPD/C/GC/5 (27 October 2017), ¶ 5.
11 This deals with empowerment and promotion of social, economic and political inclusion for all.
12 This deals with ensuring access to adequate, safe and affordable housing and affordable services for all.
13 *General Comment No. 6 (2018) on equality and non-discrimination*, CRPD/C/GC/6 (26 April 2018).
14 Ibid., ¶ 71.
15 Ibid., ¶ 72.
16 *General Comment No. 7 (2018) on the participation of persons with disabilities, including children with disabilities, through their representative organizations, in the implementation and monitoring of the Convention*, CRPD/C/GC/7 (9 November 2018).
17 Ibid., ¶ 9.
18 Ibid., ¶ 14.
19 Ibid., ¶ 32.
20 Ibid., ¶ 92. The GC also calls upon states to actively involve and consult with persons with disabilities, through their representative organizations on, *inter alia*, the monitoring of the SDGs at the national level. Ibid, ¶ 94(r). UNGA, 69/283, *Sendai Framework for Disaster Risk Reduction 2015–2030*, A/RES/69/283 (23 June 2015).
21 *Concluding observations on the initial report of the Dominican Republic*, CRPD/C/DOM/CO/1 (8 May 2015), ¶ 60, ["DR report"]. On the report of Mongolia, the Committee was concerned about the lack of information on the inclusion of perspectives of persons with disabilities on the implementation of the MDGs at the national level. *Concluding observations on the initial report of Mongolia*, CRPD/C/MNG/CO/1 (13 May 2015), ¶ 48. See also, *Concluding observations on the initial report of Uganda*, CRPD/C/UGA/CO/1 (12 May 2016), ¶ 62; *Concluding observations on the initial report of Chile*, CRPD/C/CHL/CO/1 (13 April 2016), ¶ 65 ["Chile report"]. Of the concluding observations surveyed between 2015 and 2019, a similar paragraph appears in relation to international cooperation in 45 country reports. While some recommendations are a little more detailed than others including the need to ensure the meaningful participation and consultation of persons with disability through their representative

organizations in the design and implementations of programs, most concluding observations carry almost identical language referred to above.

22 DR Report, supra note 21, ¶ 61. Similar recommendations were made on the report of Mongolia where the Committee stressed the importance of meaningful and empowered participation and inclusion of persons with disabilities in designing, implementing and monitoring disability-inclusive development projects.

23 *Concluding observations on the initial report of the European Union*, CRPD/C/EU/CO/1 (2 October 2015), ¶ 75 (emphasis added).

24 *Concluding observations on the initial report of Luxembourg*, CRPD/C/LUX/CO/1 (10 October 2017), ¶ 57. Similar recommendations were made on the reports of Seychelles, Vanuatu, Turkey, Mongolia (referred to the meaningful and empowered participation and inclusion of persons with disabilities and their organizations in designing and implementing disability-inclusive development projects).

25 Article 9 addresses accessibility.

26 Target 11.2 provides: "By 2030, provide access to safe, affordable, accessible and sustainable transport systems for all, improving road safety, notably by expanding public transport, with special attention to the needs of those in vulnerable situations, women, children, persons with disabilities and older persons" while target 11.7 provides: "By 2030, provide universal access to safe, inclusive and accessible, green and public spaces, in particular for women and children, older persons and persons with disabilities." See https://sdgs.un.org/goals/goal11. Similar recommendations were made on the reports of Guatemala, Honduras and Cyprus.

27 Chile report, supra note 21, ¶ 20. Similar recommendations were made on the report of Portugal where the Committee called upon the state party to provide access to safe, affordable, accessible and sustainable transport systems for all and to provide access to safe, inclusive and accessible, green and public spaces. CRPD, *Concluding observations on the initial report of Portugal*, CRPD/C/PRT/CO/1 (20 May 2016), ¶ 22 ["Portugal report"].

28 Article 28 deals with an adequate standard of living and social protection while target 10.2 provides: "By 2030, empower and promote the social, economic and political inclusion of all, irrespective of age, sex, disability, race, ethnicity, origin, religion or economic or other status." https://sdgs.un.org/goals/goal10.

29 See *Concluding observations on the initial report of Slovakia*, CRPD/C/SVK/CO/1 (17 May 2016). Similar recommendations were made on the reports of Portugal, Serbia, Guatemala (recommended that the state party be guided by Article 5 of the Convention when implementing targets 10.2 and 10.3 of SDGs), Colombia, Italy, Bolivia, Ethiopia, Cyprus, and Sudan.

30 Target 1.3 provides: "Implement nationally appropriate social protection systems and measures for all, including floors, and by 2030 achieve substantial coverage of the poor and the vulnerable" and Target 1.4 provides:

> By 2030, ensure that all men and women, in particular the poor and the vulnerable, have equal rights to economic resources, as well as access to basic services, ownership and control over land and other forms of property, inheritance, natural resources, appropriate new technology and financial services, including microfinance.

See https://sdgs.un.org/goals/goal1.

31 Chile report, supra note 21, ¶ 21.

32 Ibid., ¶ 22. See also Portugal report, supra note 27, ¶ 27, where reference was made to the need to provide assistance to migrant, refugees ad asylum seekers with disabilities; *Concluding observations on the initial report of Ethiopia*, CRPD/C/ETH/CO/1 (4 November 2016), ¶ 24, where the Committee referred to the need to prepare a risk and emergency strategy that integrates disability issues in compliance with the Sendai

Framework; *Concluding observations on the initial report of Montenegro* CRPD/C/MNE/CO/1 (22 September 2017), where reference was made to the Sendai Framework and the need to have an accessible and comprehensive emergency strategy for situations of risk and to develop local plans in consultation with persons of disabilities through their representative organizations.

33 *Concluding observations on the initial report of the United Kingdom of Great Britain and Northern Ireland*, CRPD/C/GBR/CO/1 (3 October 2017), ¶ 28.

34 Ibid., ¶ 29. Similar recommendations were made on the report of Nepal where the Committee noted the disproportionate impact of the 2015 earthquake on persons with disabilities and recommended the adoption of an emergency strategy in accordance with the Sendai Framework. The Committee further recommended ensuring that post-disaster recovery and rehabilitation efforts follow a human rights-based approach to protect persons with disabilities, *Concluding observations on the initial report of Nepal*, CRPD/C/NPL/CO/1 (16 April 2018), ¶ 20. Similar recommendations were made on the *Concluding observations on the initial report of Oman*, CRPD/C/OMN/CO/1 (17 April 2018).

35 *Concluding observations on the initial report of Guatemala*, CRPD/C/GTM/CO/1 (30 September 2016) ["Guatemala report"], ¶ 30. It is not clear which "climate summit" this paragraph refers to.

36 *Concluding observations on the initial report of Colombia*, CRPD/C/COL/CO/1 (30 September 2016), ¶ 27. See also *Concluding observations on the initial report of the Plurinational State of Bolivia*, CRPD/C/BOL/CO/1 (4 November 2016), ¶ 26.

37 *Concluding observations on the initial report of Honduras*, CRPD/C/HND/CO/1 (4 May 2017), ¶ 26. A similar recommendation was made on the report of Panama where the Committee also recommended that the state party include persons with disabilities in all stages of disaster management to ensure that their needs and rights are respected, *Concluding observations on the initial report of Panama*, CRPD/C/PAN/CO/1 (29 September 2017) ["Panama report"], ¶ 29.

38 *Concluding observations on the initial report of Vanuatu*, CRPD/C/VUT/CO/1 (13 May 2019), ¶ 20 ["Vanuatu report"].

39 CBM-Nossal Institute Partnership for Disability Inclusive Development, *Disability Inclusion in Disaster Risk Reduction: Experiences of people with disabilities in Vanuatu during and after Tropical Cyclone Pam and recommendations for humanitarian agencies* (July 2017), https://mspgh.unimelb.edu.au/__data/assets/pdf_file/0011/2567576/WEB-DIDRR-Report-14112017.pdf.

40 Vanuatu report, supra note 38, ¶ 21.

41 Panama report, supra note 37, ¶ 55 (emphasis added).

42 Guatemala report, supra note 35, ¶ 65.

43 Ibid. Other recommendations included ratifying the Protocol to the African Charter on Human and Peoples' Rights on the Rights of Persons with Disabilities in Africa. This recommendation was made on the reports of Niger, Rwanda, Senegal, and Algeria. However, the wording on them differed somewhat. On the reports of Niger, Rwanda and Algeria the Committee recommended that the state party *consider ratifying* the Protocol while on the report on Senegal the Committee recommended that the state party *take the measures necessary* to ratify the Protocol. The rather mandatory nature of the latter recommendation is striking.

44 Adopted by UNGA resolution 45/158 (18 December 1990), https://www.ohchr.org/sites/default/files/cmw.pdf.

45 Ibid., art. 72. Fourteen members are elected by secret ballot by state parties and serve in their personal capacity and are experts of high moral standing with recognized competence in the field. They serve for a period of four years and are eligible for re-election.

46 Ibid., art. 73.

47 Ibid., art. 73(7). In addition, the UN Secretary General may transmit reports to specialized agencies and the ILO will be invited to appoint representatives to participate in a consultative capacity in the meetings of the Committee.
48 Ibid., art. 76.
49 Global Compact for Safe, Orderly and Regular Migration (13 July 2018), https://refugeesmigrants.un.org/sites/default/files/180713_agreed_outcome_global_compact_for_migration.pdf
50 CMW, Half-day of general discussion on draft General Comment No. 6 (27 September 2022), https://www.ohchr.org/en/events/days-general-discussion-dgd/2022/half-day-general-discussion-draft-general-comment-no-6.
51 Draft General Comment No. 6 on the Convergence of the Convention and the Global Compact for Safe, Orderly and Regular Migration, Concept Note, Guiding Questions and Call for Submissions, https://www.ohchr.org/en/calls-for-input/2022/call-submissions-concept-paper-and-draft-outline-its-draft-general-comment-no.
52 CMW, Draft outline, https://www.ohchr.org/sites/default/files/documents/hrbodies/cmw/cfi-gc6–2022/2022–07–29/Outline-for-CMW-GC-No6.docx.
53 See Global Compact for Migration, supra note 49, which recognizes climate change as a driver of migration. Inter-governmental Panel on Climate Change (IPCC) has also recognized this link.
54 *Concluding observations on the second periodic report of Guatemala,* CMW/C/GTM/CO (2 May 2019), ¶ 5.
55 *Concluding observations on the initial report of Madagascar*, CMW/C/MDG/CO (15 October 2018), ¶ 56. Similar recommendations were made on the reports of Mozambique, Guatemala, Colombia, and Bosnia and Herzegovina.
56 See reports of Algeria, Madagascar, Mozambique, Guatemala, Bosnia and Herzegovina and Colombia. SDG target 17.18 states:

> By 2020, enhance capacity-building support to developing countries, including for least developed countries and small island developing States, to increase significantly the availability of high-quality, timely and reliable data disaggregated by income, gender, age, race, ethnicity, migratory status, disability, geographic location and other characteristics relevant in national contexts.

https://sdgs.un.org/goals/goal17.
57 Adopted and opened for signature by UNGA resolution 2106 (XX) of 21 December 1965 and entered into force 4 January 1969, https://www.ohchr.org/sites/default/files/cerd.pdf.
58 Ibid., art. 8.
59 Ibid., art. 9.
60 Ibid., art. 14.
61 *Mapping Human Rights Obligations Relating to the Enjoyment of a Safe, Clean, Healthy and Sustainable Environment, Individual Report on the International Convention on the Elimination of All Forms of Racial Discrimination, Report No.* 3 (December 2013) ["Mapping Report No. 3"], https://www.ohchr.org/sites/default/files/Documents/Issues/Environment/Mappingreport/3.CERD-25-Feb.docx.
62 Ibid., p. 7.
63 Ibid.
64 Ibid., ¶ 51.
65 Ibid.
66 *General Recommendation XXIII on the rights of indigenous peoples* (1997), ¶ 1, https://tbinternet.ohchr.org/Treaties/CERD/Shared%20Documents/1_Global/INT_CERD_GEC_7495_E.doc.
67 Ibid., ¶ 4©.
68 Ibid., ¶ 4(c).

69 Ibid., ¶ 5.

70 *Concluding observations on the report of Slovakia*, CERD/C/304/Add.110 (1 May 2001), ¶ 14.

71 *Concluding observations on the report of Croatia*, CERD/C/HRV/CO/8 (24 March 2009), ¶ 20.

72 *Concluding observations on the combined tenth, eleventh and twelfth periodic reports of Australia*, CERD/C/335/Add.2 (14 December 1999).

73 Ibid., ¶ 32.

74 Ibid., ¶ 41.

75 *Concluding observations on the fifteenth periodic report of Finland*, CERD/C/363/Add.2 (9 August 2000), ¶ 214. See also Chapter 5 where the UN Human Rights Committee made a similar recommendation.

76 Ibid.

77 *Concluding observations on the thirteenth and fourteenth periodic reports of Sweden*, CERD/C/362/Add.5 (22 August 2000), ¶ 338.

78 Ibid.

79 *Concluding observations on the report of South Africa*, CERD/C/ZAF/CO/3 (19 October 2006), ¶ 18, ["South Africa report"]. On the *Concluding observations on the report of Namibia*, CERD/C/NAM/CO/12 (22 September 2008), ¶ 20, the Committee expressed concern about the persistence of discrimination based on ethnicity with regard to the enjoyment of ESC rights, despite measures taken to reduce poverty and to progressively realize equal and sustainable development. It recommended undertaking studies to assess the level of enjoyment of ESC rights by different ethnic groups and requested the state party to provide in its next report information on the active involvement of targeted beneficiaries in decisions relating to their rights and interests. On the *Concluding observations on the report of China (including Hong Kong and Macau Special Administrative Regions)*, CERD/C/CHN/CO/10–13 (15 September 2009), ¶ 21, the Committee recommended creating conditions for sustainable development in the western areas and eliminating economic and social disparities in the regions. It requested the state party to provide information on the enjoyment of ESC rights by all ethnic groups.

80 South Africa report, supra note 79, ¶ 18. See also the *Concluding observations on the report of New Zealand*, where the Committee recommended holding a dialog with the Māori Community on the Foreshore and Seabed Act of 2004 to seek ways to mitigate its discriminatory effects and monitor its implementation closely, minimize its negative impacts and broaden the redress available to the Māori, in *Report of the Committee on the Elimination of Racial Discrimination*, A/62/18 (30 July–17 August 2007), ¶ 452 ["CERD report"], available at: https://undocs.org/A/62/18.

81 *Concluding observations on the report of Ethiopia*, CERD/C/ETH/CO/15 (20 June 2007), ¶ 22.

82 Ibid., ¶ 22 (emphasis added). The use of this phrase is disconcerting as it does not refer to sustainable development but rather to sustainable economic and social development. See Chapter 3 where China made similar recommendations on several reports during the UPR process.

83 Ibid., ¶ 26.

84 *Concluding observations on the report of Congo*, CERD/C/COG/CO/9 (23 March 2009) ["Congo report"], ¶ 13.

85 *Concluding observations on the report of India*, CERD/C/IND/19 (6 March 2007), ¶ 168.

86 Ibid., ¶ 177.

87 *Concluding observations on the report of Guyana*, CERD/C/GUY/CO/14 (4 April 2006), ¶ 19.

88 Ibid.

89 *Concluding observations on the report of Congo*, CERD/C/COD/15 (17 August 2007), ¶ 332.
90 Ibid., ¶ 333. Similar recommendations were made on Congo's 2009 report including establishing their forest rights and ensuring that their interests and environmental protection are taken into account in land rights and increasing efforts to consult with indigenous peoples in the administration of their lands, waters and forests. See Congo report, supra note 84, ¶ 14.
91 Stands for National Institute against Discrimination, Xenophobia and Racism, in Argentina.
92 *Concluding observations on the report of Argentina*, CERD/C/ARG/CO/19–20 (29 March 2010), ¶ 21.
93 Ibid., ¶ 26.
94 *Concluding observations on the report of Guatemala*, CERD/C/GTM/CO/12–13 (19 May 2010), ¶ 14.
95 Ibid.
96 *Concluding observations on the report of Guatemala*, CERD/C/GTM/CO/14–15 (12 June 2015), ¶ 11.
97 Ibid., ¶ 12 (emphasis added).
98 Ibid., ¶ 14 (emphasis added).
99 Ibid.
100 *Concluding observations on the combined fifteenth and sixteenth periodic reports of Colombia*, CERD/C/COL/CO/15–16 (25 September 2015), ¶ 15 ["Colombia report"].
101 Ibid., ¶ 16.
102 Ibid., ¶ 21.
103 Ibid., ¶ 22 (emphasis added). On the report of Argentina (supra note 92), the Committee expressed concern about lack of regulations governing consultations with a view to obtaining the free, prior and informed consent of indigenous peoples. When consultations have been undertaken, they have been discretionary and have failed to observe international standards. The Committee recommended, in line with GR No. 23, to establish mechanisms to ensure that prior consultations are conducted with a view to securing free, prior and informed consent, are carried out in good faith with adequate information being provided and that legislative or administrative measures that could affect indigenous peoples and all infrastructure and natural resource exploitation projects are subject to prior consultation in order to obtain their free, prior and informed consent, para 19.
104 Ibid., ¶ 23.
105 It is not clear whether sustainable *economic* development and sustainable development are the same. See also supra note 82.
106 Colombia report, supra note 100, ¶ 24.
107 Ibid., ¶ 15.
108 *Concluding observations on the combined twentieth to twenty-second periodic reports of the Russian Federation*, CERD/C/RUS/CO/20–22 (17 April 2013), ¶ 20.
109 Ibid. (emphasis added). The reference to the country visit by UN Special Rapporteur on Indigenous Peoples is noteworthy.
110 *Concluding observations on the combined fifteenth to twenty-first periodic reports of the Niger*, CERD/C/NER/CO/15–21 (25 September 2015), ¶ 19. On the report of Rwanda, the Committee noted the eviction of indigenous peoples from forest lands in order to create a national park; yet these communities were not compensated, *Concluding observations on the eighteenth to twentieth periodic reports of Rwanda*, CERD/C/RWA/CO/18–20 (10 June 2016), ¶ 11.
111 *Concluding observations on the combined twenty-first to twenty-third periodic reports of Argentina*, CERD/C/ARG/CO/21–23 (11 January 2017), ¶ 6.
112 Ibid., ¶ 7.

113 *Concluding observations on the combined nineteenth to twenty-first periodic reports of Chile*, CERD/C/CHL/CO/19–21 (23 September 2013), ¶ 13.
114 Ibid.
115 Ibid (emphasis added). The requirement to hold free, prior and informed consultations with a view to obtaining free and fully informed consent seems to go even beyond the requirements of UNDRIP.
116 In its 1999 report, US referred to President Clinton's Executive Order (EO) 12898 which directed every federal agency to consider adverse health human or environmental effects of its programs, policies and activities on minority and low-income populations and established a working group on environmental justice whose task was to coordinate, provide guidance and serve as a clearinghouse for federal agencies on their environmental justice strategies. The EO directed all agencies to address disproportionate human health or environmental effects of programs, to collect additional data on these subjects and to coordinate efforts through the inter-agency working group. While most environmental laws do not address the potential impact on low-income and minority communities, EO directs the EPA to make environmental justice part of its mission by identifying and addressing disproportionately high and adverse human health or environmental effects of its programs, policies and activities on minority and low-income populations.
117 *Concluding observations on the combined seventh to ninth periodic reports of the United States of America*, CERD/C/USA/CO/7–9 (25 September 2014) ["US report"]. See Chapter 5 for recommendations made by the Human Rights Committee.
118 Ibid., ¶ 10.
119 Ibid.
120 Ibid., ¶ 24.
121 Such as EO 14008, of 1 June 2021, on Tackling the Climate Crisis at Home and Abroad and the suspension of oil and gas leases for the Arctic National Wildlife Refuge, https://www.govinfo.gov/content/pkg/FR-2021–02-01/pdf/2021–02177.pdf.
122 *Concluding observations on the combined tenth to twelfth reports of the United States of America*, CERD/C/USA/CO/10–12 (30 August 2022), ¶ 45 ["Twelfth US report"].
123 Ibid., ¶ 46.
124 *Concluding observations on the seventeenth and eighteenth periodic reports of Costa Rica*, CERD/C/CRI/18 (15 August 2007), ¶ 300, ["Costa Rica report"].
125 Ibid., ¶ 303.
126 Ibid., ¶ 308.
127 It is noteworthy that the Committee referred to intersectionality (or as the Committee called it "double discrimination"). See also Chapter 7 on the CEDAW Committee.
128 Costa Rica report, supra note 124, ¶ 305.
129 *Concluding observations on the initial to third periodic reports of Indonesia*, CERD/C/IDN/3 (17 August 2007), ¶ 359.
130 Ibid.
131 Ibid.
132 *Concluding observations on the report of Peru*, CERD/C/PER/CO/14–17 (3 September 2009), ¶ 11.
133 Ibid.
134 Ibid., ¶ 14 (emphasis added). This is taking the FPIC principle a step further.
135 Ibid., ¶ 15.
136 Ibid., ¶ 20.
137 Ibid.
138 Ibid., ¶ 23. This recommendation does not seem to be confined to racial/ethnic minorities or Indigenous peoples and could go beyond the mandate of the Committee.
139 *Concluding observations on the eighteenth to twenty-first periodic reports of Peru*, CERD/C/PER/CO/18–21 (25 September 2014), ¶ 15.

140 Ibid. Similar recommendations were made by the CEDAW Committee especially with regard to rural women. See Chapter 7.
141 *Concluding observations on the combined twenty-second and twenty-third periodic reports of Peru*, CERD/C/PER/CO/22–23 (23 May 2018), ¶ 16.
142 Ibid., ¶ 17.
143 Ibid., ¶ 21.
144 *Concluding observations on the fifth to seventh periodic reports of Kenya*, CERD/C/KEN/CO/5–7 (8 June 2017), ¶ 3.
145 Ibid., ¶ 19.
146 Ibid., ¶ 20.
147 *Concluding observations on the combined twenty-first and twenty-second periodic reports of New Zealand*, CERD/C/NZL/CO/21–22 (22 September 2017), ¶ 20 ["New Zealand report"].
148 This Act recognized the Whanganui River as a legal person, available at: https://www.legislation.govt.nz/act/public/2017/0007/latest/whole.html.
149 New Zealand repot, supra note 147, ¶¶ 22–23. The Committee called upon the state to develop a national action plan against racial discrimination and xenophobia in line with the Durban Declaration and Program of Action and include, *inter alia*, developing a plan, in partnership with Māori and other relevant groups to meet the targets of the SDGs to reduce inequalities in social indicators among ethnic groups with a special focus on security of tenure.
150 *Concluding observations on the combined eighth to tenth periodic reports of Kyrgyzstan*, CERD/C/KGZ/CO/8–10 (30 May 2018), ¶ 24. The Committee referred to the need to reduce poverty and levels of inequality among indigenous and Afro-Honduran peoples taking account of SDGs on the report of Honduras (2019); SDG 10 on the report of Cuba (2018); requested an explanation on how demands for sustainable development, for the promotion of agriculture and for protection of rights of indigenous and peasant communities will be reconciled in the draft bill on land reform on the report of Bolivia (1996).
151 *Concluding observations on the report of Bolivia*, CERD/C/304/Add.10 (27 September 1996). Notably, the Committee referred to sustainable development as far back as 1996.
152 *Concluding observations on the seventeenth and eighteenth periodic reports of Canada*, CERD/C/CAN/18 (5 March 2007), ¶ 78.
153 Ibid.
154 US report, supra note 117.
155 Ibid., ¶ 10. The Committee expressed same concerns on the Twelfth US report, supra note 122, and recommended taking appropriate measures to prevent transnational corporations' impact on indigenous peoples and local communities.

References

CBM – Nossal Institute Partnership for Disability Inclusive Development (2017) *Disability Inclusion in Disaster Risk Reduction: Experiences of People with Disabilities in Vanuatu during and after Tropical Cyclone Pam and Recommendations for Humanitarian Agencies*, https://mspgh.unimelb.edu.au/__data/assets/pdf_file/0011/2567576/WEB-DIDRR-Report-14112017.pdf.

CERD (1997), *General Recommendation XXIII on the rights of indigenous peoples*.

CMW (2022) Draft General Comment No. 6 on the Convergence of the Convention and the Global Compact for Safe, Orderly and Regular Migration, Concept Note, Guiding Questions and Call for Submissions, https://www.ohchr.org/en/calls-for-input/2022/call-submissions-concept-paper-and-draft-outline-its-draft-general-comment-no.

CMW (2022) Draft Outline, https://www.ohchr.org/sites/default/files/documents/hrbodies/cmw/cfi-gc6–2022/2022–07–29/Outline-for-CMW-GC-No6.docx.

CMW (2022) Half-day of General Discussion on Draft General Comment No. 6, https://www.ohchr.org/en/events/days-general-discussion-dgd/2022/half-day-general-discussion-draft-general-comment-no-6.

Concluding observations on the combined eighteenth to twenty-first periodic reports of Peru (2014), CERD/C/PER/CO/18–21.

Concluding observations on the combined eighteenth to twentieth periodic reports of Rwanda (2016), CERD/C/RWA/CO/18–20.

Concluding observations on the combined eighth to tenth periodic reports of Kyrgyzstan (2018), CERD/C/KGZ/CO/8–10.

Concluding observations on the combined fifteenth and sixteenth periodic reports of Colombia (2015), CERD/C/COL/CO/15–16.

Concluding observations on the combined fifteenth to twenty-first periodic reports of the Niger (2015), CERD/C/NER/CO/15–21.

Concluding observations on the combined initial to third periodic reports of Indonesia (2007), CERD/C/IDN/3.

Concluding observations on the combined nineteenth to twenty-first periodic reports of Chile (2013), CERD/C/CHL/CO/19–21.

Concluding observations on the combined seventh to ninth periodic reports of the United States of America, CERD/C/USA/CO/7–9 (25 September 2014).

Concluding observations on the combined seventeenth and eighteenth periodic reports of Canada (2007), CERD/C/CAN/18.

Concluding observations on the combined seventeenth and eighteenth periodic reports of Costa Rica (2007), CERD/C/CRI/18.

Concluding observations on the combined tenth to twelfth periodic reports of Australia (1999), CERD/C/335/Add.2.

Concluding observations on the combined tenth to twelfth reports of the United States (2022), CERD/C/USA/CO/10–12.

Concluding observations on the combined thirteenth and fourteenth periodic reports of Sweden (2000), CERD/C/362/Add.5.

Concluding observations on the combined twentieth to the twenty-second periodic reports of the Russian Federation (2013), CERD/C/RUS/CO/20–22.

Concluding observations on the combined twenty-first to twenty-third periodic reports of Argentina (2017), CERD/C/ARG/CO/21–23.

Concluding observations on the combined twenty-first and twenty-second periodic reports of New Zealand (2017), CERD/C/NZL/CO/21–22.

Concluding observations on the combined twenty-second and twenty-third periodic reports of Peru (2018), CERD/C/PER/CO/22–23.

Concluding observations on the fifteenth periodic report of Finland (2000), CERD/C/363/Add.2.

Concluding observations on the fifth to seventh periodic reports of Kenya (2017), CERD/C/KEN/CO/5–7.

Concluding observations on the initial report of Chile (2016), CRPD/C/CHL/CO/1.

Concluding observations on the initial report of Colombia (2016), CRPD/C/COL/CO/1.

Concluding observations on the initial report of the Dominican Republic (2015), CRPD/C/DOM/CO/1.

Concluding observations on the initial report of Ethiopia (2016), CRPD/C/ETH/CO/1.

Concluding observations on the initial report of the European Union (2015), CRPD/C/EU/CO/1.

Concluding observations on the initial report of Guatemala (2016), CRPD/C/GTM/CO/1.
Concluding observations on the initial report of Honduras (2017), CRPD/C/HND/CO/1.
Concluding observations on the initial report of Luxembourg (2017), CRPD/C/LUX/CO/1.
Concluding observations on the initial report of Madagascar (2018), CMW/C/MDG/CO.
Concluding observations on the initial report of Mongolia (2015), CRPD/C/MNG/CO/1.
Concluding observations on the initial report of Montenegro (2017), CRPD/C/MNE/CO/1.
Concluding observations on the initial report of Nepal (2018), CRPD/C/NPL/CO/1.
Concluding observations on the initial report of Oman (2018), CRPD/C/OMN/CO/1.
Concluding observations on the initial report of Panama (2017), CRPD/C/PAN/CO/1.
Concluding observations on the initial report of the Plurinational State of Bolivia (2016), CRPD/C/BOL/CO/1.
Concluding observations on the initial report of Portugal (2016), CRPD/C/PRT/CO/1.
Concluding observations on the initial report of Slovakia (2016), CRPD/C/SVK/CO/1.
Concluding observations on the initial report of Uganda (2016), CRPD/C/UGA/CO/1.
Concluding observations on the initial report of the United Kingdom of Great Britain and Northern Ireland (2017), CRPD/C/GBR/CO/1.
Concluding observations on the initial report of Vanuatu (2019), CRPD/C/VUT/CO/1.
Concluding observations on the report of Argentina (2010), CERD/C/ARG/CO/19–20.
Concluding observations on the report of Bolivia (1996), CERD/C/304/Add.10.
Concluding observations on the report of China (including Hong Kong and Macau Special Administrative Regions) (2009), CERD/C/CHN/CO/10–13.
Concluding observations on the report of Congo (2007), in Report of the Committee on the Elimination of Racial Discrimination, A/62/18.
Concluding observations on the report of Congo (2009), CERD/C/COG/CO/9.
Concluding observations on the report of Croatia (2009), CERD/C/HRV/CO/8.
Concluding observations on the report of Ethiopia (2007), CERD/C/ETH/CO/15.
Concluding observations on the report of Guatemala (2010), CERD/C/GTM/CO/12–13.
Concluding observations on the report of Guatemala (2015), CERD/C/GTM/CO/14–15.
Concluding observations on the report of Guyana (2006), CERD/C/GUY/CO/14.
Concluding observations on the report of India (2007), in Report of the Committee on the Elimination of Racial Discrimination, A/62/18.
Concluding observations on the report of Namibia (2008), CERD/C/NAM/CO/12.
Concluding observations on the report of New Zealand (2007), in Report of the Committee on the Elimination of Racial Discrimination, A/62/18.
Concluding observations on the report of Peru (2009), CERD/C/PER/CO/14–17.
Concluding observations on the report of Slovakia (2001), CERD/C/304/Add.110.
Concluding observations on the report of South Africa (2006), CERD/C/ZAF/CO/3.
Concluding observations on the second periodic report of Guatemala (2019), CMW/C/GTM/CO.
Convention on the Rights of Persons with Disabilities (2006), https://www.un.org/development/desa/disabilities/convention-on-the-rights-of-persons-with-disabilities.html.
Executive Order 12898 (1994), Federal Actions to Address Environmental Justice in Minority Populations and Low-Income Populations.
Executive Order 14008 (2021), Tackling the Climate Crisis at Home and Abroad and the Suspension of Oil and Gas Leases for the Arctic National Wildlife Refuge, https://www.govinfo.gov/content/pkg/FR-2021-02-01/pdf/2021-02177.pdf.
General comment No. 2 (2014), CRPD/C/GC/2.
General comment No. 3 (2016), CRPD/C/GC/3.
General comment No. 4 (2016), CRPD/C/GC/4.

General Comment No. 5 (2017), CRPD/C/GC/5.
General Comment No. 6 (2018), CRPD/C/GC/6.
General Comment No. 7 (2018), CRPD/C/GC/7.
Global Compact for Safe, Orderly and Regular Migration (2018), https://refugeesmigrants.un.org/sites/default/files/180713_agreed_outcome_global_compact_for_migration.pdf.
Mapping Human Rights Obligations Relating to the Enjoyment of a Safe, Clean, Healthy and Sustainable Environment, Individual Report on the International Convention on the Elimination of All Forms of Racial Discrimination, Report No. 3 (2013), https://www.ohchr.org/sites/default/files/Documents/Issues/Environment/Mappingreport/3.CERD-25-Feb.docx.
Resolution 45/158 (1990), https://www.ohchr.org/sites/default/files/cmw.pdf.
Resolution 69/283 (2015), A/RES/69/283.
Resolution 2106 (XX) (1965), https://www.ohchr.org/sites/default/files/cerd.pdf.
Te Awa Tupua (Whanganui River Claims Settlement) Act (2017), (N.Z.), https://www.legislation.govt.nz/act/public/2017/0007/latest/whole.html.

10 UN Human Rights Institutions and the Environment

Reflections, Challenges, Trajectories

10.1 Introduction

The recognition of a distinct human right to a clean, healthy, and sustainable environment first by the HRC in 2021 followed by the UNGA a few months later brought a much-needed addition to the family of rights recognized under international human rights law and is the culmination of years of work by academics and civil society groups. Despite the inordinate time it took to recognize a distinct right to a healthy environment, the link between environmental issues and the enjoyment of rights has long been recognized. A degraded environment, unsustainable development, and adverse consequences of climate change affect the enjoyment of a wide array of rights including, in extreme cases, the right to life itself. After years of vacillating, the link between human rights and the environment was consolidated with the recognition of a human right to a healthy environment. The Vienna Declaration on Human Rights recognized that *all human rights* are universal, indivisible, interdependent, and interrelated. However, a gap existed in the framework of human rights due to the non-recognition of a right to a healthy environment- a gap that the international community took almost 75 years to fill since the adoption of the UDHR.

Despite being a latecomer to the family of human rights, UN human rights institutions have long discussed the link between human rights and the environment within their mandates. This volume examined how several charter-based and treaty-based institutions and mechanisms have addressed this issue. It is clear that long before the international community recognized a stand-alone right to a healthy environment, human rights bodies have discussed the relationship between human rights and the environment, and how environmental degradation, climate change, and unsustainable development affect the enjoyment of rights. Many of these bodies have adopted General Comments on the link (or are on the verge of adopting them) and an impressive number of concluding observations discuss the link.

As discussed, the mandate of the apex UN human rights body – the Human Rights Council – is all encompassing, covering *all* human rights in *all* countries. Thus, it is no surprise that many of the recommendations under the UPR process include references to environmental pollution/degradation and climate change.

DOI: 10.4324/9781003128847-13

Examples include adopting a rights-based approach to climate change, including adaptation and mitigation measures as well as building resilience to disasters, cleaning up environmental damage caused by, *inter alia*, phosphate mining and oil pollution, addressing land, water, and air pollution, deforestation, unsustainable development, and environmental and human rights damage caused by transnational corporations. More specific recommendations related to complying with commitments under the Paris Agreement, increasing the level of ambition of NDCs, moving away from investing in or supporting fossil fuel industries, and ratifying ILO Convention No. 169 on Indigenous and Tribal Peoples.

While it is impossible to summarize all the principles, recommendations, and guidelines of the human rights bodies examined in this volume, we will highlight some of the salient points.

10.2 Principles, Guidelines and Standards

There is a wide range of recommendations made to states with regard to environmental issues, ranging from the provision of information to adopting new laws. Many recommendations have referred to the need to ensure the preparation of impact assessments ranging from environmental/social/human rights/gender/child-rights (or a combination) and conducting human rights due diligence, respecting rights by transnational corporations, and ensuring compliance with environmental, human rights, labor, and health standards, especially by the business sector.

Several human rights institutions have addressed the extraterritorial reach of activities of transnational corporations registered in countries under review and the reviewing institution/s recommended that states adopt legislation and other measures to control the activities of these corporations abroad so that they do not cause damage to the environment and violate human rights. Reference was made on several occasions and across different human rights institutions to the UN Guiding Principles on Business and Human Rights and the need to adopt its "Protect, Respect and Remedy" framework.

A notable feature of many of the institutions surveyed is the extension of the free, prior, and informed consent principle to communities other than indigenous peoples. Thus, the CEDAW Committee extended this principle to rural women while the Committee on Racial Discrimination and the ESCR Committee extended it to local communities. Another feature was the application of the principle to *all stages* of a development project, especially extractive industries in the context of indigenous people and even going beyond the requirements of UNDRIP.

Another principle articulated by many of the Committees is the need to pay adequate compensation to communities when their rights are violated (sometimes formulated as "prompt and adequate compensation") and to ensure that these communities benefit from the developmental and/or extractive industries on or near their lands, especially if these lands were expropriated for these activities. On a few reports the Committees recommended returning ancestral lands to indigenous people where they had been expropriated for development projects. One of the contentious issues is the tension between the establishment of national parks near or on indigenous lands restricting or completely blocking their access, and the need to uphold the rights of indigenous peoples. In these instances, Committees have recommended holding consultations and involving affected groups in the decision-making process.

All institutions surveyed endorsed procedural rights of information, participation, and remedy. Be it women, children, people with disabilities, indigenous peoples or racial minorities, all charter-based and treaty-based institutions endorsed their right to participate in decisions, programs, legislation, strategies, and projects that affect them, the need to provide timely and accurate information and access to remedies when their rights are violated. These institutions further endorsed the need to develop data disaggregated by gender, age, disability etc., so that programs, strategies, and projects can be informed by such data.

The CRC Committee took this one step further by calling upon states to adopt *continuous child-rights impact assessments* which should complement ongoing monitoring and evaluation of laws, policies, and programs on children's rights.[1]

10.3 General Comments

Many human rights institutions are increasingly adopting General Comments on environmental issues, especially climate change. They recognize that, as an existential threat posed to humanity and their rights, they need to elaborate on states' obligations in relation to climate change. Thus, General Comments on land and ESC rights, sustainable development and ESC rights, and children's rights and climate change are in the pipeline while several existing General Comments embody provisions on environmental degradation and climate change. The CEDAW Committee is the only treaty body so far to have adopted a specific General Recommendation on gender, climate change, and disasters.

10.4 Individual Communications

In the same vein, an increasing number of individual communications addresses climate change and its impact on the enjoyment of rights and these treaty bodies' decisions are becoming increasingly bold, sophisticated, and nuanced. In its latest decision, in holding Australia accountable for violating the rights of Torres Strait Islanders, the UN Human Rights Committee held that Australia had failed to fulfill its mitigation commitments under the Paris Agreement as well as adopt adaptation measures to help this community who is already suffering the adverse consequences of climate change. Its decision in the *Teitiota case* opened the door (slightly) to claims by climate refugees by recognizing the application of the *non-refoulement* principle in the context of serious risk to life caused by climate change. It further recognized that small island states face the prospect of completely disappearing in 10–15 years' time due to sea level rise and extreme weather events associated with climate change, as did several other treaty bodies.

Likewise, the CRC Committee in *Sacchi v. Argentina* laid down a very important principle that the global nature of climate change does not absolve states of their individual responsibility to address climate change, thereby dismissing the contention often advanced by states that they cannot be held individually accountable for a collective action problem that requires concerted action by all states. It is noteworthy that the Committee relied on the international environmental law principle of common but differentiated responsibilities in this context which the Inter-American Commission of Human Rights was reluctant to do in the Inuit Petition. The Committee also endorsed the "effective control" test in relation to extraterritorial environmental damage.

10.5 Critiques

As noted throughout this book, the reference to "natural disasters" in the context of climate change is misleading as most disasters and extreme weather events are triggered by climate change. While climate change obviously is not the sole reason for these extreme weather events, the IPCC has recognized that climate change is contributing to their severity, frequency, and duration.[2] Thus, referring to them as *natural* disasters gives the impression that these are acts of God and cannot be avoided and that there is no human agency involved.[3]

Another critique is the very soft language used when making recommendations. Often the word "consider" is used before making recommendations. Thus, for example, in many concluding observations, the human rights institution in question stated, "consider ratifying ILO Convention 169," "consider adopting legislation" or "consider amending legislation" etc. As these human rights bodies have only persuasive authority, they cannot use more authoritative language. On at least one occasion, the CRC Committee recommended establishing a legally binding framework for domestic and foreign businesses, while the Committee on Racial Discrimination indicated that the state party is bound to apply the provisions of ILO Convention No. 169 even in the absence of domestic legislation. This critique thus goes to the heart of the international legal system itself, a system entrenched in the notion of sovereignty, rather than a critique specific to human rights.

10.6 Challenges

Despite these positive developments, several challenges remain. One of the biggest challenges is obtaining relief for damage caused by transnational corporations and the failure of states to control their activities abroad. Very often these corporations wield enormous power and the power asymmetry between countries in the global South and the corporations predominantly based in the global North prevent victims from obtaining relief as well as the laws and institutions that shield them from liability.

Another obstacle is the need to establish victimhood and the "imminent damage" test, especially in the context of climate change. In the *Teitiota case*, the

Human Rights Committee decided that the alleged damage was not imminent enough and referred to a time frame of 10–15 years before small island states could become submerged during which time the state could have taken measures to protect the victim's rights. In the face of climate change, how imminent is imminent?

As with most international procedures, cost, delay, and legal representation can prevent many victims from seeking relief. The causation requirement seems to have been relaxed somewhat by relying on scientific reports, especially IPCC reports detailing the consequences of climate change. Refuting States' contention that climate change is a global issue and individual states cannot be held responsible is another significant development. Yet, some of the jurisdictional challenges remain. The need to exhaust local remedies is one of them. This was the main reason why a submission by a group of children against their state for its inaction on climate change was held inadmissible (in *Sacchi v. Argentina*). While these human rights bodies do not want to become "courts of first instance," in complex issues such as climate change, domestic courts may not be able to provide proper relief to victims.

The human toll of the COVID-19 pandemic which caused untold misery worldwide as well as led to infringement of many human rights revealed cracks in our legal systems as well as systemic problems that contributed to the disproportionate impact on racial minorities, women, children, and other vulnerable and marginalized communities. With consequences of climate change looming large, we need to address these gaps and systemic problems, before it is too late.

10.7 Fragmentation, Cohesion, or Cross-Fertilization?

A concern that is often raised about diversification and expansion of law and proliferation of institutions is that they will lead to fragmentation. These concerns were such that the International Law Commission decided at its 52nd session to include the topic "Risks ensuing from the fragmentation of international law" into its long-term program of work.[4] Put simply, the concern is about the possibility of conflicting decisions by international/regional tribunals or in this case conflicting recommendations by human rights institutions, leading to fragmentation of international law.[5] While not attempting to engage in a debate about the pros and cons of fragmentation, we will make some observations based on the examination of the human rights institutions surveyed in the book.

Far from fragmentation, there seems to be considerable cohesion in the recommendations made to states in relation to environmental issues across issues and across the human rights institutions surveyed. A notable feature of many of the concluding observations is the frequent reference to reports of special mandate holders. For example, on the report of the Marshall Islands the CRC Committee recommended the implementation of the recommendations of the Special Rapporteur on Toxics, especially those relating to the impacts of nuclear tests by the USA on children's health and the right to live in a healthy environment. Likewise, the Committee on Racial Discrimination also referred to the recommendations of the Special Rapporteur on Toxics on the report of Marshall Islands.

Several concluding observations referred to the reports of the Special Rapporteur on Indigenous Rights and frequent references were made by many of the institutions to the UN Guiding Principles on Business and Human Rights, ILO Convention No. 169 and UNDRIP.

It is also noteworthy that these human rights institutions refer not just to the jurisprudence of other human rights bodies but also to other UN bodies and environmental treaties. Most concluding observations referred to Agenda 2030, SDGs and its targets predominantly as a yardstick to measure compliance with human rights obligations. Thus, often reference was made to SDG 1 on poverty, SDG 2 on hunger, SDG 4 on education, SDG 5 on gender equality, SDG 6 on water, SDG 13 on climate change as well as some of the targets under them when discussing compliance with relevant rights. Prior to 2015 several concluding observations referred to MDGs. Similarly, in relation to climate change, the human rights institutions surveyed often referred to the commitments made under the Paris Agreement, Nationally Determined Contributions (NDCs) that states made, reducing GHG emissions to meet the 1.5 degree target as well as the need to formulate adaptation plans to protect vulnerable communities. Moreover, these human rights bodies have made very specific recommendations which may seem to go beyond their mandate such as increasing ambition under their NDCs, stop investing in fossil fuel projects, or moving away from coal and investing in renewable energy.

Moreover, the human rights institutions surveyed look at the jurisprudence of other human rights bodies, especially, regional human rights bodies. The Advisory Opinion on Environment and Human Rights of the Inter-American Court of Human Rights was referred to on several instances as well as to the jurisprudence of the European Court of Human Rights. All these references seem to lead to cross-fertilization, rather than fragmentation but a more in-depth study is obviously needed.

Despite the absence of a specific mandate on environmental issues, the UN human rights institutions surveyed have addressed environmental issues within their mandate to the extent that these environmental issues impinge upon the enjoyment of rights covered in the treaty that the institution is responsible for. Now that the right to a healthy environment forms part of the rights recognized under international law, these human rights institutions will become even more important players to address environmental degradation, unsustainable development, and climate change. They can continue to develop states' human rights obligations vis-à-vis the increasingly catastrophic threat caused by environmental damage, climate change, and unsustainable development which will be useful not only for other human rights and environmental institutions, but also for civil society groups and institutions at the national level.

The sophisticated network of human rights institutions and courts will become even more relevant for victims of environmental degradation and climate change as they seek relief for the damage they suffer. The jurisprudence developed by these human rights institutions assumes greater significance in that light and indivisibility of rights becomes important in this context.

10.8 A Research Agenda for the Future

This book scratched only the surface of a rich and growing area of law covering a few UN human rights institutions. Moreover, this field is constantly evolving. An interesting exercise would be to examine to what extent environmental institutions incorporate human rights principles although nothing akin to charter-based and treaty-based bodies exist in the environment field, the closest being conference of parties (COP) established under various treaties. However, they do not have the power to receive individual complaints. Another interesting project would be to study specific countries or groups of countries to analyze how different human rights institutions have addressed environmental issues and the various recommendations made. There is rich material to be uncovered, examined, and synthesized, and it is hoped that other scholars will take up the challenge.

Notes

1 *See also* GC *No. 11 on Indigenous children and their rights under the Convention*, CRC/C/GC/11 (12 February 2009), ¶ 35, available at: https://undocs.org/CRC/C/GC/11, which states: "States parties should closely consider the cultural significance of traditional land and the quality of the natural environment while ensuring the children's right to life, survival and development to the maximum extent possible." It urges states to take all reasonable measures to ensure that indigenous children and their communities receive information and education on issues relating to, *inter alia*, environmental sanitation and the dangers of pesticides and herbicides, ¶ 53. It refers to importance of the MDGs, ¶ 36, the predecessor to the SDGs. GC *No. 9*, *The rights of children with disabilities*, CRC/C/GC/9 (27 February 2007), available at: https://undocs.org/CRC/C/GC/9, which also refers to MDGs in the context of universal education. It notes that certain disabilities result directly from human-caused or natural disasters. While it does not refer to climate change, the reference to "human-caused disasters" can include extreme weather events and disasters caused by climate change.

2 Intergovernmental Panel on Climate Change (2021) "Summary for Policymakers," 6th Assessment Report, Cambridge: Cambridge University Press., available at: https://www.ipcc.ch/report/ar6/wg1/downloads/report/IPCC_AR6_WGI_SPM.pdf

3 See Scott, M. (2020) *Climate Change, Disasters, and the Refugee Convention*, Cambridge: Cambridge University Press, doi: 10.1017/9781108784580, who advocates for a social paradigm approach to disasters, taking into account the wider social context in which disasters unfold. He argues that human agency is inherent in *all* disasters.

4 International Law Commission, *Fragmentation of International Law: Difficulties arising from the diversification and expansion of international law: Report of the Study Group of the International law Commission, finalized by Martti Koskenniemi*, A/CN.4/L.682 (13 April 2006), https://digitallibrary.un.org/record/574810. The original title was changed to this title at the 54th session (2002).

5 See Koskenniemi, M. & Leino, P. (2002) "Fragmentation of international Law? Post-modern Anxieties," *Leiden Journal of International Law*, vol. 15, no. 3, pp. 553–579, doi: 10.1017/S0922156502000262.

References

Committee on the Rights of the Child (2007) *General Comment No. 9*, CRC/C/GC/9.

Committee on the Rights of the Child (2009) *General Comment No. 11*, CRC/C/GC/11.

ILC (2006) *Fragmentation of International Law: Difficulties arising from the diversification and expansion of international law: Report of the Study Group of the International law Commission, finalized by Martti Koskenniemi*, A/CN.4/L.682, https://digitallibrary.un.org/record/574810.

Intergovernmental Panel on Climate Change (2021) "Summary for Policymakers," 6th Assessment Report, Cambridge: Cambridge University Press, https://www.ipcc.ch/report/ar6/wg1/downloads/report/IPCC_AR6_WGI_SPM.pdf

Koskenniemi, M. & Leino, P. (2002) "Fragmentation of international Law? Postmodern Anxieties," *Leiden Journal of International Law*, vol. 15, no. 3, pp. 553–579, doi: 10.1017/S0922156502000262.

Scott, M. (2020) *Climate Change, Disasters, and the Refugee Convention*, Cambridge: Cambridge University Press, doi: 10.1017/9781108784580.

Index

1.5 degrees 7, 53, 173, 246, 302

Aarhus Convention (UN Economic Cooperation for Europe) 25
Abidjan 238
aboriginal rights 126, 129
access to electricity 162, 210
access to genetic resources 153
access to information 14, 20, 34, 36, 38, 64, 71, 81, 122, 125, 153, 255
access to remedies 20, 76, 78, 148, 161, 281, 299
accessible homes 60
accountability 51, 55, 57, 88, 93, 168, 193, 197; mechanisms 88, 93
actio popularis 246–247
adaptation: measures 46, 60, 104, 134–137, 153, 155, 196–197, 204, 216, 231, 246, 299; policies 91, 103, 153
adequate compensation 50, 137, 158, 160, 164, 179, 185, 207, 210, 219, 276–277, 298
affirmative action 134, 280
Afghanistan 66, 218, 223
African Charter on Human and Peoples' Rights 28
African Commission and Court of Human and Peoples' Rights 18
Afro-descendants 84; Afro-Colombian peoples 278–279; Afro-Peruvian communities 282
Agenda 2030 4, 7–8, 20, 48–49, 54–56, 62, 67, 69–70, 94, 97, 154, 168, 173, 206–207, 217–218, 243–244, 269–270, 273, 284, 302
agriculture 22, 84, 88, 90–91, 94, 151, 153, 160, 163–164, 167, 177, 193–194, 198, 200, 209, 214, 239, 284, 293; agribusinesses 84, 94, 110; agricultural wastes 89; agro-biodiversity 193; agrochemicals 160–161, 179, 193, 209, 211, 236–237, agroecology 91, 95; agrofuel 89; industrial agriculture 91, 153, 194; and trade policies 92; traditional agriculture 88
air pollutant emissions 236, 238
air travel levy 83
Alliance of Small Island States 48; *see also* small island states
Amazon 51, 123, 127, 282
ambient air quality 82
amici curiae brief 135, 246–247, 261, 263
ancestral lands 8, 49, 62, 101–102, 151–152, 170–171, 186, 276–277, 280, 282, 284–285
Andorra 206, 218, 223
Andre Brun v. France 130
Ángela Poma Poma v. Peru 131; River Uchusuma 131
Angola 49, 65–66, 72, 174, 184–185, 188, 207, 219, 223
anthrax 162
Antigua 202, 214, 221, 231, 255, 264
apartheid 99
Arab Charter on Human Rights 29, 35, 37
archaeological sites 164
Arctic Ocean 154; environment 88
Argentina 10, 69, 79, 123, 139, 143–145, 155–157, 159, 169, 172, 177–178, 183–184, 186, 188, 218, 220, 223, 236, 241, 245, 248, 253, 258, 260–263, 266–267, 277, 280, 291, 294–295, 300–301; General Law on the Environment 248; National Institute of Indigenous Affairs 169; Salinas Grandes 172
armed conflict 92, 121, 213, 226, 240
artisanal and small-scale mining 162
asbestos 243

Asia 90, 95, 173
Asian Human Rights Charter 29, 35
Association of Southeast Asian Nations (ASEAN) Human Rights Declaration 29; ASEAN Intergovernmental Commission on Human Rights (AICHR) 29
asylum seekers 121, 287
Australia 5, 10, 50, 134–135, 138, 142, 156, 170, 231, 241, 252, 275, 299; Native Title Act 1993 170
Austria 124, 231; Code of Criminal Procedure 124
autonomy 102
Aymara people 131

Bahamas 202; National Emergency Management Agency 202
Bangladesh 67, 103, 157, 183–184; Bangladesh Climate Change Resilience Fund 157; Bangladesh Climate Change Trust 157; Climate Change Strategy and Action Plan of 2009 157
Barbuda 202, 221, 231; Department of Environment 202; National Office of Disaster Services 202
Barents Sea 154
Belarus 43, 260, 263
Belgium 124, 140, 236, 241, 244
Belize 65, 67, 123
benefit sharing 50, 219
best interests of the child 229, 245–247
best practices 52–53, 78, 80, 103, 216
Bhutan 45, 65, 205, 217–218, 232
biodiversity 4, 28, 53, 69, 76, 79–81, 83, 86–87, 91–94, 99, 101, 106, 110, 127, 152, 154, 165, 173, 193, 211, 229–230; and ecosystems 76, 80–81; loss 69, 83, 93, 154, 229; terrestrial biodiversity 86
biofuel 89–91; biodiesel 89; bioenergy 91, 153; bioethanol 89; first-generation biofuels 91; production 90–91
biological hazards 197
biopiracy 193
birth defects 208, 210, 239
"blue economy" 210
Bolivia 43, 46, 49, 63, 102, 174–175, 284, 287–288, 293
boron 159
Bosnia and Herzegovina 44, 65, 237, 263, 289
Brazil 43, 49, 67, 99, 102, 111, 178, 237, 245; Belo Monte dam 237
Brunei Darussalam 49, 65, 67
build back better 97
Burkina Faso 161, 167, 179, 218, 220; Environmental and Social Management Plan 161; Essakane community 161; Mining and Environment Code 161; Strategy for Accelerated Growth and Sustainable Development 167
business enterprises 51, 75, 77, 97–98, 106, 111, 115, 228, 243, 260; sector 51, 228, 240–244, 253–254, 298

Cabo Verde 125, 156, 183, 200, 210, 232, 241, 256
Cambodia 50, 67, 111, 157, 165, 200, 210–211; National Strategic Development Plan 165
campesinos 164; *see also* peasant farmers
Canada 25, 67, 102, 127, 129–130, 157, 185, 284; Budget bill C38 (2012) 157; Port Hope 127, 130; Port Hope Environmental Group 127; *see also EHP et al v. Canada*
cancer 127, 159, 209, 229, 239, 281
carbon 5, 45, 65, 90, 96, 155–157, 201, 234, 250–252; carbon-neutral technologies 234; credits 157; footprint 234; neutral 90
causation 250, 301
Central America 102, 165
Centre International de Droit Compare de l'Environnement (International Center of Comparative Environmental Law) 27
Charter on Inclusion of Persons with Disabilities in Humanitarian Action 271
charter-based institutions 18–20, 41, 297, 299
Chernobyl accident 124
chikungunya 251
child-rights impact assessment 228
children 4, 6, 10, 18, 22–23, 44–48, 57, 77, 79, 81–82, 84, 89, 92–93, 99, 121, 134, 143, 151–152, 157, 173, 181, 210, 221, 225–226, 228–240, 242–252, 254–258, 260, 262, 269, 272, 286–287, 299, 301, 303; children in poverty 229; children on the move 229; children with disability 229; children's environmental health 237; children's health 227, 234, 236–238, 240, 258–259, 301
Children's Rights and Business Principles 82, 241

Chile 44, 50–51, 65, 67, 102, 180, 185, 203, 219, 256, 260, 270, 280, 286–287; National Indigenous Development Corporation 280
China 44, 51, 64–66, 68, 72, 163, 166, 181–182, 188–189, 290, 295
civil and political rights 10, 16–17, 21, 25, 31, 37, 96, 119, 138–139, 145, 147, 225, 229; *see also* first generation rights
civil society 31, 41, 43, 51, 56, 60, 62, 95, 105, 111, 154, 157, 163–164, 170, 269, 297, 302
clean air 21, 81–82, 86
clean environment 242
clean, healthy and sustainable environment 4, 11, 14, 21, 30, 32, 37–38, 42, 45, 51–53, 55, 63, 68, 71, 73, 78, 104–108, 113–115, 138, 145, 175, 190, 212, 223, 254, 266, 289, 296–297
climate: adaptation 46–47, 87, 201–202, 205, 233–234, 236, 256; anxiety 25; crisis 82, 94, 104–105, 229, 245–248, 292, 295; displacement 87, 103, 199; equity 196; finance 59, 83, 204; mitigation 96, 173, 196, 204, 231, 233; negotiations 47–48, 203; refugees 12–13, 57, 70, 92, 132, 134, 138, 142, 144, 254, 263, 299; science 198; vulnerability assessments 88
climate change 4–7, 9–14, 19–23, 26–27, 33–34, 37–38, 41–48, 51–66, 68–73, 75, 79, 82–83, 85–93, 95–96, 99, 103–105, 110, 112, 115, 120–125, 127, 132–139, 142–147, 153–157, 170, 173–174, 176–177, 185–187, 191–193, 195–218, 220, 225–238, 240, 244–259, 261–263, 266, 270–271, 273, 281, 285, 289, 297–304; climate change-induced disasters, extreme weather events 13, 57, 92–93, 205, 227, 230; human-induced climate change 90, 196; *see also* global warming
Le Club des Juristes 27
coal-fired power plants 204, 233, 240
Coalition of Low Lying Atoll Nations on Climate Change 48
coastal resiliency master plans 104
Colombia 102, 111, 123, 126, 139–141, 144–146, 160, 178, 183–184, 189, 237, 258, 260, 264, 271, 278, 287–289, 291, 294; diversion of the Rancheria River 279; El Cercado dam 279; El Cerrejon mine 279
colonialism 85
Committee on Economic, Social and Cultural Rights 5, 9–10, 147, 149, 151, 153, 155, 157, 159, 161, 163, 165, 167, 169, 171, 173, 175, 177, 179, 181, 183, 185, 187, 189–190; 2018 statement on climate change 155–156; Draft GC No 26 on land and ESC rights 153; Draft GC on sustainable development and ESC rights 153; GC No 12 90, 149, 151, 167; GC No 14 55, 150; GC No 15 on the Right to Water 151, 160–163
Committee on the Elimination of All Forms of Discrimination against Women (CEDAW Committee) 9–10, 191–192, 199, 211, 298–299; GR No 21 192; GR No 23 192–193; GR No 27 192; GR No 34 192–195, 198, 208–209, 211; GR No 37 195, 198, 200–204
Committee on the Elimination of Racial Discrimination 9–10, 50, 268, 273, 285, 298, 300–301; GR No 23 276, 283–284; GR No 25 282
Committee on the Protection of Rights of All Migrant Workers and Members of their Families 9–10, 227, 268, 272–273; GC on the Convergence of the Convention and the Global Compact for Migration 272–273
Committee on the Rights of Persons with Disabilities 9–10, 18, 268–272, 285; GC No 2 268, 270; GC No 3 on women and girls with disability 269; GC No 5 269; GC No 6 269; GC No 7 269
Committee on the Rights of the Child 10, 225–263, 299–301; GC on children's rights and the environment 229; GC No 12
commodification of land 94
common heritage of humankind 45
community forestry 157
Comoros 47, 66, 72
compensation 50, 87–88, 126, 129, 137, 158–161, 164–165, 171, 174, 179, 182, 185–186, 198, 207–208, 210–211, 219–220, 238, 242, 274–277, 279, 283, 298
compound injustice 109
Concluding Observations 5, 8–10, 12–13, 18–19, 32, 119–120, 123, 139–140, 144, 154–155, 158, 168, 176–189, 199, 206, 213–223, 230, 240–241, 244, 255–261, 263–266, 270, 273, 275, 284, 286–295, 297, 300–302

Conference of Parties (COP) 303
contamination of underground water supplies 235
continuous environmental impact assessments 228
Convention against Torture and other Cruel, Inhuman or Degrading Treatment or Punishment (CAT) 18
Convention on Biological Diversity 67, 103, 176, 207, 223
Convention on the Elimination of All Forms of Discrimination Against Women (CEDAW) 10, 18, 23, 59, 61, 191, 196–197, 199, 206, 225; Optional Protocol 191
Convention Relating to the Status of Refugees 58, 132
Convention on the Rights of the Child (CRC) 18–20, 23, 225–226, 229, 244, 246–247; optional protocol on child prostitution and child pornography 226, 244; optional protocol on children in armed conflict 226, 244; optional protocol on a communications procedure 226, 244, 248, 250–251
conventional crops 130
Cook Islands 215, 222, 230, 255, 264
Copenhagen Declaration 7, 34, 38
coral bleaching 135
corporate: due diligence 100; Guiding Principles on Business and Human Rights 43–44, 51, 82, 93, 98, 100, 106, 124, 228, 238, 241–242, 253, 298, 302; "respect, protect and remedy" framework 100; responsibility 98; UN Global Compact 98
corruption 97–98, 111, 273; Convention Against Corruption 97, 111
Costa Rica 45, 47, 64–66, 72, 184, 187, 292, 294
Côte d'Ivoire 43, 200, 214, 223, 259, 266
country reports 4, 10, 18, 119, 175, 191, 268, 286
COVID-19 pandemic 8, 21, 55, 60, 83, 86, 96, 168, 301
crimes against humanity 128
Croatia 275, 290, 295
crop and disaster insurance 93
culture: cultural assets 238; cultural expressions 88; cultural extinction 87–88; cultural heritage 87–89, 104, 109, 164, 186, 228, 282; cultural identities 87, 104; cultural interaction 152; cultural losses 88; cultural policies 87; cultural practices 87; cultural rights 4–5, 9–11, 14, 17, 37, 61, 75, 84–89, 109, 112–113, 119, 122, 147, 149, 151, 153, 155, 157, 159, 161, 163, 165, 167, 169, 171, 173, 175–177, 179, 181, 183, 185–187, 189–190, 245, 282; cultural sites 87, 104, 109, 281; cultural survival 87–88, 101; cultural ties 130; culture apartheid 87; intangible cultural heritage 88; natural and cultural heritage 87
cyber attacks 121
Cyclone Idai 199–201
Cyclone Kenneth 201
Cyprus 67, 184, 189, 203, 216, 222, 287
Czechia 237, 258, 263

decent work 4, 60–61
deforestation 6, 29, 45, 69, 74, 99, 152, 158, 160, 165, 172, 178, 201, 210, 239, 242, 257, 298; and forest degradation 45, 69, 74
democracy 7, 96
Democratic People's Republic of Korea 47, 66, 72, 218, 222, 256, 265
Democratic Republic of the Congo 44, 64, 67, 72, 165, 182, 187, 207, 219, 222, 238, 258–259, 264–265, 276–277, 290–291, 295; Frontier Mine 207; Ndoki National Park 276
dengue 251
deportation 133
desertification 69, 152, 167, 176, 193, 232, 257
development projects 44, 49, 97, 104, 126, 153, 159, 161, 164, 166, 171–173, 179, 182, 185–186, 194–195, 209, 219, 240, 277, 283, 287, 299
diarrheal diseases 22, 229, 258
dignity 3, 15–17, 24, 27, 30, 36, 55, 60, 76, 82, 88, 108, 110, 121, 126, 134, 136–137, 142, 144, 150–151, 173, 199, 254, 263
disability 6, 8, 53, 59–60, 70, 83, 206, 229, 268–272, 286–289, 293, 299
disaggregated data 198, 206, 230–232, 235, 252, 255, 257, 289, 299
disasters 7–8, 10, 13, 43, 45–48, 57–58, 65–66, 70, 84, 88, 92–94, 103, 107, 110, 114, 121–122, 125, 149–151, 156, 161, 183, 192–193, 195–198, 200–205, 209–212, 214–217, 220–221, 227, 229–236,

252, 254–255, 269–271, 273, 281, 285, 298–300, 303–304; management plans 121; man-made disasters 110, 121, 227; natural disasters 13, 43, 45–48, 58, 66, 70, 92, 110, 122, 125, 149, 151, 156, 192–193, 201–202, 204, 209, 212, 214, 216, 220–221, 230–232, 234–235, 254, 269, 271, 273, 281, 285, 300, 303; post-disaster recovery 46, 202, 288; preparedness 124, 271; response 200, 202–204, 209, 215, 217, 233; risk management 48, 199–200, 203, 215, 230–233, 235–236, 238, 240, 252, 257; risk reduction 10, 20, 44, 46–48, 65, 70, 93–94, 156–157, 192–193, 196–206, 212–215, 217, 230–231, 233–236, 256, 269–272, 285, 288, 293
Djibouti 210
domestic remedies 120, 135, 192, 226, 247, 251, 262
Dominica 67, 124–125, 139–140, 144; 2018 Climate Resilience Act 124
Dominican Republic 67, 165, 180, 182, 184, 188–189, 270, 286, 294; Economic Partnership Agreements 165; US-Dominican Republic-Central America Free Trade Agreement (CAFTA-DR) 165
Draft International Covenant on Human Right to the Environment 27
drought 157, 163, 177, 203, 209, 229, 251
due diligence regulation 98
Durban Accord and Action Plan 102
duty bearers 15, 79

early warning systems 234, 238, 256, 271
eco-guards 276
economic crisis 96
economic development 3, 26, 54, 123, 159, 161, 244, 279, 291
Economic and Social Council (ECOSOC) 3, 16, 18, 27, 41, 147, 191, 226, 268; Commission on the Status of Women 16; Procedure 1503 18, 31; Sub-Commission on the Discrimination and Protection of Minorities 16, 26, 75; Sub-Commission on the Promotion and Protection of Human Rights 16
economic, social and cultural (ESC) rights 10, 16–22, 25, 28–29, 44, 49, 55, 61, 81, 84, 87–88, 93, 95–97, 119, 135, 147–174, 225–226, 229, 275, 277, 298–299; *see also* second generation rights
ecosystem services 25, 79–80, 173
Ecuador 31, 37, 45, 51, 64, 67–68, 72, 123, 140, 146, 155, 166, 168, 172, 175–177, 181–184, 186, 188–189, 238, 259, 263; Environmental Management Act 238; Yasuní National Park 168
effective control 35, 248–250, 252–253, 262, 300
effective remedy 57, 98, 120, 129, 137, 142
Egypt 49, 65–66, 72, 167, 183, 188
EHP et al v. Canada 127, 130
electric power plants 129
eliminating subsidies 231
emergency assistance 124
emergency protocols 47, 231, 234–235
emission reduction strategies 91
empowerment 20, 49, 90, 95, 196–197, 202, 286
energy efficient heating systems 162
environmental: conservation 45, 240; crisis 4, 97; degradation 4–6, 9–10, 20–21, 23–25, 28, 31, 41, 44–45, 51, 55, 62, 69, 75–76, 78, 81, 93–94, 101, 111, 120–123, 132–133, 136–138, 141, 153–154, 158, 160, 164, 169, 178, 181, 184, 191–192, 201, 210–211, 225, 227–229, 238, 242, 252, 261, 274, 297, 299, 302; determinants of health 90; hazards 44, 122, 125, 158, 181, 228, 242; health 8, 33, 38, 227, 234, 236–241, 244, 26; and health damage 207; human rights defenders 79, 81, 111; impact assessments 8, 97, 122, 125, 154, 157, 159, 194, 207, 228, 242, 276, 278, 280, 283; injustices 82, 85–86; justice movement 25; preservation projects 161; resources 92, 211; rights 4–5, 8–9, 11, 13, 15, 18, 23–31, 34–36, 41, 43, 51–52, 62, 78, 81, 127, 140–141, 143, 172, 174–175, 187, 189, 253, 269; sanitation 163, 226, 230, 239, 254, 303; standards 44, 81, 122, 125, 140, 241, 253; tribunals 228
EP et al v. Colombia 129
equality 17, 20, 22, 24, 26, 46, 56–58, 60, 75, 98–100, 129, 142, 147, 152–154, 192, 196–198, 201, 204, 206–207, 212–214, 217–218, 268, 284, 286, 302
equity 56–57, 154, 196, 230, 245, 255
Escazú Agreement 25, 43, 65
Estonia 159, 178, 182–183, 188–189
Eswatini 238, 257, 259, 264

Ethiopia 49, 65, 67, 72, 164, 179, 182, 187, 276, 287, 290, 294–295; Gilgel Gibe III hydroelectric dam 164; Sustainable Development and Poverty Reduction Program 276
ethnic protection plans 278
ethnic, religious or linguistic minorities 138
ethnicity 198, 206, 287, 289–290
European Convention on Human Rights 28, 35–36
European Court of Human Rights 18, 130, 249, 262, 302
European Union (EU) 90, 159, 165, 270
evacuation plans 124
excessive consumption 92
exploitation of natural resources 158–160, 169, 171, 179–180, 185, 199, 277, 279–281, 283–285
expropriation 179, 184, 195, 207, 238
extractive activities 50, 102, 155, 158–159, 168–170, 172, 179, 184, 282
extractive industries 8, 97, 111–113, 157–158, 165, 171, 174, 185, 193, 195, 197, 201, 219–220, 232, 241, 243, 279–280, 298
Extractive Industries Transparency Initiative 97, 111, 113, 165
extractivist projects 99; extractivism economy 98–100, 111–112; global extractivism 98–99
extraterritorial obligations 153–155, 174, 230, 255; extraterritorial violations 95

family farming 89
fauna and flora 127, 152–153
Federated States of Micronesia 215, 222, 230, 255, 266; Climate Change Law 230
female headed households 157
Fiji 46, 65–66, 72, 203, 214, 216, 219, 223, 232, 256, 258, 264; National Climate Change Coordinating Committee 232; National Climate Change Policy 232
financial and technical support 156
Finland 126, 141, 145, 168, 177, 184–185, 189, 275, 290, 294
first generation rights 17; *see also* civil and political rights
floods 92, 153, 193, 196, 199, 201, 205, 229, 245, 248
fluoride 159
food 4, 7–9, 15, 19–20, 22, 24, 46, 48, 54–56, 58–61, 64, 69, 71, 75–76, 79–81, 83–84, 86, 89–95, 97, 100, 104, 107, 109–110, 112–115, 128, 132, 136–137, 147, 149–151, 153, 157–160, 162–164, 166–167, 172–173, 177, 184, 194–196, 199–201, 203, 209–210, 213, 227–229, 233, 236–240, 242, 244, 249, 256–257, 276; chain 128, 166, 237; contamination 163–164, 237; crops 89, 194; globalization of food systems 94; healthy and sustainable food 83–84, 107, 114; industrial food systems 84; insecurity 59–60, 92–95, 132, 157, 167, 199, 201, 238, 244; production systems 92; security 8, 22, 91–92, 95, 112, 150, 158, 164, 172, 200, 210, 213, 227–228; systems 83–84, 90–91, 93–94, 107, 110, 113–114; waste 92
Food and Agriculture Organization (FAO) 92, 94, 165; FAO global forest survey 165
forced displacement 23, 102–103, 238, 241; forced evictions 32, 90, 166, 182, 277; forced migration 95, 273
foreseeable risks 247
forest carbon stocks 45, 65
forest conservation 93, 97
Former Yugoslavia Republic of Macedonia 45, 64, 67, 72, 200, 214, 223
fossil fuels 82, 85, 155, 234, 244
fracking 99, 184, 197, 208; *see also* hydraulic fracking
fragmentation of international law 6, 11, 110, 301, 303–304
Free, Prior and Informed Consent (FPIC) 8, 140, 153, 174, 185–186, 194–195, 199, 207, 211, 213, 219, 280, 285, 292
French Guinea *see* Guinea
French Polynesia 127
freshwater 84, 232, 284
full reparation 137–138
fundamental human rights 3, 16, 76
future generations 7, 21, 24, 26–29, 51–52, 54, 84–85, 87, 90, 121–122, 127, 133, 136, 150–151, 155, 225, 229–230, 238, 245

Gabon 239, 259, 266
Gambia 65, 239, 259, 264
gas extraction 98, 158, 161
Gaza 158
gender 10, 20, 22, 33, 46–48, 52–53, 57–61, 70, 74, 77, 83, 89, 95, 100, 110–111, 150, 154, 192–194, 196–207,

209, 211, 213–217, 219–220, 274, 282, 289, 298–299, 302; balance 197; and environmental impact assessments 194; equality 20, 22, 46, 154, 196–198, 201, 206–207, 213–214, 217, 302; impacts 194; inequality 59, 193; gender-based discrimination 58–59; gender-based violence 22, 59, 65, 121, 196, 199, 201, 211–213; gender-sensitive disaster risk reduction policies 197; mainstreaming strategies 200; *see also* LGBTQ+ persons; women
General Comments 4–5, 8–11, 18–19, 32, 61, 120, 148–149, 153, 212, 227, 268, 272, 297, 299
genetic defects 127
genetically modified organisms (GMOs) 130; seeds and breeds 91
Geneva Pledge for Human Rights in Climate Action 53
Ghana 43
Global Compact on Migration 58, 273
Global Dryland Alliance 201
global energy consumption 82
Global North 85, 99, 300
Global Pact for the Environment 27–28
Global political economy 98
Global South 99–100, 300
global warming 34, 37, 57, 90, 114, 132, 154–155, 186, 190; *see also* climate change
global water crisis 84, 107, 114
glyphosate 160
gold mines 43, 99, 182, 208
governance mechanisms 98
Greece 43
greenhouse gases (GHGs) 5–6, 21, 45, 48, 82–84, 90–91, 94–96, 134, 136, 154–157, 197, 201, 203–204, 231–234, 236, 246, 248, 252, 302
"greening" human rights law 25
grievance mechanisms 84
gross national product (GNP) 167
groundwater contamination 99, 237
Guatemala 111, 209, 211, 218–221, 270–273, 277, 285, 287–289, 291, 295; Environmental Protection and Enhancement Act 278; Regulations on Environmental Impact Assessments 278
Guinea 44, 51, 64–65, 67–68, 72–73, 123, 139, 144, 158, 167, 178, 183, 188, 218, 220, 222, 242, 256, 260, 265; 2013 Mining Code 158; Accelerated Food and Nutrition Security and Sustainable Agriculture Programme 2016–2020 167
Guyana 123, 139, 144, 185, 187, 201, 207, 214, 219, 223, 276, 290, 295; Green State Development Strategy Vision 2040 201, 207

Haiti 216, 220–221
hazards 44, 92, 122, 125, 151, 158, 181–182, 195, 197, 204, 228, 235, 242; hazardous activities 238, 251, 262, 266; hazardous conditions of work 199, 207; hazardous waste 19, 22, 33, 38, 162, 259
health care inequalities 60; health disparities 227; health shocks 95
healthy aquatic systems 85
healthy environment 4–5, 8, 11–13, 20–21, 23–26, 28–30, 34, 41, 43, 48, 52, 54–56, 62–63, 71, 76, 78–79, 81, 83–86, 104–108, 113, 115, 124, 128–130, 150, 158, 175, 193, 195, 212, 237, 248, 259, 297, 301–302; distinct right to a healthy environment 5, 24–25, 28, 76, 297; healthy and sustainable environment 4, 11, 14, 21, 30, 32, 37–38, 42, 45, 51–53, 55, 63, 68, 71, 73, 78, 84, 104–108, 113–115, 138, 145, 147, 175, 190, 212, 223, 254, 266, 289, 296–297
heat waves 229, 245, 248, 251
herbicides 158
high-efficiency particulate air (HEPA) 240
HIV/AIDS 92
human mobility 59, 61
human rights 3–21, 23–38, 41–63, 65–73, 75–89, 91–108, 110–115, 119–123, 125, 127–145, 147–148, 150–153, 155, 157, 159, 161, 163, 169–170, 172–175, 186–187, 189–192, 196–197, 208–209, 211–212, 219, 223–229, 238–239, 241–250, 252–255, 260–263, 266–269, 273–274, 279–280, 285, 289–290, 292, 296–303; defenders 52, 59, 77, 79, 81, 95, 111, 121, 163, 172, 174, 186, 190, 211; due diligence 84, 98, 239, 298; education 41, 76; human rights-based approach to climate change 58; human rights-based development 57; human rights, labor, and environmental standards 241, 253; impact assessment 92
Human Rights Commission 9, 34–35
Human Rights Committee 9–10, 12, 21, 29, 79, 119, 121, 123, 125, 127–129, 131–133, 135, 137, 139–141, 143–145,

147, 228, 248–249, 261, 290, 292, 299, 301; Conciliation Commission 119; GC No 31 133; GC No 36 21, 120, 125, 133–134
Human Rights Council (HRC) 5, 9, 16, 19–21, 26–27, 30, 41–63, 78–80, 83, 85, 103, 105, 242–243, 297; Advisory Committee 42, 110; Complaint Procedure 42; Resolution 7/23 79; Resolution 8/7 243; Resolution 10/4 53; Resolution 16/11 54; Resolution 19/10 52, 78; Resolution 28/11 52; Resolution 34/20 52; Resolution 38/4 58; Resolution 41/21 53, 59; Resolution 44/7 60; Resolution 47/24 54; Resolution 48/13 51, 54; subsidiary expert mechanisms 42; Universal Periodic Review (UPR) 9, 18–20, 42–43, 110, 137, 297; Universal Periodic Review Working Group 42–43
hurricanes 121, 196, 202, 272; Hurricane Irma 202; Hurricane Katrina 25, 61, 124; Hurricane Matthew 202; Hurricane Pam 272
hydraulic fracking 99, 184; hydraulic fracturing 155, 159–160; *see also* fracking
hydrocarbon activities 168
hydropower projects 91

illegal logging 67, 102
illegal settlements 102
imminent damage test 300
impunity 85, 185
Independent Expert on Human Rights and the Environment 5, 9, 78–79, 148, 274
India 48, 51, 66, 72, 111, 276, 290, 295
indigeneity 60–61; Declaration on the Rights of Indigenous Peoples (UNDRIP) 23, 50, 62, 94, 101, 170–171, 174, 279, 281–282, 285, 298, 302; Expert Mechanism on the Rights of Indigenous Peoples 50; indigenous and native communities 282; indigenous lands 49, 100–102, 111, 170, 186, 278, 282–283, 299
individual: communications 4–5, 9–11, 19, 21, 31, 120, 127, 138, 226, 299; complaints 8, 18–19, 31, 104, 244, 303; complaints procedure 18, 244
Indonesia 43, 162, 181, 188, 217–218, 221, 282, 292, 294; Kalimantan Border Oil Palm Mega-Project 282; Law on Plantations 282; National Policy for Community-Based Water Supply and Environmental Sanitation 163; Presidential Regulation on Procurement of Land 163
Indonesia-Malaysia border 282
industrial waste 44–45, 151, 181
infrastructure projects 48, 102, 211, 278
insecticides 160
Inter-American Commission and Court of Human Rights 18; Advisory Opinion OC-23/17 28, 249–250, 252; Inter-American Court of Human Rights 11, 28, 50, 172, 209, 249, 252–253, 302
intergenerational equity 230, 245, 255; health impacts 208; justice 245
Intergovernmental Panel on Climate Change (IPCC) 54, 56, 59, 83, 87, 173, 196, 246, 300–301; 1.5 Degree report 173, 246
Intergovernmental Science-Policy Platform on Biodiversity and Ecosystem Services (IPBES) 173
internally displaced persons 4, 92, 112, 173
International Bill of Rights 16–17, 191
international certification 239, 242–243, 253, 260
International Convention for the Protection of All Persons from Enforced Disappearance (CED) 18–19
International Convention on the Elimination of All Forms of Racial Discrimination (CERD) 18–19, 23, 273–274, 280
international cooperation 45, 47, 57, 149, 157, 226, 230–231, 233–235, 246, 269–270, 286
International Covenant on Civil and Political Rights (ICCPR) 10, 17–19, 22, 61, 119–120, 122–123, 127–132, 138–139, 147–148; First Optional Protocol to the ICCPR 10, 17, 120, 122, 127–129, 133, 135, 138
International Covenant on Economic, Social and Cultural Rights (ICESCR) 17–20, 22, 44, 59, 61, 81, 119, 135, 147–148, 154, 173, 226
international displacement 205; *see also* forced displacement, mobility
International Labor Organization (ILO) 44, 49–50, 62, 169–171, 174, 207, 228, 276, 282, 285, 298, 300, 302;

Convention No 107 276; Convention No 169 on Indigenous and Tribal Peoples 44, 49–50, 62, 169–171, 207, 275, 277, 282, 285, 298, 300, 302
international law 5–6, 8, 11–12, 15–16, 25, 30, 34–37, 48, 57–58, 76, 80, 92, 102, 104, 106, 108, 110, 113, 122, 132, 135, 143–144, 152, 175, 262, 267, 276, 301–304; crime under international law 122; international environmental law 11–12, 14, 24, 26–27, 33–34, 36–37, 58, 121–122, 249, 252, 300; international human rights law 4, 9, 11, 14–15, 17–19, 21, 23–25, 27, 29–31, 33, 35–37, 80, 88, 105, 108, 135, 137, 174, 189, 254, 267, 297; international refugee law 133
International Law Commission 110, 175, 301, 303–304
international organizations 47, 246
international solidarity 87
International Sustainable Development Law (ISDL) 7
intersectionality 109, 213, 292; intersectional discrimination 272
Inuit petition 53, 83, 107, 300
investment and trade agreements 159
Iran 46, 65, 72, 164, 181, 189, 239, 259, 265; river diversion program 164, 239
Iraq 44, 64–65, 72, 206, 209, 218, 220–221, 223, 239, 259, 264
irrigation facilities 158–159
Israel 123–124, 139, 144, 158, 178, 183, 188, 218, 223
Ivory Coast *see* Côte D'Ivoire

Jamaica 233, 256, 258, 265; Office of Disaster Preparedness and Emergency Management 233
Japan 67, 233, 239, 256, 259, 264; Comprehensive Support Projects for Health and Life of Afflicted Children 239; Fukushima Citizen Health Management Fund 239; Nuclear Accident Victimized Children Support Act 239
Johannesburg Declaration on Sustainable Development 54
jurisdictional hurdles 138
just transition 59, 96

Kazakhstan 43, 49, 64–66, 72, 159, 178, 181, 183, 188–189, 210, 220, 223
Kenya 47, 49, 64–65, 67, 72, 161, 180, 187, 216, 222, 240, 259–260, 265, 284, 293–294; 2010 Constitution of Kenya 240; National Adaptation Plan 240; Vision 2030 284
Keystone XL pipeline 101
Kuwait 164, 181, 189; Failaka Island 164
Kyoto Protocol 5, 156–157, 201
Kyrgyzstan 43, 46, 65, 72, 162, 180, 187, 284, 293–294; National Sustainable Development Strategy 284

labor standards 208, 253
land: claims 283; dispossession 100; disputes 132–133, 163, 275; governance 97; land-grabbing 161–163, 165; management 97, 153, 173, 244; rights 8, 88, 102, 171, 208, 211, 275, 29; trafficking 164, 18; use 100, 153, 171, 173, 211, 277; and water pollution 45
Lao People's Democratic Republic 64, 72, 203, 215, 221; National Committee for Natural Disaster Prevention and Control 203
large infrastructure development projects 97, 153, 166, 185, 209, 219
large-scale movement of people 96, 101, 227
Latin America 34, 36, 64, 71, 90
League of Arab Nations 29
Lesotho 46; National Climate Change Policy 46
lex ferenda 76
LGBTQ+ persons 4, 59
licensing investments 159
life of dignity 24, 121, 134, 137, 151
livelihoods 45, 49, 59–61, 65, 87–88, 94, 97, 129, 136, 158, 161, 193, 197, 199, 204, 220, 243, 245, 280; livelihood resources 102
livestock 84, 91–92, 94, 131, 157, 162, 193, 256
llama-raising 131
locus standi 29
loss and damage 83, 103, 142
low-lying areas 87, 232; low-lying islands 23
Lubicon Lake Band v. Canada 129–130
Luxembourg 218, 222, 270, 287, 295

Magna Carta of 1215 16
malaria 229, 237, 251, 276
Malawi 233, 257, 260, 265

Maldives 45, 64, 73, 217–218, 221
Mali 159, 167, 178, 183, 188; Framework for Economic Recovery and Sustainable Development 167; National Action Plan on Achieving the Sustainable Development Goals 167
malnutrition 22, 94, 177, 226, 256
mangrove swamps 129
Māori claims 125, 170, 284
marine ecosystem 22; resources 137, 158–159
Marshall Islands 203, 208, 216, 219, 222, 233, 245, 251, 257, 259, 265, 301; Bikini Atoll 204; Joint National Action Plan for Climate Change Adaptation and Disaster Risk Management 233; National Climate Change Policy Framework 233
Mauritius 65, 156, 167, 177, 183, 187–188, 203, 210, 215–216, 220, 222
maximum available resources 57, 149, 153, 246
means of subsistence 22, 126, 131, 133, 141, 152, 158, 164, 283
mental health 54, 59, 61, 85, 147, 161, 251, 266; impacts 251
mercury 99, 162, 208, 210, 238–239, 276; contamination 239; mercury-free processing plants 162; poisoning 99
methane 99
Mexico 111, 123, 139, 144, 215, 219, 223
Middle East 90
migrants 4, 49, 57–58, 69, 73, 92, 152, 173
migration with dignity 88, 199
Millennium Declaration 54; Millennium Development Goals (MDGs) 52, 206–207, 270, 302
minimum core obligation 149
mining projects 94, 159, 186, 211, 283
minorities 4, 6, 16, 23, 26, 50, 61, 75, 81, 121–123, 127, 138, 142, 152–153, 275, 280–281, 284–285, 292, 299, 301
miscarriages 130, 210
mitigation policies 46, 48, 53, 91–92, 103, 233
mobility 59–61, 97, 199
Mongolia 43, 157, 162, 177, 180, 188, 209, 216, 220–221, 240, 256, 259, 265, 286–287, 295
monocultures 193
Montenegro 162, 180, 188, 288, 295
Mossville case 25
Mozambique 201, 210, 214, 220–222, 260, 265, 289
multinational companies 158, 174

Nansen Initiative 69, 71
National Human Rights Institutions (NHRIs) 62
Native Americans 100–101, 111
natural resources 6, 8, 29, 44, 48, 52, 55, 59, 87, 92, 97, 99–100, 103, 111, 122, 125–127, 150, 152–154, 158–161, 165–166, 168–169, 171–172, 179–180, 185–186, 193–195, 198–199, 204–205, 219–220, 228, 230, 238, 274, 277–285, 287
Nauru 43–44, 64, 73, 209, 216, 218, 220–221
net zero emissions 83, 135, 234
Netherlands 161, 179, 189
New Zealand 31, 37, 50, 67, 73, 125, 132, 140–142, 144–146, 170, 178, 183–184, 188, 201, 214, 222, 256, 261, 265–266, 284, 290, 293–295; 2016 Matike Mai Aotearoa report 170; Climate Commission 170; Foreshore and Seabed Act of 2004 290; Immigration and Protection Tribunal of New Zealand 132; Ko Aotearoa Tēnei report 170; Marine and Coastal Area (Takutai Moana) Act 2011 125; National Plan of Action for Human Rights 50; Supreme Court of New Zealand 132; Te Awa Tupua (Whanganui River Claims Settlement) Act 284, 296; Treaty of Waitangi 50, 125, 170
Nicaragua 47, 49, 64–65, 67, 73
Niger 44, 67, 179, 183, 188, 244, 257, 261, 265, 270, 279, 288, 291, 294
Nigeria 38, 43–44, 64, 73, 100, 166, 182, 188, 210, 218, 220, 222; Ogoniland 33, 38, 166
nomadic pastoralism 88
non-exhaustion of local remedies 127, 129
non-GM fields 130
non-regression 36
non-toxic environments 81
Nordic Sami Convention 171
North Siberia 125, 183
Norway 49, 67, 73, 154, 176, 189, 204, 207, 216, 218, 223, 244, 261, 263
nuclear weapons 67, 101, 122, 127–128, 138–139, 141; disarmament 192; testing 101

obtaining views of the child 229
occupational health and safety 160, 166, 193
occupied territory 158
ocean acidification 135
oil 43–44, 89, 98–100, 104, 111, 123, 129, 154–155, 158–159, 161, 166, 179, 182, 184–186, 201, 204–205, 238–239, 242–243, 282–283, 292, 295, 298; exploration 43, 100, 166, 186; and gas extraction 98, 158; and mineral sectors 239; reserves 123; spills 44, 99, 166
oil palm plantations 89, 282
Office of the High Commissioner for Human Rights (OHCHR) 5, 8–9, 12, 14–15, 20, 23, 33, 37, 41–63, 69–73, 79, 137, 229; Eliminating Discrimination and Inequalities in Access to Water and Sanitation 56
older persons 4, 44, 60–61, 64, 70–71, 81, 152, 157, 272, 287
organic farming 195
Organization for Economic Cooperation and Development (OECD) 98, 228
"Our Common Future," *see* World Commission on Environment and Development (WCED)

Pacific Islands Forum 208
Palau 234, 251; Climate Change Policy for Climate and Disaster Resilient Low Emissions Development 234
Palestinians 158
palm oil companies *see* oil palm plantations
Papua New Guinea 51, 67–68, 73, 218, 220, 222
Paris Agreement 5, 7, 12, 14, 24, 27, 48, 54, 56, 58, 62, 65, 83, 91, 94, 103, 135, 137, 142, 145, 153–156, 174, 176, 197, 231, 236, 250, 298–299, 302; intended nationally determined contribution (INDC) 247; nationally determined contribution (NDC) 5, 46, 48, 83, 155, 298, 302; Task Force on Displacement 58
participation in decision-making 20, 34, 38, 77, 123, 192, 198, 215, 274
patriarchy 85; patriarchal structures 59
peasant 89–90, 100, 273, 284, 293; Declaration on the Rights of Peasants and Other People Working in Rural Areas 167; farmers 90, 100, 273; *see also* campesino
pensioners 157
permanent injunction 129
permanent sovereignty over natural resources 100
persons with disabilities 4, 18, 45, 47, 49, 52–53, 59–60, 70–71, 75, 152, 173, 176, 268–272, 285–288, 295
Peru 43, 111, 124, 140–141, 145–146, 219, 222, 242, 260, 264, 282–283, 292–295; Cajamarca 242; Cerro de Pasco 242; La Oroya 242; Penal Code 124
pesticides 91, 94, 158, 160, 162, 166, 172, 182, 193, 195, 209, 237, 254, 303; burial sites for pesticides 162
Philippines 29, 67, 73, 111, 161, 170, 180, 184–188; 1974 Forest Reform Code 171; Comprehensive Agrarian Reform Program Extension 161; Indigenous Peoples' Rights Act 171; National Land Use Bill 171; Supreme Court of the Philippines 29–30
pipelines and hydroelectric dams 126; Keystone XL pipeline 101
planetary boundaries 25, 97
plastic bags 158; waste 22
pledge to leave no one behind 168, 173, 187, 190
Poland 124, 140, 146, 180, 184, 189, 210, 220, 236, 258, 261, 264; Youth Climate Council 236; Youth Ecological Council 236
pollution 4, 6, 20, 22, 24, 32–33, 38, 44–46, 48, 69, 76–77, 81–86, 94, 99–101, 105, 107, 114, 120–121, 123–124, 137–138, 147, 149, 151, 154, 158–164, 166–167, 173–174, 179–181, 191–193, 199, 201, 207–208, 211, 220, 226–229, 231, 234, 236–240, 242–243, 251, 257–258, 260, 274–277, 279–281, 283, 285, 297–298; air pollution 22, 44–45, 81–82, 85, 99, 158–159, 162, 164, 181, 229, 234, 236–238, 240, 258, 298; atmospheric pollution 158, 166, 231, 234; electromagnetic pollution 237; hotspots 85; indoor and outdoor pollution 82; industrial pollution 45, 162–164, 238–239; persistent organic pollutants 160, 162; of rivers 160, 277, 279; soil and groundwater pollution 160; soil pollution 69, 83, 99, 257–258; surface water pollution 99; water pollution 45, 83–84, 99, 107, 114, 160, 180, 193, 201, 229, 277, 280, 283; of wells 199

poor working conditions 162
poverty 4, 9, 19, 49, 53–55, 57–59, 62, 66–67, 75, 77, 81–82, 85, 91, 94–97, 100, 110, 112–113, 115, 131, 152, 167, 173, 195–196, 201, 204, 208, 213, 217, 229, 233, 244, 246, 255–256, 269–270, 272–273, 276, 290, 293, 302; alleviation 217, 270; eradication 54, 97, 213; extreme poverty 75, 82, 94–96, 110, 112–113, 115, 208, 272, 276; people living in poverty 4, 97, 152, 173
precaution 86; precautionary approach 122, 125, 140
prevention 18, 26, 31, 38, 41, 46, 75, 86, 93, 124, 197, 203, 250, 256, 262, 266
principles: of common but differentiated responsibility (CBDR) 23, 250, 300; of non-discrimination 17, 56–58, 60, 76, 80, 86, 90, 99, 134, 147–148, 152, 206, 229, 231, 268; of *non-refoulement* 10, 121, 133–134, 227–228, 299; of proportionality 130–132; of sovereign equality 100
prior consultation 50, 103, 125, 139, 169, 171–172, 242, 277–278, 291; prior, informed consent 100, 186, 219, 276, 282; *see also* FPIC
private and family life 130
procedural rights 20, 23–25, 76, 78, 86, 150, 299
progressive realization 148
protected areas 94, 101–103, 168
protection of aliens 133; ecosystems 230
public emergencies 121
public interest litigation 29
pygmy communities 239

Qatar 201, 214, 223, 244, 261, 265; National Vision 2030 244

radiation 21, 150, 162, 239; exposure 239; radioactive accidents 121; radioactive and toxic waste 127, 281; radioactivity 127; radiological hazards 151
radon 159
rapid urbanization 227, 240
reconstruction plans 124
refugees 12–13, 57, 70, 74, 81, 92–93, 109, 121, 132, 134, 138, 142, 144, 173, 205, 254, 263, 274, 287, 299; High Commissioner for Refugees 205; political refugees 70, 132; refugee status 10, 132–133
regional human rights tribunals 133, 136
reindeer husbandry 126, 141, 275
relief measures 124, 203
relocation 104, 156, 210, 277
remedies 20, 24, 27, 29, 76–78, 88, 120, 127, 129, 135, 148, 161, 163, 172, 176, 181, 184, 192, 197, 226, 228, 236, 239, 242–243, 247, 251, 262, 274, 281, 299, 301
renewable energy 5, 8, 66, 154–158, 199–200, 210–211, 232, 236, 238, 252–253, 302; production 5, 66, 157; projects 158
reparations 46, 63, 169, 211
Republic of Kiribati 48, 132–134, 199, 206; Joint Implementation Plan on Climate Change and Disaster Risk Management 48; National Adaptation Programme of Action (NAPA) 132; South Tarawa 132–133
Republic of Moldova 160, 178, 189, 262, 270
resettlement 50, 60, 104, 164–165, 182, 195, 207, 276
resilience 46, 48, 54, 91, 124–125, 157, 167, 183, 202, 230, 235–236, 238, 255, 271, 298
resource extraction 101, 179, 281, 285; scarcity 88
respect, protect, and fulfill 8, 17, 81–82, 153, 227–228; *see also* typology of rights
respiratory diseases 236, 240; pneumonia 240, 258
responsible business conduct 98
retrogressive measures 84, 149
rights: to a clean, healthy and sustainable environment 11, 14, 21, 32, 38, 42, 51–52, 68, 73, 105, 107, 115, 297; to food 8, 19, 46, 55–56, 69, 71, 75, 84, 89–95, 109–110, 112–113, 115, 149–151, 177, 195; to health 20–21, 24, 44, 55, 69, 71, 75, 124, 150–151, 159, 169, 181, 208, 227, 231, 234, 245, 274; to a healthy environment 4–5, 11–13, 20–21, 23–26, 28–30, 41, 43, 48, 52, 54, 56, 62–63, 76, 78, 81, 83–86, 105–108, 115, 128–130, 195, 212, 248, 297, 302; holders 15, 79, 168; indigenous rights 8, 19, 25, 49, 62, 126, 137–138, 154, 168, 174, 302; labor rights 43–44, 51, 100, 241, 243, 253, 298; to life 10, 15, 21, 23–24, 28–29, 57, 82, 84–85, 120–125, 127–130, 132–134, 136, 138, 140–141,

228–229, 245–246, 254, 297, 303; minority rights 129, 245; of nature 5, 26, 28, 31, 37, 45, 105, 147; of persons with disabilities 49, 53, 59–60, 70–71, 75, 268, 270, 285–286, 288, 295; to physical and mental health 147; to self-determination 57, 100, 102, 104, 122, 142; to take part in cultural life 147; to water and sanitation 8, 21, 56, 180; to work 147
rights-based approach to governance 95; rights-based deep decarbonization plans 83
Rio Declaration on Environment and Development 25, 154, 197
Roma communities 162; settlements 275; Serb and Roma minorities 275
Royal Dutch Shell oil spill 99
rule of law 96, 197
rural development projects 193–195
Russian Federation 67, 125–126, 140, 144, 159, 177–178, 183–185, 187, 189, 242, 260, 264, 279, 291, 294

Sacchi v. Argentina 10, 79, 245, 253, 300–301
sacred sites 101, 104, 281; of indigenous peoples 281
sacrifice zones 85–86
safe and potable water 55, 150; and sanitation 45, 53–56, 59, 75, 103, 112, 162, 167, 178–180, 228, 230, 235–236, 239, 257, 260, 266, 279
safe food, nutrition and housing 55, 150
safe, clean, healthy and sustainable environment 4, 32, 37, 45, 52–53, 63, 71, 78, 104, 106–108, 113–115, 138, 145, 175, 190, 212, 223, 254, 266, 289, 296; *see also* healthy environment
Sami 49–50, 126, 171, 185, 275; culture 126, 275; lands 50, 171, 275; people 49–50, 126, 171, 185, 275; reindeer-breeding areas 275; Sami Parliament 126
San Salvador Protocol to the American Convention on Human Rights 28
Saramaka people 50
sea level rise 7, 23, 109, 132–135, 153, 156, 196, 232, 248, 299
sea walls 132
second generation rights 17; *see also* economic, social and cultural (ESC) rights
Sendai Framework for Disaster Risk Reduction 46, 94, 197, 269–271
Senegal 158; Constitutional Act of 2016 158; Dakar 158
sex-disaggregated data 200
sexual harassment 22; violence 112, 195, 203
Seychelles 200, 214, 223, 287; Disaster Risk Management Act 200; National Vulnerability Assessment Committee 200
silicosis 99
Slovakia 33, 36, 177, 189, 254, 263, 270, 275, 287, 290, 295
slow-onset impacts 58
small herders 157
small island states 7, 12, 22–23, 48, 54, 57, 62, 87–88, 132, 134, 289, 299, 301; *see also* Alliance of Small Island States
smallholder farmers 59, 91–92, 94, 179
social and environmental impact assessments 283; social impact assessments 165–166, 186, 238
social inclusion 26, 269; justice 26; pillar 26
social protection 46, 61, 92, 97, 167, 183, 233–234, 236, 270, 272, 287; social protection floors 167, 183
social security 147, 167, 240
social transformation 96–97
sockeye salmon fisheries 101
soft law 17, 21, 33, 36, 52, 68, 71, 106
soil quality 91, 160
Solomon Islands 43, 45, 51, 65, 68, 73, 209, 218, 220, 222, 256, 264
South Africa 99, 167, 169, 172, 183–184, 186, 188, 244, 261, 266, 275, 290, 295; Traditional and Khoi-San Leadership Bill 169
South Asia 95
South Pacific 127
South-South cooperation 48
soya bean production 89, 172, 236
Spain 43, 46, 65–66, 73, 183, 189, 240, 259, 264
Special Rapporteur on Climate Change and Human Rights 27, 103
Special Rapporteur on Contemporary Forms of Racism, Racial Discrimination, Xenophobia, and Related Intolerance 49, 98–100, 111, 115
Special Rapporteur on Cultural Rights 86
Special Rapporteur on Extreme Poverty 75, 82, 95–96, 110, 115

Special Rapporteur on Hazardous Substances 77–78
Special Rapporteur on Housing 53
Special Rapporteur on Human Rights and the Environment 4, 79; Draft Principles on Human Rights and the Environment 26, 30, 76, 105; Framework Principles on Human Rights and the Environment 4, 9, 30, 80, 84–85; Mapping Report No 2 78, 138; Mapping Report No 6 43
Special Rapporteur on Right to Water and Sanitation 235
Special Rapporteur on Toxics 75, 208, 301
Sri Lanka 128–129, 141, 145, 160, 177, 179, 183, 187; Central Environment Authority 129; Road Development Authority of Sri Lanka 128; Supreme Court of Sri Lanka 129
statelessness 67, 199, 205; stateless persons 121
Stockholm Conference on the Human Environment 26, 51, 105; Stockholm Declaration 24, 76
storm surge 135
structural exploitation 99
Subcommittee on Prevention of Torture and other Cruel, Inhuman or Degrading Treatment or Punishment (SPT) 18
Sub-Saharan and North Africa 90, 100
subsidies 33, 38, 82–83, 92, 94, 231
substantive equality 197–198, 204, 218, 284
sugar-cane farming 164, 239
supermarketization 94
supply chains 98, 213
Suriname 45, 50, 64, 67, 73, 204, 208, 216, 218, 220–221, 244, 256, 261, 265; Mining Decree 208; Moiwana community 50
survival and development 228–231, 235, 254, 303
Susila Malani Dahanayake 128, 141
sustainable development 4, 6–10, 12–14, 20, 22–23, 25–26, 30, 32, 34–36, 38, 41, 43–44, 48–49, 51–52, 54–55, 61–62, 64, 66–71, 73, 76, 82, 85, 94, 97, 105, 108, 112, 120, 123, 125, 137, 147, 153–154, 166–168, 174–176, 183, 187, 190–193, 196, 198–199, 205–207, 211, 213, 215, 217, 226, 243–244, 252, 261, 266, 268–270, 273–276, 278–279, 284–286, 290–291, 293, 299; Addis Ababa Action Agenda 62; Conference on Housing and Sustainable Urban Development (Habitat III) 269; Conference on Sustainable Development 52; Development Coordination Office (DCO) 68, 73; Development Programme (UNDP) 45, 68, 73–74, 184; "Future We Want" outcome document 197; High-Level Political Forum on Sustainable Development (HLPF) 62, 269; sustainable consumption and production patterns 160; sustainable economic and social development 66, 192, 274, 276, 290; sustainable food production 92; sustainable forest management 45; sustainable management of natural resources 238, 283
Sustainable Development Goals (SDGs) 4–7, 22, 27, 41, 43, 49, 52, 54–56, 61–62, 82, 85, 94, 97, 103, 105, 120, 125, 137, 147, 154, 167–168, 173–174, 192–193, 196–197, 199, 206–207, 211–212, 226, 230–239, 243–244, 252, 269–270, 272–273, 280, 285, 302; SDG 1 on poverty 302; SDG 2 on hunger 22, 302; SDG 3 on health 22, 197; SDG 4 on education 197, 302; SDG 5 on gender equality 22, 125, 197, 206–207, 211, 302; SDG 6 on water and sanitation 56, 85, 197, 302; SDG 7 on energy 211; SDG 13 on climate change 103, 197, 211, 272, 302; SDG 14 on oceans 103; Target 1.3 230, 232–235, 238, 255–259, 270; Target 1.4 270; Target 1.5 167, 231; Target 3.9 234, 236–237, 239, 255; Target 5.1 206; Target 5.A 206; Target 13.1 235; Target 13.3 235–237; Target 13.5 232, 235; Target 13.b 232–235, 238; Target 17.18 273
Sweden 123, 126, 171, 275; Environmental Code 171; Minerals Act 171; Minerals Strategy 171
Switzerland 155, 174, 176, 183, 188, 234, 257, 261, 264
Syrian Arab Republic 181, 189, 240, 259, 261, 265

Tajikistan 234–235; National Climate Change Adaptation Strategy 235
Taskforce on Displacement, *see* Paris Agreement
technology development and transfer 47

Teitiota case 12, 127, 136, 138–139, 143, 228, 299–300
terminator seeds 195
Thailand 111, 162, 180, 185, 187, 211, 214, 218–220, 222
thermal coal 135
third generation rights 17
time-bound benchmarks 155–156, 177
Tonga 235, 257, 266; Joint National Action Plan on Climate Change and Disaster Risk Management 235
Torres Strait Islander 50, 143, 145; dispute 127
tourism 45, 103, 215, 241, 281
toxic chemicals 85, 208, 237; toxic waste dumps 162; toxification of planet Earth 85
traditional and eco-friendly agricultural knowledge 195
traditional foods and medicines 88
traditional livelihoods 87, 129
transboundary harm 249, 251–252, 262, 266; environmental damage 27, 30, 248–249
transgenic crops 130
transnational corporations 8, 51, 62, 75, 77, 97–100, 111, 115, 153, 236, 241, 260, 280, 284–285, 293, 298, 300; extraterritorial reach of 298; Working Group on Human Rights and Transnational Corporations 75, 97
transparency 90, 97, 111, 113, 153, 165, 170, 197
tribal communities 276; tribal nations 104
tsunamis 121
Tunisia 167–168; Ninth Economic and Social Development Plan 167
Turkmenistan 167–168; National Programme on Socioeconomic Development 167–168
Tuvalu 45, 48, 205, 210, 235; National Adaptation Programme of Action (NAPA) 45; Water and Sanitation Policy 235
typology of rights 8, 17; *see also* respect, protect, fulfill

Uganda 100, 286, 295
Ukraine 124, 157, 177, 180, 189
UN Framework Convention on Climate Change (UNFCCC) 5–7, 12, 14, 45, 47, 53–54, 58–60, 69–70, 74, 82–83, 91–92, 103, 142, 145, 157, 176, 197, 201, 203–204, 213, 224, 246, 250, 252; 1.5 degree report 173, 246; Adaptation Committee 60; Clean Development Mechanism 157; Collaborative Programme on Reducing Emissions from Deforestation and Forest Degradation in Developing Countries (UN-REDD) 57, 65, 69, 91, 157; Doha work program 60; Gender Action Plan 59; Lima Work Program on Gender 59, 197
UN Guiding Principles on Business and Human Rights 43–44, 51, 82, 93, 98, 100, 106, 114, 228, 238, 241, 253, 298, 302
undernutrition 229
UNESCO Universal Declaration on Cultural Diversity 87, 109
United Kingdom (UK) & Northern Ireland 51, 208, 271
United Nations: Charter 3, 8, 11, 13–16, 18–19, 28–29, 31–32, 34–35, 37–38, 119, 149, 173, 271, 288; Environment Programme 102; General Assembly (UNGA) 16, 21, 24, 26–28, 30, 41–42, 56, 81, 103, 105, 119, 191, 226, 268, 272–273, 297; Green Climate Fund 154; Resolution 41/120 81; Secretary General (UNSG) 27, 41, 54, 191, 272, 289; Special Envoy for Climate Change 41, 63, 73; special mandate holders 9, 19, 26, 53, 68, 75, 85, 91, 103, 105, 253, 301; treaty bodies 4–5, 10, 18–20, 23, 31–32, 37–38, 42–43, 55–56, 58, 63, 82–83, 85, 89, 110, 123, 128, 137–138, 190, 230, 233, 268–269, 271, 273, 275, 277, 279, 281, 283, 285, 287, 289, 291, 293, 295, 299
United States (US): Alaska 25, 101, 103–104, 113, 245; Alaska Institute for Justice 103, 113; Executive Order issued by President Clinton 25; Flint, Michigan 25; Kivalina 25, 113; Louisiana 34–35, 104, 112, 281; nuclear testing program 204, 208; nuclear trust fund 208; Pacific Northwest 100; Pebble Mine Project 101; Shishmaref 25
Universal Declaration of Human Rights (UDHR) 3, 15–17, 24, 81, 119, 297
universal health coverage 49
universality 86, 109, 113
unlawful encroachment 151
unsustainable development 21, 51, 55, 62, 121–122, 133, 136, 174, 240, 297–298, 302

uranium 101, 111, 162, 165, 179, 237, 239, 279–280; extraction 165; groundwater contamination 237; mining 101, 111, 280; pollution 239; tailings 162
Uruguay 44, 50, 64–65, 67, 73, 179, 183–184, 187
Uzbekistan 163, 181, 189, 220, 223

Vaihere Bordes and John Temeharo v. France 127–128
Vanuatu 43, 47, 66–67, 73, 217–218, 221, 236, 258, 265, 271, 287–288, 293, 295; Ambae Island 271; Climate Change and Disaster-Induced Displacement Policy 47; Disaster Risk Management, Climate Change Adaptation and Energy 236; National Advisory Board on Climate Change and Disaster Risk Reduction 47; National Policy on Climate Change and Disaster Risk Reduction 47; Strategic Plan of 2016–2020 47
vector-borne diseases 59, 251
Venezuela 67, 186, 189, 258, 265
victims 10, 25, 44, 46, 50, 76–77, 83, 87, 89, 93, 120–122, 128–129, 131, 135, 138, 149, 161, 177, 191–192, 196, 209, 212, 217, 226, 236, 238, 248, 250, 254, 274, 278, 281, 300–302
Vienna Declaration on Human Rights 297
Viet Nam 44, 47, 64, 66, 73
vulnerable people 48

waste-recycling programs 160
water 4, 6–9, 12, 19–22, 24–25, 32–33, 44–46, 48, 53–56, 58–61, 64, 69, 71, 73, 75–76, 79–86, 89–92, 99–100, 103–104, 107–108, 110, 112, 114, 123–124, 129, 131–134, 136, 149–153, 158–164, 166–168, 170–173, 176–181, 184, 191, 193–196, 198–199, 201, 203, 207–210, 220, 227–233, 235–240, 242, 249, 251, 257–258, 260, 266, 272, 275–277, 279–280, 282–283, 298, 302; depletion 83, 194; governance 85; pollution 45, 83–84, 99, 107, 114, 160, 180, 193, 201, 229, 277, 280, 283; resources 151–152, 160, 179, 237, 242, 283; scarcity 7, 84–85, 107, 114, 195; waterborne diseases 84, 163; water-related disasters 84, 107, 114; water-related ecosystems 152
watersheds 152; management 93
weapons of mass destruction 122
wetlands 131
widespread, long-term, and severe damage 152
wildfires 229, 251
women: breastfeeding or pregnant women 88; discrimination against women 9–10, 18, 191, 193, 195, 197–199, 201, 203, 205–207, 209, 211–213, 215, 217, 219–221, 223–224; empowering women and girls 20, 95; farmers 91–92, 151, 161, 194–195; full and equal participation of women 197; human rights defenders 59, 211; indigenous women 57, 88, 194, 199, 204, 208, 211–212, 219–220, 272, 282; Māori and Pasifika women and girls 125; rural women 47, 140, 191–195, 199–201, 204–212, 214–217, 219–220, 293, 298; *see also* gender
Working Group on Transnational Corporations 97
World Bank 33, 38, 95
World Commission on Environment and Development (WCED) 7, 26; WCED Experts Group on Environmental Law 26, 34; "Our Common Future" 7, 26
World Environment Day 4, 12, 14
World Food Program 89, 256
World Health Organization (WHO) 21; WHO guidelines 81, 151; WHO standards 44, 139, 159, 174; Health Principles of Housing 149
World Heritage sites 101, 103
World Human Rights Day 16
World Parks Congress 102
World War II 3, 119

xenophobia 57, 98, 111, 115, 291, 293

Zambia 67, 243, 261, 265
Zimbabwe 213, 218, 223
zoonotic diseases 83, 229